SEMA SIMAI Springer Series

Volume 33

As of 2013, the SIMAI Springer Series opens to SEMA in order to publish a joint series aiming to publish advanced textbooks, research-level monographs and collected works that focus on applications of mathematics to social and industrial problems, including biology, medicine, engineering, environment and finance. Mathematical and numerical modeling is playing a crucial role in the solution of the complex and interrelated problems faced nowadays not only by researchers operating in the field of basic sciences, but also in more directly applied and industrial sectors. This series is meant to host selected contributions focusing on the relevance of mathematics in real life applications and to provide useful reference material to students, academic and industrial researchers at an international level. Interdisciplinary contributions, showing a fruitful collaboration of mathematicians with researchers of other fields to address complex applications, are welcomed in this series.

THE SERIES IS INDEXED IN SCOPUS

Gabriella Bretti • Roberto Natalini •
Pasquale Palumbo • Luigi Preziosi
Editors

Mathematical Models and Computer Simulations for Biomedical Applications

Springer

Editors
Gabriella Bretti
Istituto per le Applicazioni del Calcolo (IAC) "M. Picone"
Consiglio Nazionale delle Ricerche (CNR - National Research Council)
Roma, Italy

Roberto Natalini
Istituto per le Applicazioni del Calcolo (IAC) "M. Picone"
Consiglio Nazionale delle Ricerche (CNR - National Research Council)
Roma, Italy

Pasquale Palumbo
Department of Biotechnology and Biosciences
University of Milano-Bicocca
Milano, Italy

Luigi Preziosi
Department of Mathematical Sciences
Politecnico di Torino
Torino, Italy

ISSN 2199-3041 ISSN 2199-305X (electronic)
SEMA SIMAI Springer Series
ISBN 978-3-031-35714-5 ISBN 978-3-031-35715-2 (eBook)
https://doi.org/10.1007/978-3-031-35715-2

This Springer imprint is published by the registered company Springer Nature Switzerland AG
The registered company address is: Gewerbestrasse 11, 6330 Cham, Switzerland

Preface

Mathematical modelling and computer simulations are playing a crucial role in the solution of the complex problems arising in the field of biomedical sciences and provide a support to clinical and experimental practices in an interdisciplinary framework. Indeed, the development of mathematical models and efficient numerical simulation tools is of key importance when dealing with such applications.

Moreover, since the parameters in biomedical models have peculiar scientific interpretations and their values are often unknown, accurate estimation techniques need to be developed for parameter identification against the measured data of observed phenomena. In the light of the new challenges brought by the biomedical applications, computational mathematics paves the way for the validation of the mathematical models and the investigation of control problems. This series hosts high-quality selected contributions containing original research results as well as comprehensive papers and survey articles including prospective discussion focusing on some topical biomedical problems:

- Drug diffusion in pharmacokinetic compartmental models
- In silico modelling immunocompetent behaviour in lab-on-chip experiments
- Particle models for cell migration and aggregation
- HIF-PHD dynamics related to oxygen availability
- Mechanical modelling of brain tumours
- Multi-scale immune system simulator for diabetes
- Machine learning techniques for biological tissues
- COVID variants modelling and pandemic waves
- Multifractal spectrum-based classification for breast cancer

These topics constitute the chapters of the book according to the contributions collected.

Roma, Italy — Gabriella Bretti
Roma, Italy — Roberto Natalini
Milano, Italy — Pasquale Palumbo
Torino, Italy — Luigi Preziosi
January 2023

Contents

An Application of the Grünwald-Letinkov Fractional Derivative to a Study of Drug Diffusion in Pharmacokinetic Compartmental Models

Tahmineh Azizi

Abstract In this study, we present the application of fractional calculus (FC) in biomedicine. We present three different integer order pharmacokinetics models which are widely used in cancer therapy with two and three compartments and we solve them numerically and analytically to demonstrate the absorption, distribution, metabolism, and excretion (ADME) of drug or nanoparticles (NPs) in different tissues. Since tumor cells interactions are systems with memory, the fractional order framework is a better approach to model the cancer phenomena rather than ordinary and delay differential equations. Therefore, the nonstandard finite difference analysis or NSFD method following the Grünwald-Letinkov discretization may be applied to discretize the model and obtain the fractional order form to describe the fractal processes of drug movement in body. It will be of great significance to implement a simple and efficient numerical method to solve these fractional order models. Therefore, numerical methods using finite difference scheme has been carried out to derive the numerical solution of fractional order two and tri-compartmental pharmacokinetics models for oral drug administration. This study shows that the fractional order modeling extends the capabilities of integer order model into the generalized domain of fractional calculus. In addition, the fractional order modeling gives more power to control the dynamical behaviors of (ADME) process in different tissues because the order of fractional derivative may be used as a new control parameter to extract the variety of governing classes on the non local behaviors of a model, however, the integer order operator only deals with the local and integer order domain. As a matter of fact, NSFD may be used as an effective and very easy method to implement for this type application, and it provides a convenient framework for solving the proposed fractional order models.

T. Azizi (✉)
Department of Mechanical Engineering, Florida State University, Tallahassee, FL, USA
e-mail: tazizi@fsu.edu

G. Bretti et al. (eds.), *Mathematical Models and Computer Simulations for Biomedical Applications*, SEMA SIMAI Springer Series 33,
https://doi.org/10.1007/978-3-031-35715-2_1

1 Introduction

Pharmacokinetics have been defined as the flow and rate of distribution and removal of a drug or nanoparticle (NP) inside body. Pharmacokinetics may be represented as mathematical models to determine the process of administered drug movement throughout body. Pharmacokinetics models are the main piece of modeling based drug development and can be performed by non-compartmental or compartmental methods. Compartmental modeling helps to find the most efficient route of drug administration based on time of uptake and elimination. These models provide a theoretical and mathematical framework to demonstrate the transmit of molecules biochemistry and transport phenomena in the body, and it has been done by dividing body into two main compartments based on pharmacokinetic and pharmacodynamic of different tissues. Therefore, compartmental modeling defines a comprehensive framework which makes an effective drug delivery toward targeted tissue and has been attracted by different researchers to find the most optimized therapy for different diseases such as cancer. These models receives information regarding route of administration such as intravenous or intramuscular injection, and or oral and combine them by different assumptions related to single or multiple doses to demonstrate drug traveling states inside body, starting from the absorbing by tissue and distributing from one organ to the other organ, and then chemical alteration of the specific tissue, and finally declining drug or NPs concentration because of elimination of chemical or biochemical drug by all removal paths [1–6].

Another approach to simulate distribution of drug in body that involves different organs and their interactions with other tissues called physiologically based pharmacokinetics models (PBPK). This computer based modeling method receives biological information such as physiological and chemical parameters for administered drug, blood flow rates to different tissues, volume of different organs, absorption and metabolism parameters, exposure and dose parameters and depends on different administration route, predicts the concentration of drug in different tissues using computational and numerical tools [1–3].

To understand the absorption, distribution, metabolism, and excretion (ADME) of drugs or NPs in different tissues mathematical and statistical models have been used by many different researchers [7–10]. These models use differential equation framework which are well-defined based on frequency responses, but should be appropriately scaled because of sensitivity of drugs and NPs to body mass of adults which is an index of body fat measurement in terms of weight or height of different persons. However, to build a comprehensive computational model, we need to obtain different experimental data which demonstrates the (ADME) process in different body organs [5, 6, 11–13]. Basically, these models provide us more flexibility in analysis and then modification and update by converting given biological system including all parameters into a mathematical model which can be written as the form of deterministic or stochastic differential equations (ODE, PDE). For example, a PBPK model is a mathematical model and structural framework which explains the (ADME) process of drugs or NPs in different tissues such as kidney, liver, lung, muscle, plasma, and so on. One of the advantages of this type modeling is preserving the same physiological parameters as the original system during modeling process.

During the recent decades, the area of fractional calculus has been attracted by many researchers in biology, physics, chemistry and biochemistry, hydrology, medicine, finance, and engineering. Due to wide applications of fractional modeling in science, and engineering, its importance and popularity is increasing day by day and it has allocated many attentions [14–19]. Fractional calculus is a new approach for modeling biological and physical phenomena with memory effects. Fractional calculus uses differential and integral operators including non-integer orders to study the non-linear behavior of physical and biological systems with some degrees of fractionality or fractality. While the integer order models provide a small class of non-integer order models, using fractional order operators, we can study different classes of the same model through changing the fractional order. Therefore, these models provide us more flexibility to analyze and control the behavior of a system.

There are many studies that proved the fractional order differential equations (FODEs) and or models with integral operators provide more precise results in real world applications since there are some biological or physical systems which display memory in their long term behavior and the traditional ODEs of integer-order disregard this fact [15, 19]. However, analytical solutions of these equations cannot explicitly be obtained. Therefore, to find the dynamical behaviors of solutions, we require to use approximation and numerical schemes. The finite difference method, Adomian decomposition method, extrapolation method, multistep method, iterative methods, and predictor corrector techniques can be used to obtain the numerical solutions for linear and nonlinear FODEs.

Fractional calculus as a field of mathematics can be considered as an old and novel topic. Old because it has been started by some works of G. W. Leibniz (1965, 1697), and L. Euler (1730), important contribution by P.S. Laplace (1812), J. B. J. Fourier (1822), N. H. Abel (1823–1826), J. Liouville (1832–1873), B. Riemann (1847), H. Holmgren (1865–1867), A. K. Grünwald (1867–1872), A. V. Letnikov (1868–1872), H. Laurent (1884), P. A. Nekrassov (1888), A. Krug (1890), J. Hadamard (1892), O. Heaviside (1892–1912), S. Pincherle (1902), G.H. Hardy and J. E. Littlewood (1917–1928), H. Weyl (1917), P. Levy (1953), A. Marchaud (1927), H. T. Davis (1924–1936), A. Zygmund (1935–1945), E. R. Love (1938–1996), A. Erdelyi (1939–1965), H. Kober (1940), D. V. Widder (1941), M. Riesz (1949). It can be considered as a novel area since we have H. T. Davis (1936), Erdelyi (1965), M. Caputo (1969), Yu. I. Babenko (1986), R. Gorenflo and S. Vessella (1991) [20–23].

To introduce some of the widely used fractional derivatives and integrals of a function $f(x)$ of arbitrary order α, we start with definition of fractional integral in the sense of Riemann and Liouville. According to Riemann-Liouville, fractional integral of order α ($\alpha > 0$) may be defined as a natural consequence of Cauchy formula:

$$J^n f(t) := f_n(t) = \frac{1}{(n-1)!} \int_0^t (t-s)^{n-1} f(s) ds, \quad t > 0, \quad n \in \mathbb{N} \tag{1}$$

where $f_n(t)$ vanishes at $t = 0$ with its derivatives of order $1, 2, \ldots, n-1$. Using the Gamma function:

Definition 1 The gamma function is defined as follows:

$$\Gamma(x) = \int_0^{\infty} y^{x-1} e^{-y} dy \tag{2}$$

where for convergence of the integral, $x > 0$

we can extend (1) from positive integer values to any positive real values and for $\alpha > 0$ write it as:

Definition 2 (Riemann Liouville integral, Riemann (1953) and Liouville (1832)) The fractional integral of order $\alpha \in \mathbb{R}^+$ of the function $f(t)$, for $t > 0$ where $f : \mathbb{R}^+ \to \mathbb{R}$ has been defined by

$$I_a^{\alpha} f(t) = \frac{1}{\Gamma(\alpha)} \int_a^t (t-s)^{\alpha-1} f(s)\, ds \qquad t > 0 \tag{3}$$

Definition 3 (Riemann-Liouville Fractional Derivative [24]) The fractional derivative of order $\alpha \in (n-1, n)$ of $f(t)$ is defined by taking fractional integral of order $(n-\alpha)$, and then take nth derivative as follows:

$$D_*^{\alpha} f(t) = D_*^n I_a^{n-\alpha} f(t) \tag{4}$$

where

$$D_*^n = \frac{d^n}{dt^n}, \qquad n = 1, 2, \ldots \tag{5}$$

Then, we define the Grünwald-Letinkov definition of fractional derivative:

Definition 4 (Grünwald-Letinkov Fractional Derivative, Grünwald and Letinkov 1872 [25]) The Grünwald-Letinkov definition of fractional derivative of a function generalizes the notion of backward difference quotient of integer order. In this case $\alpha = 1$ if the limit exists the Grünwald-Letinkov fractional derivative is the left derivative of the function. The Grünwald-Letinkov fractional derivative of order α of the function $f(x)$ is define as

$$D_x^{\alpha} f(x) = \lim_{N\to\infty} \frac{(x-a)^{-\alpha}}{\Gamma(-\alpha)} \sum_{j=0}^{N-1} \frac{\Gamma(j-\alpha)}{\Gamma(j+1)} f(x - j[\frac{x-a}{N}]) \tag{6}$$

If $\alpha = -1$, we have a Riemann sum. If $\alpha = 1$, then we have

$$\lim_{N\to\infty} \frac{f(x) - f(x - [\frac{x-a}{N}])}{[\frac{x-a}{N}]} \tag{7}$$

which is left derivative of the function f at x.

In 2015, Caputo and Fabrizio proposed a new derivative with fractional order [26] with two different forms, however, the first representation was proposed by Joseph Liouville in 1832 [24, 27]. Caputo's representation which is a modification of the Riemann Liouville definition and can be used for initial value problems, has the following definition:

Definition 5 (Caputo-Fractional Derivative, Caputo (1967) [26]) The fractional derivative of order $\alpha \in (n-1, n)$ of $f(t)$ is defined by taking nth derivative, and then take a fractional integral of order $(n - \alpha)$

$$D^{\alpha} f(t) = I_a^{n-\alpha} D_*^n f(t), \qquad n = 1, 2, \ldots \tag{8}$$

Therefore, Caputo derivative of order α has the form:

$$D^{\alpha} f(t) = \frac{1}{\Gamma(n-\alpha)} \int_0^t \frac{f^n(s)}{(t-s)^{\alpha-n+1}} ds \tag{9}$$

where $n - 1 < \alpha < n$, n is an integer, and f^n is nth derivative of $f(s)$.

Definition of time-fractional derivative of a function $f(t)$ at $t = t_n$ includes an integration and calculating time-fractional derivative that requires all the past history, that is, all the values of $f(t)$ from $t = 0$ to $t = t_n$.

A common approach to solve the problems with Caputo fractional derivative is using the Laplace transform. The Laplace transform formula for Caputo fractional derivative has the following form

$$\mathcal{L}\{D^{\alpha} f(t)\} = z^{\alpha} F(z) - \sum_{k=0}^{n-1} z^{\alpha-k-1} f^{(k)}(0) \tag{10}$$

where $f^{(k)}(0)$ is the initial conditions of $f(t)$. If we neglect the initial conditions, we get

$$\mathcal{L}\{D^{\alpha} f(t)\} = z^{\alpha} F(z) \tag{11}$$

From (11) and using the definition of Laplace transform of Riemann Liouville fractional integral:

$$\mathcal{L}\{I^{\alpha} f(t)\} = z^{-\alpha} F(z) \tag{12}$$

we get the definition of general differintegral fractional operator as:

$$\mathcal{L}\{D^{\pm\alpha} f(t)\} = z^{\pm\alpha} F(z) \tag{13}$$

It has been proved that for the most analytic functions, Grünwald-Letnikov fractional derivative is identical to Caputo fractional derivative. The difference between the Grünwald-Letnikov definition and Caputo definition appears when dealing with constant function. For example, for a constant function, the Caputo fractional derivative is zero while its Riemann–Liouville fractional derivative is not. In the application, Caputo fractional derivative has been used for initial value fractional ordinary differential equations. For the case α is an integer, the fractional derivative would be identical to the integer derivative and we can conclude that fractional calculus is a kind of interpolation of the integer calculus. One important fact about fractional operators such as fractional integral and fractional derivative is that there has not been developed any acceptable physical and or geometrical interpretation for these operators during 300 years [27].

Here, we want to explore the behaviors of the solutions of three pharmacokinetic compartmental models and their corresponding fractional order forms. We organize this study as follows: to drive the fractional-order compartmental models, we apply the nonstandard finite difference (NSFD) method since they provide better results compare to traditional standard finite difference (SFD) methods and then we discretize these pharmacokinetic compartmental models using the Grünwald-Letinkov discretization method. We use the numerical methods to demonstrate (ADME) process for any drug or NPs in different compartments in both original and fractional order models. By considering the fractional order framework in pharmacokinetic modeling, we can explain all the possible geometric mechanisms underlying drug distribution in different tissues.

2 Pharmacokinetic Two Compartmental Model

To demonstrate the uneven transition of drug and NPs in body, we use a bicompartmental model which follows a biexponential distribution to describe disposition of drugs and NPs. We start with the simplest case when we assume that drugs and NPs distribute from second compartment or capillary bed into the third compartment or tissue compartment but they would be eliminated from the second compartment (see the schematic diagram in Fig. 1). Here, transfer rate constants k_{23} and k_{32} illustrate the reversible transfer of drugs and NPs between compartment two and compartment three.

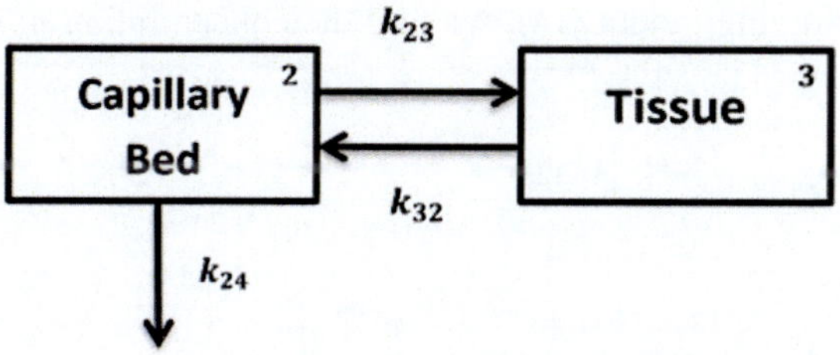

Fig. 1 Schematic diagram of a simple bi-compartmental model for drug distribution, where k_{23} and k_{32} represent the rate of drug distribution between compartment two or capillary bed and three or tissue, k_{24} is the rate of elimination from vascular capillary bed to the venous

In general, this is a simplification of absorption, distribution, metabolism, and excretion procedure in the whole body. In this case, we consider that drug or NPs concentration initially decreases very fast and they distribute rapidly into the tissue compartment. Here, tissue compartment may be contained several organs in body. After this phase, there would be an equilibrium phase for drug and NPs concentration in which concentration of them decreases so slowly, and therefore we only have elimination of these particles.

In this bicompartmental model, we consider a transfer rate of first order between second and third compartments and also we assume that elimination of drug from the second compartment or capillary bed follows a rate of first order, and lastly, we ignore any other metabolism in capillary bed.

To demonstrate distribution of drugs and NPs in Fig. 1, we write the following kinematic equations,

$$\begin{cases} \frac{1}{V_2}\frac{dA_2}{dt} = \frac{-k_{23}\,A_2}{V_2} + \frac{k_{32}\,A_3}{V_3} - \frac{k_{el}\,A_2}{V_2} \\ \frac{1}{V_3}\frac{dA_3}{dt} = \frac{-k_{32}\,A_3}{V_3} + \frac{k_{23}\,A_2}{V_2} \end{cases} \tag{14}$$

where, A_2 and A_3 are the drug or NPs concentrations in compartment two or capillary bed and compartment three or tissue respectively, and V_2 and V_3 represent the volume of compartment two and three respectively.

We assume the following boundary conditions for the Eqs. (14) at $t = 0$,

$$A_2 = \frac{\text{Dose}}{V_2}, \quad A_3 = 0$$

Therefore, after solving model (14), we find the concentration of drug in capillary bed, A_2, and tissue, A_3, as the form,

$$\begin{cases} A_2 = \dfrac{\text{Dose}\left[(k_{32} - r_1)\, e^{-r_1 t} - (k_{32} - r_2)\, e^{-r_2 t}\right]}{V_2(r_2 - r_1)} \\ \\ A_3 = \dfrac{\text{Dose}\, k_{23}\left[e^{-r_2 t} - e^{-r_1 t}\right]}{V_2(r_1 - r_2)} \end{cases} \tag{15}$$

where,

$$\begin{cases} r_1 = \dfrac{1}{2}\left[(k_{23} + k_{32} + k_{el}) + (\sqrt{(k_{23} + k_{32} + k_{el})^2 - 4\, k_{32}\, k_{el}})\right] \\ \\ r_2 = \dfrac{1}{2}\left[(k_{23} + k_{32} + k_{el}) - (\sqrt{(k_{23} + k_{32} + k_{el})^2 - 4\, k_{32}\, k_{el}})\right] \end{cases}$$

The first order constants r_1 and r_2 determine if drug distribution and removal take place slow or fast and from (16) we can see that they depend on constant rates k_{23}, k_{32}, and k_{el}.

The following equality describes the relationships between r_1 and r_2 and three rates k_{23}, k_{32} and k_{el};

$$\frac{r_1 + r_2}{r_1\, r_2} = \frac{k_{23} + k_{32} + k_{el}}{k_{32}\, k_{el}}$$

Using mathematical modeling not only helps us to find the concentration of drug or NPs in different compartments, but also helps to approximate distribution rates between different compartments. The concentration of drug in capillary bed of model (14) can be simply obtained by adding the amount of drug that is sent to the tissue and the amount of drug that is eliminated from capillary:

$$A_2 = A(\text{Distribution}) + A(\text{Elimination})$$

Thus,

$$A_2 = \eta_1\, e^{-r_1 t} + \eta_2\, e^{-r_2 t}$$

where, A_2 consists of two phases, first the exponential term $\eta_1\, e^{-r_1 t}$ related to distribution phase and $\eta_2\, e^{-r_2 t}$ related to elimination phase of drug from compartment two. In model (14), because distribution is faster than removal, therefore, we assume that the exponential term $\eta_1\, e^{-r_1 t}$ is zero and we rewrite A_2 as

$$A_2 = \eta_2\, e^{-r_2 t}$$

If we convert it to log form, we have

$$\log A_2 = \log \eta_2 - \frac{r_2 t}{2.303}$$

Moreover, for this simple case, we can easily evaluate pharmacokinetic parameters, k_{23}, k_{32} and k_{el}.

2.1 Grünwald-Letinkov Approximation for Bicompartmental Model (14)

We define the fractional differential as the following form [28, 29]

$$D^{\gamma} A(t) = f(t, A(t)), \quad A(t_0) = A_0$$

where $\gamma > 0$ represents the order of derivative and D^{γ} denotes the fractional derivative which is given by:

$$D^{\gamma} A(t) = J^{k-\gamma} D^k A(t)$$

where $\gamma \in (k-1, k]$, for $k = 1, 2, \ldots$ and integral operator J^k called the Riemann-Liouville of kth-order which is obtained by the following formula

$$J^k A(t) = \frac{1}{\Gamma(k)} \int_0^t (t-\tau)^{(k-1)} A(\tau)\, d\tau, \quad t > 0$$

where $\Gamma(.)$ denotes the gamma function.

To apply the Micken's (NSFD), we need to find the fractional order derivative using the Grünwald- Letinkov (G-L) approximation for model equations (14) as the form [30–32]

$$D^{\gamma} A(t) = \lim_{s \to 0} s^{-\gamma} \sum_{i=0}^{T} (-1)^i \binom{\gamma}{i} A(t - i\, s) \tag{16}$$

where $T = [t]/s$ and [.] used to show the integer value and s represents the step size. Thus, Eq. (16) would be discretized as

$$\sum_{i=0}^{T} C_i^{\gamma} A(t_{k-i}) = f(t_k, A(t_k)) \tag{17}$$

where $t_k = k\,s$ and C_i^γ are the coefficients for (G-L) approximation written as

$$C_i^\gamma = \left[\frac{i-1-\gamma}{i}\right] C_{i-1}^\gamma, \quad C_0^\gamma = s^{-\gamma} \quad i = 1, 2, \ldots$$

Next we introduce the non-standard finite difference schemes.

2.2 Non-standard Discretization of Bicompartmental Model (14)

To discretize a system of differential equations, ordinary differential equations (ODEs) and or partial differential equations (PDEs), one may apply the Mickens NSFD discretization method which is more flexible in construction rather than standard finite difference method and therefore has better performance. This method checks the positivity of solutions and is concerned about boundedness and monotonicity of them. Another advantage of using NSFD schemes is their ability to preserve the structure and properties of the system of differential equations and therefore, we apply NSFD schemes on the general compartmental model of the form:

$$\frac{dA}{dt} = f(A) \tag{18}$$

However, to apply the non-standard scheme we need to check that if non-local approximation is used and or we need to have a non traditional discretization of derivatives and also we may need to use a non-negative function $\Phi(h) = s + O(s^2)$. To apply NSFD scheme, we consider a grid $t_k = t_0 + k\,s$, such that $s > 0$, and we approximately write the discretized function A as $A_k \approx A(t_k)$. Next, we discretize (18):

$$\frac{dA}{dt} = \frac{A_{k+1} - A_k}{\Phi(s)} + O(\Phi(s)), \quad \text{when} \quad s \to 0 \quad \Longrightarrow \quad \frac{dA}{dt} \approx \frac{A_{k+1} - A_k}{\Phi(s)} \tag{19}$$

where real valued $\Phi(s)$ as a function of the step size s needs to satisfy the following properties:

$$\Phi(s) = s + O(s^2), \quad \Phi(s) \in (0, 1), \quad \forall s \in (0, \infty) \tag{20}$$

Here, the equality (19) is equivalent with the integer order derivative because:

$$\begin{aligned}\frac{dA}{dt} &= \lim_{s\to 0}\left[\frac{A(t+s)-A(t)}{\Phi(s)} + O(\Phi(s))\right]\\ &= \lim_{s\to 0}\left[\frac{A(t+s)-A(t)}{s}\right]\lim_{s\to 0}\left[\frac{s}{\Phi(s)}\right] + \lim_{s\to 0} O(\Phi(s)) = \dot{A}(t)\end{aligned}$$

As $s \to 0$ the discrete form in (19) converges to its associated continuous derivative. NSFD methods are convergent without any restriction related to step size s but this is not always true for SFD methods which depend on the step size s. Moreover, when we discretize a system using NSFD method, if the original system is persistent, and solutions are stable and convergent, these properties remain the same after discretization, but not for the case we use SFD to discretize the system of differential equations.

2.3 *Fractional Bicompartmental Model*

Finally to apply the Mickens NSFD method, we substitute the step size s by $\Phi(s)$. Next, we use the G-L technique to discretize the Eqs. (16). We consider $A_2(t_k) = A_2^k$, $A_3(t_k) = A_3^k$, we have:

$$\begin{cases} \dfrac{1}{V_2}\sum_{i=0}^{k+1} C_i^{\gamma} A_2^{k+1-i} = \dfrac{-k_{23}A_2}{V_2} + \dfrac{k_{32}A_3}{V_3} - \dfrac{k_{el}A_2}{V_2} \\ \dfrac{1}{V_3}\sum_{i=0}^{k+1} C_i^{\gamma} A_3^{k+1-i} = \dfrac{-k_{32}A_3}{V_3} + \dfrac{k_{23}A_2}{V_2} \end{cases} \tag{21}$$

After simplification, the fractional-order system which is linear and time-invariant has the following form:

$$\begin{cases} A_2^{k+1} = \left[\dfrac{k_{32}A_3}{V_3} - \dfrac{\sum_{i=1}^{k+1} C_i^{\gamma} A_2^{k+1-i} + k_{23}A_2 + k_{el}A_2}{V_2}\right]\dfrac{V_2}{C_0^{\gamma}} \\ A_3^{k+1} = \left[\dfrac{k_{23}A_2}{V_2} - \dfrac{\sum_{i=1}^{k+1} C_i^{\gamma} A_3^{k+1-i} + k_{32}A_3}{V_3}\right]\dfrac{V_3}{C_0^{\gamma}} \end{cases} \tag{22}$$

We have demonstrated the solutions of the fractional order system (22) for different orders in Figs. 2 and 3.

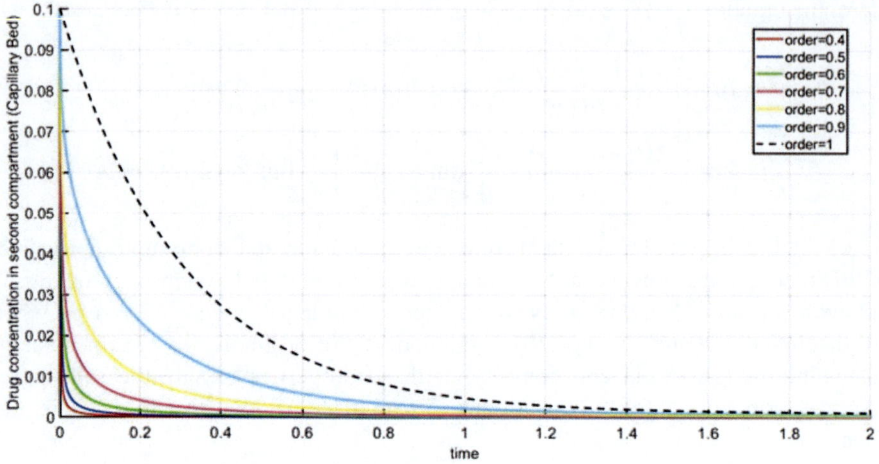

Fig. 2 Evolution of drug concentration in capillary bed in fractional-order bi-compartmental model (22) for orders $\gamma = 0.4$–1. In this model, the constant rates k_{23}, k_{32}, and k_{el} determine the fast drug distribution and slow drug elimination

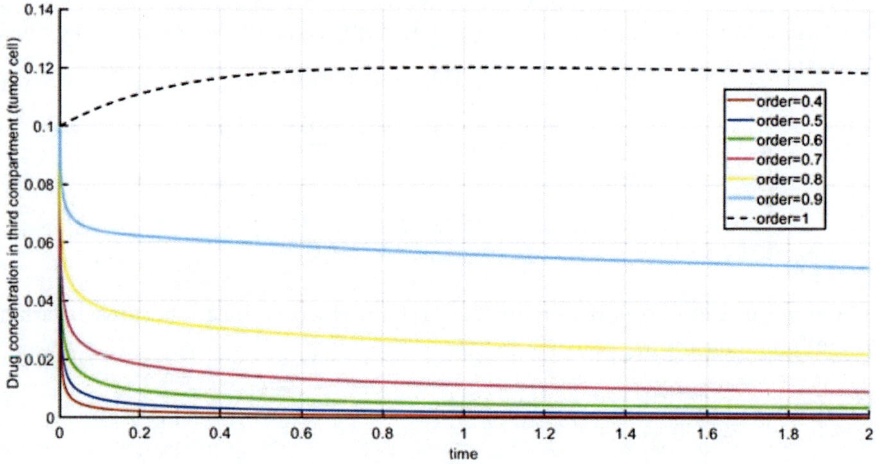

Fig. 3 Evolution of drug concentration in tissue in fractional-order bi-compartmental model (22) for orders $\gamma = 0.4$–1

3 Bicompartmental Model with NPs Infusion

For this case, we have the following two compartmental model (see Fig. 4)

$$\begin{cases} \frac{1}{V_2}\frac{dA_2}{dt} = \frac{k_{12}\,A_1}{V_1} + \frac{-k_{23}\,A_2}{V_2} + \frac{k_{32}\,A_3}{V_3} - \frac{k_{el}\,A_2}{V_2} \\ \frac{1}{V_3}\frac{dA_3}{dt} = \frac{-k_{32}\,A_3}{V_3} + \frac{k_{23}\,A_2}{V_2} \end{cases} \quad (23)$$

where, A_1 is the concentration of NPs in compartment one, V_1 is the volume of compartment one, and k_{12} is NPs infusion rate. For model (23), we can derive A_2 as the form,

$$A_2 = \frac{k_{12}}{V_2\,k_{el}}\left[1 + \frac{k_{el} - r_2}{r_2 - r_1}\,e^{-r_1 t} - \frac{k_{el} - r_1}{r_1 - r_2}\,e^{-r_2 t}\right] \quad (24)$$

At steady state, when $t \to \infty$, $e^{-r_1 t}$ and $e^{-r_2 t}$ converge to zero and we have,

$$A_2 = \frac{k_{12}}{V_2\,k_{el}} = \frac{\text{infusion rate}}{\text{clearance}}$$

We define the Grünwald-Letinkov approximation as before to apply the non-standard finite difference Micken's scheme and finding the following fractional

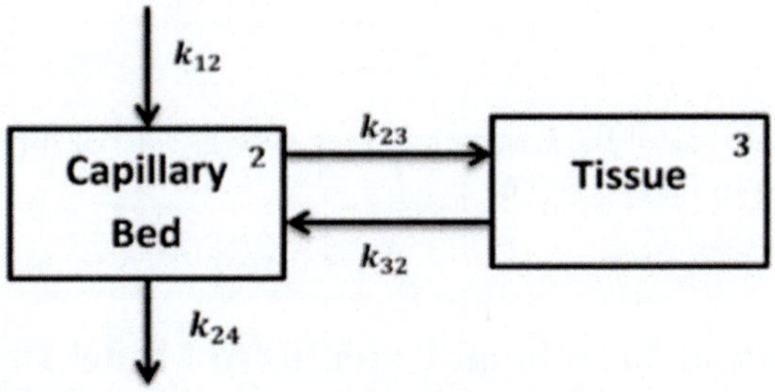

Fig. 4 Schematic diagram of a simple bi-compartmental model for drug or NPs distribution, where k_{12} describe the transition rate from artery to capillary, k_{23} and k_{32} represent the rate of drug distribution between compartment two or capillary bed and three or tissue, k_{24} is the rate of elimination from vascular capillary bed to the venous

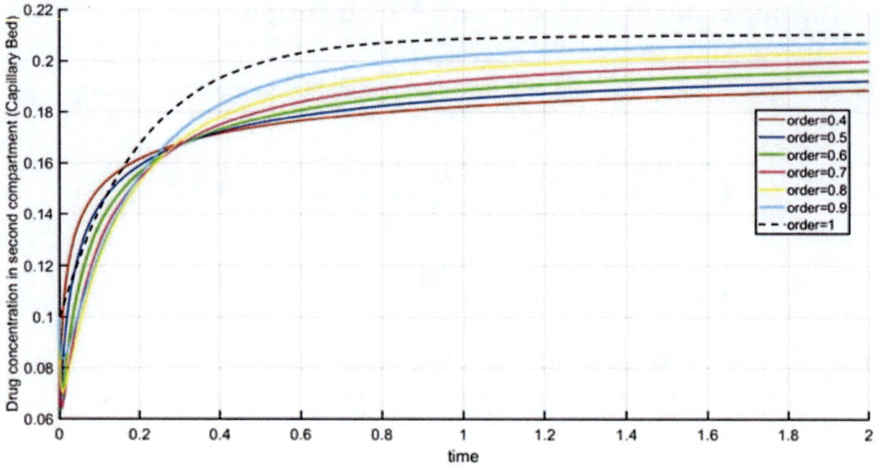

Fig. 5 Evolution of drug concentration in capillary bed in fractional-order bi-compartmental model (25) for orders $\gamma = 0.4$–1

order system for (23):

$$\begin{cases} A_2^{k+1} = \left[\dfrac{k_{12}\,A_1}{V_1} + \dfrac{k_{32}\,A_3}{V_3} - \dfrac{\sum_{i=1}^{k+1} C_i^{\gamma}\,A_2^{k+1-i} + k_{23}\,A_2 + k_{el}\,A_2}{V_2} \right] \dfrac{V_2}{C_0^{\gamma}} \\ \\ A_3^{k+1} = \left[\dfrac{k_{23}\,A_2}{V_2} - \dfrac{\sum_{i=1}^{k+1} C_i^{\gamma}\,A_3^{k+1-i} + k_{32}\,A_3}{V_3} \right] \dfrac{V_3}{C_0^{\gamma}} \end{cases} \tag{25}$$

The numerical solutions of the fractional order system (25) of different orders have been demonstrated in Figs. 5 and 6.

4 Applications of Fractional Calculus to Model Drug Diffusion in a Three Compartmental Pharmacokinetic Model

Since all blood vessels in body are covered with Endothelial cells, we have added a new compartment to the previous two compartmental model to have a more realistic model. These cells based on their size and location play different roles in our body. For example, some of them carry small size molecules and or specific hormones such as insulin, the others are effective in regulation of blood pressure. However, all of these cells have a common role that is building a wall between blood cells and

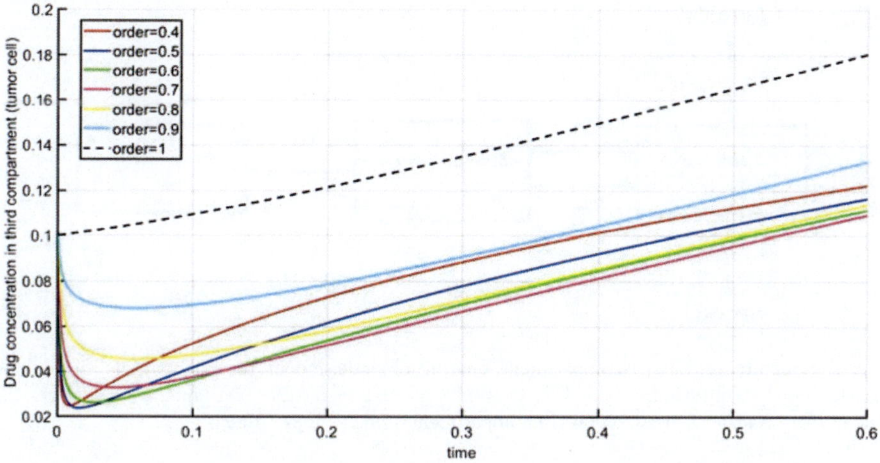

Fig. 6 Evolution of drug concentration in tissue in fractional-order bi-compartmental model (25) for orders $\gamma = 0.4$–1

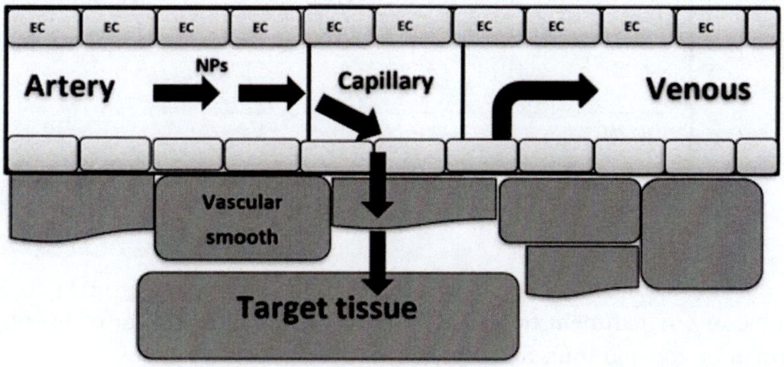

Fig. 7 EC or endothelial cells as a barrier between body vessels and different tissues facilitate the transportation of NPs from capillary to tissue

other tissue cells [33]. For simplicity, we ignore their interactions with surrounding cells.

We consider a new compartmental model with three compartments to represent (ADME) of drugs or NPs as it is demonstrated in Fig. 7. In this case, we assume that we have three phases; absorption and distribution, and leakage or elimination. Drugs enter from artery which is the first compartment to the capillary or second compartment with absorption rate k_{12}. Next, we have distribution to the third compartment or Endothelial cells with a rate k_{23}. Then, we have transition of drugs

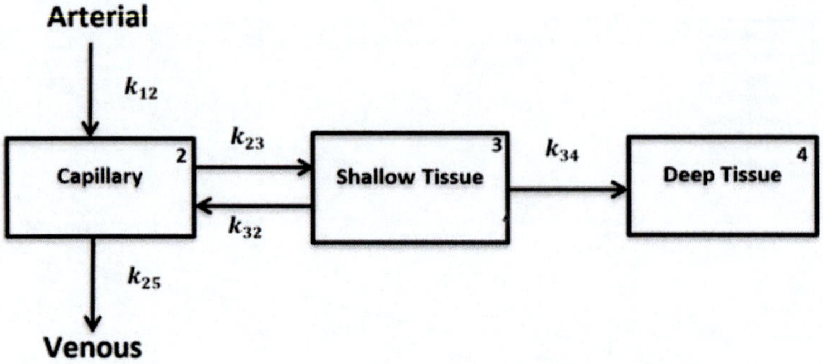

Fig. 8 Structure of the tri-compartment pharmacokinetic model for nanoparticle disposition considering NPs infusion, where k_{23}, k_{32} and k_{34} are transfer rate constants, k_{25} describes the rate of mass transfer from vascular (2) compartment to the venous effluent

or NPs into tissue with transfer rate k_{34}:

$$\begin{cases} \dfrac{1}{V_2}\dfrac{dA_2}{dt} = \dfrac{k_{12}\,A_1}{V_1} + \dfrac{-k_{23}\,A_2}{V_2} + \dfrac{k_{32}\,A_3}{V_3} - \dfrac{k_{el}\,A_2}{V_2} \\ \\ \dfrac{1}{V_3}\dfrac{dA_3}{dt} = \dfrac{-k_{32}\,A_3}{V_3} + \dfrac{k_{23}\,A_2}{V_2} - \dfrac{k_{34}\,A_3}{V_3} \\ \\ \dfrac{1}{V_4}\dfrac{dA_4}{dt} = \dfrac{k_{34}\,A_3}{V_3} \end{cases} \tag{26}$$

where, A_4 is the concentration of NPs in deep tissue or compartment four, V_4 is the volume of compartment four, and k_{34} is the reversible transfer of NPs between compartment three and four. See Fig. 8.

After solving model (26) for A_3 and A_4 we have:

$$A_2 = M_2 \tag{27}$$

$$A_3 = \frac{M_2\,k_{23}}{k_{32}+k_{34}}\left[1 - e^{-(k_{32}+k_{34})t}\right] + A_3(0)\,e^{-(k_{32}+k_{34})(t-\Gamma)} \tag{28}$$

$$A_4 = \frac{M_2 k_{23} k_{34}}{k_{32}+k_{34}}\left[t - \frac{1-e^{-(k_{32}+k_{34})t}}{k_{32}+k_{34}}\right] + \frac{k_{34}A_3(0)}{k_{32}+k_{34}}\left[1 - e^{-(k_{32}+k_{34})t(t-\Gamma)}\right] + A_4(0) \tag{29}$$

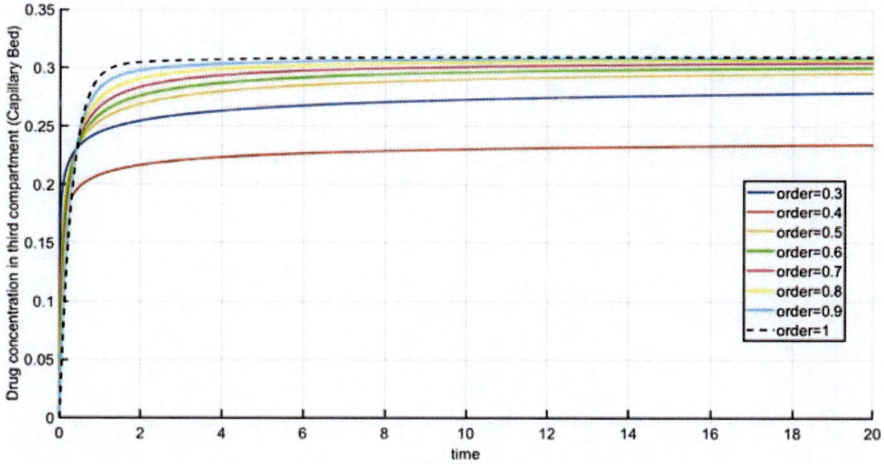

Fig. 9 Evolution of drug concentration in capillary bed in fractional-order tri-compartmental model (30) for orders $\gamma = 0.3$–1

As before, we apply the NSFD Micken's method and we find the following fractional order system for (26):

$$\begin{cases} A_2^{k+1} = \left[\dfrac{k_{12}\, A_1}{V_1} + \dfrac{k_{32}\, A_3}{V_3} - \dfrac{\sum_{i=1}^{k+1} C_i^{\gamma}\, A_2^{k+1-i} + k_{23}\, A_2 + k_{el}\, A_2}{V_2} \right] \dfrac{V_2}{C_0^{\gamma}} \\ A_3^{k+1} = \left[\dfrac{k_{23}\, A_2}{V_2} - \dfrac{\sum_{i=1}^{k+1} C_i^{\gamma}\, A_3^{k+1-i} + k_{32}\, A_3 + k_{34}\, A_3}{V_3} \right] \dfrac{V_3}{C_0^{\gamma}} \\ A_4^{k+1} = \left[\dfrac{k_{34}\, A_3}{V_3} - \dfrac{\sum_{i=1}^{k+1} C_i^{\gamma}\, A_4^{k+1-i}}{V_4} \right] \dfrac{V_4}{C_0^{\gamma}} \end{cases} \tag{30}$$

We have demonstrated the numerical solutions of the fractional order system (30) of different orders in Figs. 9, 10, and 11.

5 Discussion

Pharmacokinetic compartmental models play an important role to study the chemical or biochemical reactions of drugs and their distribution entire body and to measure the rates of distribution in different tissues. Pharmacokinetic compartmental models have been used widely to demonstrate how chemical compounds such as

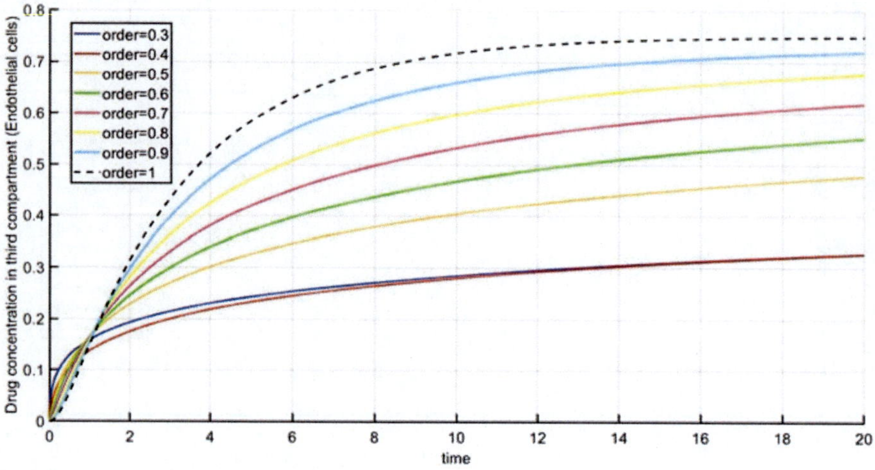

Fig. 10 Evolution of drug concentration in endothelial cells in fractional-order tri-compartmental model (30) for orders $\gamma = 0.3$–1

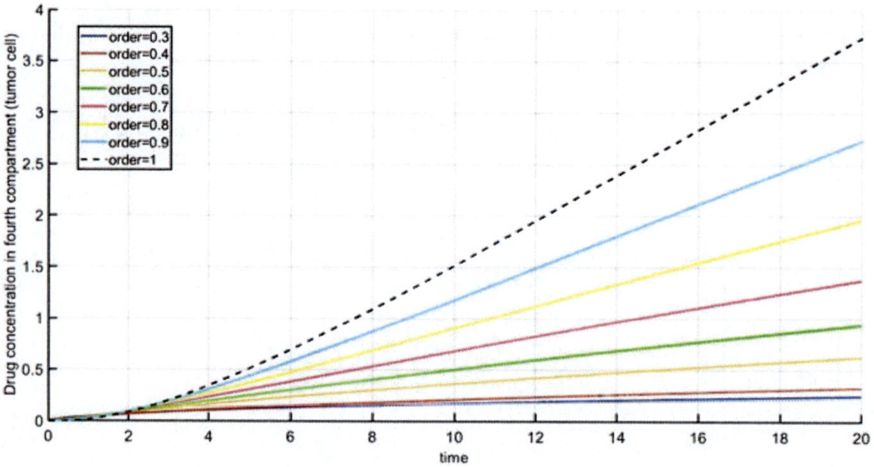

Fig. 11 Evolution of drug concentration in tissue in fractional-order tri-compartmental model (30) for orders $\gamma = 0.3$–1

drugs interact with biological processes in different tissues. Pharmacokinetic models are usually expressed as systems of differential equations. There are many studies related to cancer therapy that use the compartmental models to approximate the concentration of drugs and NPs in tumor cells and other tissues. These models are concerned about different pathways to administer drug and compare them to find the concentration of chemicals in different tissue that helps to explore the best route of drug administration for different diseases and finally better treatment. Fractional calculus is a new approach for modeling physical and biological phenomena with memory or aftereffects. Fractional calculus uses differential and integral operators with non-integer orders to study the complexity and behavior of non-linear physical and biological systems which display some degrees of fractionality or fractality. Using fractional derivatives and integrals we are able to explain the memory and patrimonial effect in different events and successive processes in real world phenomena. Since the analytical and explicit solutions of linear and nonlinear fractional ordinary differential equations (FODEs) cannot be easily obtained, someone needs to use computational methods and computer simulations to approximate the solutions of these systems. Fractional order modeling in biology and specifically physiology helps to explore the complexity of different processes inside human body and to characterize and measure different organs complication using fractal geometry and fractional calculus tools. Moreover, since tumor cells interactions are systems with memory, therefore FODEs are better candidates to model cancer phenomena rather than ordinary and delay differential equations. In the current research, we have explored the process of drug absorption, distribution, and elimination inside the body using three different pharmacokinetic compartmental models. These models include physiological parameters which describe the process of ADME of drug or NPs in different body organs. At first, we have considered integer order compartmental models to display the process of drug distribution in different tissues tissue. We tried to write the mathematical models using the schematic diagrams that makes it easier to study these biological processes. Next, we have used the Grünwald-Letinkov discretization method to discretize these pharmacokinetic compartmental models. Finally, to obtain the fractional-order compartmental model for each case, we have applied the nonstandard finite difference method (NSFD) Micken's technique. To compare the solutions of fractional order systems of different orders to the corresponding traditional integer order models, we have carried out some numerical methods (by preserving the same biological parameters) since the solutions of fractional order models may not be explicitly obtained. The graphical presentations of numerical results showed that NSFDM is easy and convenient to implement, and effective enough for solving efficiently the proposed models. Therefore, these numerical results would be a good starting point to find out the best mathematical model to demonstrate the absorption, distribution, and excretion of drug or NPs in different body organs. By considering the fractional order modeling approach, we could improve the capabilities of the integer order modeling into the generalized domain of fractional calculus and we could derive all the possible geometric mechanisms underlying the absorption, distribution, metabolism, and excretion (ADME) of drug or nanoparticles (NPs) inside body. Likewise, we have used the order of fractional derivative as a new control parameter to extract the variety of governing classes on the non-local behaviors of these models, while, the integer order framework only could study the local behaviors in the integer order domain. The fractional-order approach proposed in this study required to be developed a bit further before its general adoption in the pharmacokinetic modeling context. In principle, it is worth noting the limitations of this work. Due to the non availability of real data, future work is required to validate and verify the reliability of the proposed fractional order models. Using the real data would considerably increase the credibility of this framework for drug diffusion modeling. It is straightforward to apply fractional calculus on the pharmacokinetic compartmental model representations proposed here, however we may require to find an explicit agreement on the exact physiological relevance of the new parameter, the fractional differentiation order γ. Although it is evident from numerical simulations that different classes of these compartmental models and their dynamics have mainly been presented and captured by different fractional-order γ value, it would be of great significance to find the ranges of γ value using experimental data for clinical applications.

Acknowledgments The author would like to thank the reviewers, whose comments really helped to improve the manuscript.

References

1. Lin, Z., Gehring, R., Mochel, J.P., Lave, T., Riviere, J.E.: Mathematical modeling and simulation in animal health–Part II: principles, methods, applications, and value of physiologically based pharmacokinetic modeling in veterinary medicine and food safety assessment. J. Vet. Pharmacol. Therap. **39**, 421–438 (2016)
2. Brown, R.P., Delp, M.D., Lindstedt, S.L., Rhomberg, L.R., Beliles, R.P.: Physiological parameter values for physiologically based pharmacokinetic models. Toxicol. Ind. Health **13**, 407–484 (1997)
3. Azizi, T., Mugabi, R.: Global sensitivity analysis in physiological systems. Appl. Math. **11**, 119–136 (2020)
4. Azizi, T.: Mathematical Modeling with Applications in Biological Systems, Physiology, and Neuroscience. Kansas State University (2021)
5. Pitchaimani, A., Nguyen, T.D.T., Marasini, R., Eliyapura, A., Azizi, T., Jaberi-Douraki, M. and Aryal, S.: Biomimetic natural killer membrane camouflaged polymeric nanoparticle for targeted bioimaging. Adv. Funct. Mater. **29**, 1806817 (2019)
6. Riviere, J.E., Jaberi-Douraki, M., Lillich, J., Azizi, T., Joo, H., Choi, K., Thakkar, R. and Monteiro-Riviere, N.A.: Modeling gold nanoparticle biodistribution after arterial infusion into perfused tissue: effects of surface coating, size and protein corona. Nanotoxicology **12**, 1093–1112, (2018)
7. Marino, S., Hogue, I.B., Ray, C.J. and Kirschner, D.E.: A methodology for performing global uncertainty and sensitivity analysis in systems biology. J. Theor. Biol. **254**, 178–196 (2008)
8. Blower, S.M., Dowlatabadi, H.: Sensitivity and uncertainty analysis of complex models of disease transmission: an HIV model, as an example. Int. Stat. Rev./Revue Internationale de Statistique, JSTOR **62**(2), 229–243 (1994)
9. Zi, Z.: Sensitivity analysis approaches applied to systems biology models. IET Syst. Biol. **5**, 336–346 (2011)
10. Dalberg, J., Gimenez, H., Keeley, A., Azizi, T., Xi, X. and Jaberi-Douraki, M.: Local and global dynamics of discrete type 1 diabetes model (2019)
11. Zhao, P., Zhang, L., Grillo, J.A., Liu, Q., Bullock, J.M., Moon, Y.J., Song, P., Brar, S.S., Madabushi, R., Wu, T.C., et al.: Applications of physiologically based pharmacokinetic (PBPK) modeling and simulation during regulatory review. Clin. Pharmacol. Therap. **89**, 259–267 (2011)

12. Barrett, J.S., Della Casa Alberighi, O., Läer, S., Meibohm, B.: Physiologically based pharmacokinetic (PBPK) modeling in children. Clin. Pharmacol. Therap. **92**, 40–49 (2012)
13. Wagner, C., Zhao, P., Pan, Y., Hsu, V., Grillo, J., Huang, S.M., Sinha, V.: Application of physiologically based pharmacokinetic (PBPK) modeling to support dose selection: report of an FDA public workshop on PBPK. CPT: Pharmacom. Syst. Pharmacol. **4**, 226–230 (2015)
14. Hilfer, R., et al.: Applications of Fractional Calculus in Physics. World Scientific Singapore, pp. 497–528 (2000)
15. Rihan, F.A., Baleanu, D., Lakshmanan, S. and Rakkiyappan, R.: On fractional SIRC model with salmonella bacterial infection. In: Abstract and Applied Analysis. Hindawi (2014)
16. Rihan, F.A., Lakshmanan, S., Hashish, A.H., Rakkiyappan, R., Ahmed, E.: Fractional-order delayed predator–prey systems with Holling type-II functional response. Nonlinear Dynam. **80**, 777–789 (2015)
17. Rihan, F.A., Hashish, A., Al-Maskari, F., Hussein, M.S., Ahmed, E., Riaz, M.B., Yafia, R.: Dynamics of tumor-immune system with fractional-order. J. Tumor Res. **2**, 109–115 (2016)
18. Gómez-Aguilar, J.F., López-López, M.G., Alvarado-Martínez, V.M., Baleanu, D., Khan, H.: Chaos in a cancer model via fractional derivatives with exponential decay and Mittag-Leffler law. Entropy **19**, 681 (2017)
19. Zeinadini, M., Namjoo, M.: Approximation of fractional-order Chemostat model with nonstandard finite difference scheme. Hacettepe J. Math. Stat. **46**, 469–482 (2017)
20. Gorenflo, R., Mainardi, F.: Fractional Calculus, Fractals and Fractional Calculus in Continuum Mechanics. Springer, pp. 223–276 (1997)
21. Mainardi, F.: Fractional Calculus, Fractals and Fractional Calculus in Continuum Mechanics. Springer, pp. 291–348 (1997)
22. Machado, J.T., Kiryakova, V., Mainardi, F.: Recent history of fractional calculus. Commun. Nonlinear Sci. Numer. Simul. **16**, 1140–1153 (2011)
23. Carpinteri, A., Mainardi, F.: Fractals and Fractional Calculus in Continuum Mechanics. Springer (2014)
24. Liouville, J.: Memoire sur quelques questiona de geometrie et de mechanique, et sur un nouveau genre de calcul pour resoudre ces questions. J. Ecole Polytech. **13**, 16–18 (1831)
25. Oldham, K., Spanier, J.: The Fractional Calculus Theory and Applications of Differentiation and Integration to Arbitrary Order. Elsevier (1974)
26. Caputo, M., Fabrizio, M.: A new definition of fractional derivative without singular kernel. Progr. Fract. Differ. Appl. **1**(2), 1–13 (2015)
27. Atangana, A., Baleanu, D.: Application of fixed point theorem for stability analysis of a nonlinear Schrodinger with Caputo-Liouville derivative. Filomat, JSTOR **31**, 2243–2248 (2017)
28. Miller, K.S., Ross, B.: An Introduction to the Fractional Calculus and Fractional Differential Equations. Wiley (1993)
29. Podlubny, I.: Fractional Differential Equations: An Introduction to Fractional Derivatives, Fractional Differential Equations, to Methods of Their Solution and Some of Their Applications. Elsevier (1998)
30. Mickens, R.E.: Nonstandard Finite Difference Models of Differential Equations. World Scientific (1994)
31. Mickens, R.E.: Nonstandard finite difference schemes for reaction-diffusion equations. Numer. Methods Partial Differential Equations Int. J. **15**, 201–214 (1999)
32. Mickens, R,.E.: A nonstandard finite difference scheme for a Fisher PDE having nonlinear diffusion. Comput. Math. Appl. **45**, 429–436 (2003)
33. Lee, H.A., Imran, M., Monteiro-Riviere, N.A., Colvin, V.L., Yu, W.W., Riviere, J.E.: Biodistribution of quantum dot nanoparticles in perfused skin: evidence of coating dependency and periodicity in arterial extraction. Nano Lett. **7**, 2865–2870 (2007)

Merging On-chip and In-silico Modelling for Improved Understanding of Complex Biological Systems

Francesca Romana Bertani, Gabriella Bretti, Luca Businaro, Adele De Ninno, Annamaria Gerardino, and Roberto Natalini

Abstract In recent years an increasing interest is registered in the direction of developing techniques to combine experimental data and mathematical models, in order to produce systems, i.e., in silico models, whose solutions could reproduce and predict experimental outcomes. Indeed, the success of informed models is mainly due to the consistent improvements in computational abilities of the machines and in imaging techniques that allow a wider access to high spatial and temporal resolution data. Here we present an interdisciplinary work in the framework of Organs-on-chip (OoC) technology, and, more precisely, in Cancer-on-Chip (CoC) technology.

1 Introduction

The observation and measurement of processes occurring in our body are at the core of biological and medical applications but obtaining data at the right time and spatial scale is a formidable challenging task. Simplifying, three main factors play against the obtainment of a full description:

(1) gathering information inside a body is difficult, implying invasive approaches (biopsies, blood drawing, etc.) or indirect measurements (i.e. radiological/magnetic resonance imaging, etc.);
(2) the number of agents (proteins, single cells, populations, ...) and parameters describing the system is overwhelming and, often, they are not stable in time;
(3) many biological processes can be schematized as the result of interactions in a network of networks (genes, proteins, cell-cell interactions, tissues and

F. R. Bertani · L. Businaro · A. De Ninno · A. Gerardino
Istituto di Fotonica e Nanotecnologie – CNR, Rome, Italy

G. Bretti (✉) · R. Natalini
Istituto per le Applicazioni del Calcolo – CNR, Rome, Italy
e-mail: g.bretti@iac.cnr.it

G. Bretti et al. (eds.), *Mathematical Models and Computer Simulations for Biomedical Applications*, SEMA SIMAI Springer Series 33,
https://doi.org/10.1007/978-3-031-35715-2_2

organs, ...), with the consequent appearance of emergent behaviors (immune system collective responses, brain activity, etc..) which originating phenomena is difficult to circumscribe.

Despite the stated difficulties, our knowledge of human biology at different scales is vast. At the base of the giant progresses made there is the availability of a plethora of different models recapitulating the different biological situations and allowing precise measurements. For the sake of comprehension, we here state that with "model" it is commonly intended a biological system used for experiments, which could mean an animal (typically mice or rats), and we speak of *in-vivo models*, or one or more cell or cell populations cultured on a petri dish—*in-vitro models*.

Animals do not always recapitulate the heterogeneity of human response while conventional cultures in plastic plates fail to simulate and to control in a flexible way physical and chemical cues of in vivo cellular, tissue and organs landscapes. To overcome these limits the past decade saw the appearance of more advanced in-vitro cellular models realized merging micro-engineering, microfluidics strategies, advanced culturing, and biomaterials science strategies, which go under different names depending on implemented technologies and methods used, such as Micro Physiological Systems (MPS) [32, 48], Organ-on-Chip (OoC) i.e. microfluidic devices engineered to reproduce mechanics and surrounding physiochemical environment of entire living organisms with modular degree of complexity [16, 20, 21, 29, 50], etc.

The general idea is to obtain physiologically relevant models able to include both physical and chemical aspects of the target biological environments, interfaceable with advanced microscopy stations. They will be more deeply discussed in Sect. 2.

In this scenario, we started to explore possible methods to bridge the in-vitro and in-silico worlds in a synergistic way. A plethora of biological data can be extracted from the OoC models to feed numerical simulations, which can then be exploited to improve our interpretation ability of the phenomena occurring inside the chip models. With this goal in mind, we consider as scientific case what has been at the core of our research in the past years, the interaction of tumors and immune system in microfluidic co-culture systems, and use extracted migration patterns data in a reference experiment showing effective activation of immune cells in response to dying cancer cells exposed to a chemotherapy drug. In the experiment, immune cells (IC) migration and interaction with cancer cells is driven by the chemical gradient released by Tumor Cells (TC). However, spatiotemporal distribution of this complex gradient is highly dynamic and demand rigorous analysis for interpretation. Moreover, from experimental point of view is really demanding to realize local, specific, and dynamic description of tumor microenvironment. For this reason, we followed the approach to simplify experimental assays using on-chip approach, to derive cell migration data and implement mathematical and engineering principles to allow the understanding of complex mechanisms of gradient sensing and response, with the objective of building descriptive approach towards multiscale models on one side and aiming at exploring the possibility to build predictive hybrid (biologically informed mathematical platforms) models on the other. In this respect, we can imagine that mathematics can represent biology's next microscope [10], able to reveal hidden mechanisms.

2 The Organs-on-Chip Technology

Bidimensional cell cultures have been used for more than a century as in vitro model to study human physiology and pathology and to test therapeutic approaches, bringing to a great deal of knowledge and results in different fields including cell biology, biochemistry, pharmacokinetics. These techniques are currently a fundamental step in preclinical drug studies. However they cannot recapitulate the complexity of cell behavior in physiological condition of living tissues, characterized by three-dimensional architecture and high level of interaction of cells among them, coexistence of different populations and signal exchange with the extracellular matrix [33]. In the research for new, more physiologically reliable models for drug testing and personalized disease treating, one of the most promising is OoC technology, that belongs to the family of MicroPhysiological Systems (MPS). This approach resides at the interface between tissue engineering and microfluidic technology, whose definition "a fit-for-purpose microfluidic device, containing living engineered organ substructures in a controlled microenvironment, that recapitulates one or more aspects of the organ's dynamics, functionality and (patho)physiological response in vivo under real-time monitoring", as confirmed during ORCHID (Horizon 2020 FET-Open project "Organ-on-Chip In Development") workshop, also represents an objective [30]. OoC technology aims at reconstructing at a microscale organ architecture, thanks to the design of a specific fluidic and mechanical environment that reproduces shape, surface pattern and stiffness of organ-specific contexts. The use of microfluidic flow control ensure oxygenation and proper culture condition on one side, and the precise control over stimuli administration shaping spatiotemporal chemical gradients on the other. This set up allows long-term viability, observation of differentiation/regeneration processes, efficient circulation of immune system cells, antibodies, biochemical signaling molecules and metabolites, and the possibility to extract tiny volumes for biochemical analyses. Continuous perfusion causing controlled shear stress, and finely tuned mechanical cues help to build dynamic tissue models, this in addition to the possibility to insert electrodes to induce/detect electrical stimuli or biosensors enabling real-time monitoring of chemical landscape. All these techniques can be combined to reach a rich description about effects of the microenvironment on the cellular fate. Furthermore OoCs allow the observation at a single cell level of mutual interactions and other biological processes that cannot be monitored in animals or human patients: in fact, another feature of this approach is the compatibility with advanced optical microscopy at high spatial and temporal resolution. In addition, the great diversity and range of complexity of OoCs (and combination of single organ models in a more complex body-on-chip approach) offer the opportunity to optimize or even customize the design of devices and experiments representing a promising tool for drug discovery and personalized medicine. In particular, CoC platforms consisting by cancer cells alone [40] or co-cultured with stromal and or immune cells [38], provide a physiologically relevant 3D microenvironment because of the ability to compartmentalize and better control gradients. Therefore, these models have the potential for a more in-depth investigation of ICs interactions with TCs or of the transport processes of drugs, including therapeutic proteins, into and around tumor tissues, and allow for a quick evaluation of novel formats with respect to their transport behavior. Importantly, the highly controlled nature of experiments with microfluidic models facilitates the application of modeling approaches for the description of migration and transport processes, which, so far, have been limited to the description of processes in either static spheroids or in vivo models [39].

Although the concept of OoC is relatively new, mathematical approaches for designing and interpreting the OoC can be adapted from mathematical techniques that have been used in biological and pharmacological sciences, which have been developed and validated for several decades [46]. These models can be built on different scales (macro, meso, micro, nano), depending on the purpose of the research. Macro-models, using finite difference (FD) [45], finite volume (FV) or finite element methods (FEM) [43], are applied to explain the behavior of the whole system at the organ level; meso-models use mesoscopic methods such as dissipative particle dynamics (DPD) and explain behavior at the level of molecular clusters; micro- and nanomodels use molecular dynamics method and explain behavior at the molecular or atomistic level [14].

2.1 Setting of the Laboratory Experiments

The in vitro microfluidic experiment we refer here is described in [12, 47], where details about cells types involved and loading procedures are reported. Briefly, human breast cancer cells, previously treated with anthracycline-based chemotherapy, are then cultivated in the left compartment of the microfluidic chip. This treatment triggers the process of immunogenic cell death [24], according to which TCs release danger signals. The right chamber is loaded with unlabeled human peripheral blood mononuclear cells (PBMCs) from healthy donors. PBMCs start migrating biased by the detection of chemical signal produced by TCs and after crossing microchannels enter in contact with dying cancer cells, engaging in stable interactions leading to TCs killing events.

In the chip, the chosen culture medium is neutral, which means that no exogenous substance is introduced. Timelapse imaging was performed using a microscope placed directly inside the CO_2 incubator for all the duration of the recordings. Images were taken every 2 min over a period of 72 h of migration. Immune cells tracks (<400) were extracted in left chamber (interval time: 24–48 h) using TrackMate available in the Fiji/ImageJ software (https://imagej.nih.gov/ij/, accessed on 2020).

Our aim consists of designing a basic mathematical model able to capture the main aspects related to the migratory movement of the ICs with respect to TCs. Therefore, we only focus on the case of efficient interaction between dying cancer cells exposed to ICD inducers and immune cells from healthy donors. At this first stage, we introduce multiscale mathematical models able to reproduce qualitatively the experimentally observed behavior. In the future, we aim at developing an efficient algorithmic calibration procedure to derive model parameter values as close as possible to empirical observations.

3 Mathematical Modeling of OoC

Regarding the state-of-the-art in modeling of dynamics and interactions in microfluidic chip environment, it is worth mentioning the study in [28] based on macroscopic models, the microscopic approaches proposed in [22, 26, 27, 51] based on particle models and the cellular automata approach proposed in [9].

The present Section is devoted to the description of the mathematical framework recently introduced by some of the authors to build in silico models of OoC and, more specifically, of CoC technology. The main aim of this modeling is to investigate qualitatively the effects of chemical gradients released by TCs on immune system dynamics. Here we mainly refer to the recent papers [3] and [8], where we propose a macroscopic and a hybrid discrete-in-continuous approach, respectively.

Notice that a rather general model based on cellular automata approach was proposed by one of the author in [5]. Such an approach revealed to be promising in terms of the ability of mimicking the experimentally observed behavior and the easy implementation; however, it still needs to be improved from the point of view of the running time in order to allow accurate statistical estimates of key model parameters.

In the experimental setting here considered, TCs are experiencing apoptosis during which they liberate damaged-associated molecular patterns together with tumor antigens that elicit an immune response. For the sake of simplicity, the alarm concentrations are not differentiated, and we indicate them by means of the chemoattractant φ, whose evolution is described by a PDE. At the cellular level, the model is discrete and includes the equation of the motion of each IC. Notice that since we refer to experiment of treated cancer cells, according to the experimental setting we do not need to include TCs migration or duplication in our model. We underline that for both the macroscopic and hybrid approaches we assume as model parameters those available from literature, since a calibration against real data will be carried out in forthcoming works.

3.1 Macroscopic Model for CoC Experiment [3]

The macroscopic model for CoC was introduced in the paper [3], where the reaction-diffusion system with chemotaxis (1) was proposed to describe the behavior of two populations of cells coexisting together: the density of TCs, denoted in the system by T, and the density of ICs, denoted in the system by M and considered as a single species. The model describes death processes, interactions with chemoattractants, interactions and competition between different cell species. The microfluidic chip is schematized as a network of channels connecting two boxes (the microfluidic chambers). Following the ideas in [6] we introduce ad hoc transmission conditions in order to ensure mass conservation.

In this work we consider only the 2D case since the experimental data do not take into account the height of the chip. However, in principle, our framework (model and numerical algorithm) could be easily extended to the third dimension and we remark that, analogously to the 2D-1D case here considered, mass-preserving and positivity preserving numerical conditions at the external boundaries and at the interfaces between 3D and channels still hold. We also mention here that a multigrid parameter estimation method coupled with appropriate regularization estimation techniques for identifying macro-model parameters was proposed in [4] and tested on synthetic data to provide an estimate of the diffusion rate and the chemotactic coefficients involved in the chemotaxis term representing the velocity field of ICs. However, accurate estimates of key model parameters against real data will be done in a forthcoming paper.

TCs produce chemical signals, called φ, activating the immune response of ICs and influencing their behavior. Moreover, we take into account the presence of molecules and cytokines ω (produced by ICs), acting as a chemical killer of cancer cells. The model here considered in the 2D chambers is in the framework of Keller-Segel model [19]:

$$\begin{cases} \partial t T = D_T \Delta T - \lambda_T(\omega) T, \\ \partial t M = D_M \Delta M - \nabla \cdot (\chi(\varphi) M \nabla \varphi), \\ \partial t \varphi = D_\varphi \Delta \varphi + \alpha_\varphi T - \beta_\varphi \varphi, \\ \partial t \omega = D_\omega \Delta \omega + \alpha_\omega M - \beta_\omega \omega. \end{cases} \tag{1}$$

The system above describes the dynamics of the two cell species and the diffusion of the chemoattractant, and it needs to be complemented with suitable initial conditions and boundary conditions for the cells and the chemoattractant concentrations, as will be specified in the following.

Since here we consider the experiment where tumor is treated with chemotherapy drug, the system above describes the following situation: TCs produce a chemical substance φ attracting ICs, that start migrating towards them. In such a case the tumor does not proliferate, thus we do not include this feature in the model. ICs do not seem to proliferate during the experiment, thus also this feature is neglected in the model. Note that the proliferation of TCs, observed experimentally in the untreated case, can be easily added to the model by putting a linear or logistic source term in the equation for the tumor.

ICs also produce a chemical substance ω which should be responsible for tumor killing. Therefore, in the first equation of the system (1), besides the diffusion term we can find $-\lambda_T(\omega)T$, representing the tumor suppression operated by immune cells. We underline that at the moment we have no information from the biologists about the real killing rate in the microchip environment induced by ω, but we decided to introduce it in order to include this phenomenon in the model qualitatively.

In order to define the action of the cytotoxin ω produced by ICs (which determines the death of TCs), we introduce the function $\lambda_T(\omega)$:

$$\lambda_T(\omega) = \frac{S\omega}{\eta + \omega}, \tag{2}$$

where S is the maximum secretion rate of the cytotoxin by the ICs and η the equivalent Michaelis constant associated with the production, as described in [34]. In the second equation, besides the diffusion term we have the chemotactic term $f = \chi(\varphi)\nabla\varphi$ due to the presence of the chemical substance φ produced by the tumor. The chemotaxis function $\chi(\varphi)$ represents the chemotactic sensitivity of ICs and we refer the form suggested in [25]:

$$\chi(\varphi) = \frac{k_1}{(k_2 + \varphi)^2}, \tag{3}$$

where k_1 represents the cellular drift velocity, while k_2 is the receptor dissociation constant, which says how many molecules are necessary to bind the receptors. Note that we mainly refer to [34] for the values of the parameters k_1, k_2, and all the parameters of the macroscopic model are reported in Table 1.

Now, in order to describe in a realistic way the dynamics of cells in the narrow microchannels connecting the two boxes, we introduce the following 1D (finite speed) hyperbolic model for the dynamics in the framework of Greenberg-Alt model

Table 1 Physical parameter values for the macro and the hybrid models

Parameter	Description	Units	Value	Ref.
L	Length of the corridor	μm	500	Datum
L_x	Horiz. size of the box	μm	100 (macro model); 200 (hybrid model)	Datum; design
L_y	Vert. size of the box	μm	1000 (macro model); 200 (hybrid model)	Datum; design
D_M	Diffusivity of cells	μm²/s	9×10^2	[34]
D_T	Diffusivity of T—scenario 2	μm²/s	5.6×10^2	Empirical
D_T	Diffusivity of T—scenario 1	μm²/s	5.6×10^1	[34]
D_φ, D_ω	Diffusivity of chemoattractants	μm²/s	2×10^2	[34]
α_φ	Growth rate of φ in scenario 2	s^{-1}/cell	0	Empirical
α_φ	Growth rate of φ for scenario 1	s^{-1}/cell	0.5×10^{-1}	[11]
β_φ	Consumption rate of φ	s^{-1}	10^{-4}	[11]
α_ω	Growth rate of ω for scenario 1	s^{-1}/cell	10^{-1}	[11]
α_ω	Growth rate of ω for scenario 2	s^{-1}/cell	0	Empirical
β_ω	Consumption rate of ω	s^{-1}	10^{-4}	[11]
k_1	Cellular drift velocity	mol cm^2 s^{-1}	$3.9 \cdot 10^{-9}$	[34]
k_2	Receptor dissociation constant	mol	$5 \cdot 10^{-6}$	[34]
S	Maximum secretion rate of the chemicals	g/μm³	$1.97 \cdot 10^{-11}$	[34]
γ	Equivalent Michaelis constant	cells/μm³	10−4	[34]
K	Kedem-Katchalsky constant	–	500	Empirical
$N_{tot,I}$	N. of ICs in Ω during 24 h	–	97	Exp. Data
$N_{tot,C}$	N. of TCs in Ω during 24 h	–	4	Exp. Data
R_I	IC radius	μm	4	[49]
R_T	TC radius	μm	10	[2]

(continued)

Table 1 (continued)

Parameter	Description	Units	Value	Ref.
$\bar{R}$	Chem. Detection radius	μm	7	Biol. Assumption
R_3	Radius of repulsion for ICs	μm	8	Biol. Assumption
R_4	Radius of adhesion for ICs	μm	10	Biol. Assumption
R_1	Radius of ICs/TCs repulsion	μm	14	Biol. Assumption
ω_{adh}	Adhesion coeff. for ICs	s^{-2}	1e−7	Biol. Assumption
ω_{rep}	Repulsion coeff. for ICs	$\mu\text{m}^2\,\text{s}^{-2}$	5e−4	Biol. Assumption
ω_{repTI}	Repulsion coeff. for ICs/TCs	$\mu\text{m}^2\,\text{s}^{-2}$	8.5e−3	Biol. Assumption
μ	Damping Coeff.	s^{-1}	2.1e−3	Biol. Assumption
a	Exchange rate of chemicals	$\mu\text{m}\,\text{s}^{-1}$	10	Biol. Assumption
γ	Chemotactic Coeff. Scenarios 1/2	μm^{-1}	5e−3	Biol. Assumption
b_1	Flux cond. ($y = 0$). Scenarios 1–2	$\text{mol}\,\text{s}^{-1}\,\mu\text{m}^{-1}$	22	Biol. Assumption
b_2	Flux cond. ($x = L_x$)	$\text{mol}\,\text{s}^{-1}\,\mu\text{m}^{-1}$	12	Biol. Assumption
b_3	Flux cond. ($y = L_y$)	$\text{mol}\,\text{s}^{-1}\,\mu\text{m}^{-1}$	0	Biol. Assumption
b_4	Flux cond. ($y = 0$). Scen. 1–2	$\text{mol}\,\text{s}^{-1}\,\mu\text{m}^{-1}$	18	Biol. Assumption
σ_x	Std. Dev. of x-trajectories	$\mu\text{m}\,\text{s}^{-1/2}$	1.5e−1	Exp. Data
σ_y	Std. Dev. of y-trajectories	$\mu\text{m}\,\text{s}^{-1/2}$	1.7e−1	Exp. Data

[17], with the label c indicating the channels:

$$\begin{cases} \partial_t T_c + \partial_x v_c^T = -\lambda_T(\omega) T_c, \\ \partial_t v_c^T + D_T \partial_x T_c = -v_c^T, \\ \partial_t \omega_c = D_\omega \partial_{xx} \omega_c + \alpha_\omega T_c - \beta_c \omega_c, \\ \partial_x M_c + \partial_t v_c^M = 0, \\ \partial_t v_c^M + D_M \partial_x M_c = M_c f_c - v_c^M, \\ \partial_t \varphi_c = D_\varphi \partial_{xx} \varphi_c + \alpha_\varphi T_c - \beta_\varphi \varphi_c, \end{cases} \tag{4}$$

with T_c and M_c, respectively, as the density of TCs and ICs; $f_c = \chi(\varphi_c)\partial_x\varphi_c$ and v_c^T and v_c^M, respectively, as the average flux of TCs, called T_c and ICs, namely M_c, in the channels.

3.1.1 Interface Between 2D-1D Models in (1)–(4)

We remark that, for simplicity, the following description focuses on the left part of the domain, i.e. Ω_l and its connection with a single microchannel represented by the interval $[0, L]$. However, the numerical treatment holds analogously on the right side of the domain and also to multiple channels. In the 2D left box Ω_l, the position of node $1L$ is at $x = L_y$, $y \in [a, b]$ and for the 1D domain, represented by a given number of channels C, node $1L$ is placed at $x = 0$ (left endpoint of the channel), see Fig. 2.

Let us use the notation $F = M\ \chi(\varphi)\nabla\varphi = Mf$, and, from now on, the two components of the drift term f will be indicated as:

$$f(x, y, t) := \begin{pmatrix} f^x(x, y, t) \\ f^y(x, y, t) \end{pmatrix}.$$

Then for the mass conservation of M we impose the condition:

$$\begin{aligned} 0 &= \frac{d}{dt}\int_{\Omega_l} M(x, y, t)\, d\Omega_l + \frac{d}{dt}\int_0^L M_c(x, t)\, dx \\ &= \int_{\Omega_l} (D_M \Delta M(x, y, t) - \nabla \cdot F(x, y, t)) d\Omega_l + \int_0^L -\partial_x v(x, t) dx \\ \Longrightarrow &\oint_{\partial\Omega_l} (D_M \nabla M(x, y, t) - F(x, y, t) n)\, dS + v(0, t) = 0. \end{aligned}$$

Note that in the above formula we have $v(L, t) = 0$ because we are looking at left interface (node $1L$). Then, we finally get:

$$\int_{a_1}^{b_1} (D_M \partial_x M(L_x, y, t) - M(L_x, y, t) f^x(L_x, y, t)) dy = -v(0, t). \tag{5}$$

Now we impose the KK-condition at the interface:

$$\begin{aligned} &D_M \partial_x M(L_x, y, t) - M(L_x, y, t) f^x(L_x, y, t) \\ &\quad = K(M_c(0, t) - M(L_x, y, t)) \qquad \text{for } y \in [a, b] \end{aligned}$$

and then (5) reads as:

$$v(0, t) = -K(b - a) M_c(0, t) + K \int_a^b M(L_x, y, t)\, dy. \tag{6}$$

The next Section deals with the hybrid macro-micro approach for the same problem introduced in [8].

3.2 *Hybrid Macro-Micro Model for CoC Experiment [8]*

In the framework of hybrid deterministic PDE–ODE models, in [8] we propose a discrete–in-continuous approach previously applied in [13], where immune cell motion is governed by ODE and the diffusion of chemoattractant released by cancer cells in the environment is described by PDE. The present work is complementary to the fully macroscopic approach presented in Sect. 3.1, since it allows a zooming on immune system dynamics. Moreover, we added a Brownian component to the equations of cell dynamics in order reproduce zigzag pathways of ICs.

Let X_i be the position of the i-th IC and $f(x,t)$ the total chemoattractant concentration, the following equations are introduced:

$$\partial_t \varphi = D\Delta\varphi + \alpha_\varphi F_4(Y) - \beta_\varphi \varphi \tag{7}$$

$$\ddot{X}_i = \gamma F_1(\chi(\varphi)\nabla\varphi) + F_2(X, Y) + F_3(X) - \mu \dot{X}_i, \tag{8}$$

where γ, μ, α_φ and β_φ are given constants and $F_n(\cdot)$, $n = 1, 2, 3, 4$ are suitable functions described below.

3.2.1 Function F_1: Chemotactic Term

The term F_1 is related to the detection of a chemical signal by i-th cell in its neighborhood and it is taken to be a weighted average over a ball of radius $\bar{R}$ and centered in X_i:

$$F_1(g(x,t)) = \frac{1}{\mathcal{W}} \int_{B(X_i,\bar{R})} g(x,t) w_i(x) dx, \tag{9}$$

where

$$B(X_i, \bar{R}) := \{x : ||x - X_i|| \leq \bar{R}\},$$

$\|\cdot\|$ being the Euclidean norm,

$$w_i(x) = \begin{cases} 2\exp\left(-||x - X_i||^2 \dfrac{\log 2}{\bar{R}^2}\right), & \text{if } ||x - X_i|| \leq \bar{R}, \\ 0, & \text{otherwise,} \end{cases} \tag{10}$$

is a truncated Gaussian weight function, and

$$\mathcal{W} := \int_{B(X_i,\bar{R})} w_i(x)dx,$$

independently on i. A similar definition holds for the vector quantity F_1.

In addition, the function F_1 contains the gradient of the chemical substance and a chemotaxis function $\chi(\varphi)$, representing the chemotactic sensitivity of ICs, is the function already defined in (3).

In the integral, $\bar{R}$ will be chosen larger than the IC radius R_I, thus relation (9) describes a chemical signal that is sensed more in the center of the cell and less at the edge of the cell extensions.

3.2.2 Function F_2: ICs/TCs Repulsion

Function F_2 includes a repulsion effect among TCs and ICs. In particular, repulsion occurs at a distance between the centers of two cells less than R_1, where R_1 is defined as the sum of the two radii, R_I and R_T, of the immune and tumor cells, respectively. With this formulation, repulsion occurs when an IC and a TC start being effectively overlapped. We assume

$$F_2(X,Y) = \sum_{j:Y_j \in B(X_i,R_2)\setminus\{X_i\}} K(Y_j - X_i), \tag{11}$$

where the function K depends on the relative positions $Y_j - X_i$, namely:

$$K(Y_j - X_i) = -\omega_{repTI}\left(\frac{1}{||Y_j - X_i||} - \frac{1}{R_1}\right)\frac{Y_j - X_i}{||Y_j - X_i||}, \text{ if } ||Y_j - X_i|| \leq R_1, \tag{12}$$

where ω_{repTI} is a constant. The quantity $R_2 > R_1$ indicates the radius of the ball centered in X_i, in which the centers of the tumor cells can fall. Thus, we are considering all the tumor cells in proximity of the center of an immune cell.

3.2.3 Function F_3: ICs Adhesion/Repulsion

Function F_3 includes adhesion/repulsion effects between ICs. In particular repulsion occurs at a distance between the centers of two ICs less than R_3 and takes into account the effects of a possible cell deformation. Conversely, adhesion occurs at a distance greater than R_3 and less than $R_4 > R_3$, and it is due to a mechanical interaction between cells via filopodia. We assume

$$F_3(X) = \sum_{j:X_j \in B(X_i, R_4)\setminus\{X_i\}} K(X_j - X_i), \tag{13}$$

where the function K depends on the relative positions $X_j - X_i$, namely:

$$K(X_j - X_i) =$$

$$-\omega_{rep}\left(\frac{1}{||X_j - X_i||} - \frac{1}{R_3}\right)\frac{X_j - X_i}{||X_j - X_i||}, \text{ if } ||X_j - X_i|| \leq R_3, \tag{14}$$

$$\omega_{adh}(||X_j - X_i|| - R_3)\frac{X_j - X_i}{||X_j - X_i||}, \text{ if } R_3 < ||X_j - X_i|| \leq R_4, \tag{15}$$

where $\omega_{rep}, \omega_{adh}$ are constants. Note that we assume $R_3 = 2R_I$ so that the repulsion occurs when two cells start being effectively overlapped. The expression (14) represents a repulsion term which goes as $1/r$, r being the distance between the centers of two cells as we can find for instance in [44]. The expression (15) represents instead the Hooke's law of elasticity. The last term in the first equation is due to the cell adhesion to the substrate (see for example [1, 15]).

3.2.4 Friction

The addend $\mu\dot{X}_i$ in the equation of the motion is due to cell adhesion to the substrate, with damping coefficient equals μ.

3.2.5 Function F_4: Production of Chemical Signal

In the diffusion equation, only cancer cells are responsible for the production of the chemoattractant, so that

$$F_4(Y) = \sum_{j=1}^{N_{tot,c}} \chi_{B(Y_j, R_T)}, \tag{16}$$

where $N_{tot,c}$ is the total number of cancer cells, and

$$\chi_{B(Y_j,R_T)} = \begin{cases} 1, & \text{if } x \in B(Y_j, R_T), \\ 0, & \text{otherwise.} \end{cases} \tag{17}$$

In the previous formula R_T is the radius of a cancer cell, considering that the source of chemoattractant is defined by the dimension of a single cell.

3.2.6 Initial Conditions

Initial data for Eq. (8) are given by the position and velocity of each IC:

$$X_i(0) = X_{i,0}\ ,\ V_i(0) = V_{i,0}.$$

Moreover, since TCs do not migrate and maintain their initial position Y during the whole time, the initial data for Eq. (7) is provided by the chemoattractant produced by TCs at time $t = 0$:

$$\varphi(x, 0) = F_4(Y).$$

3.2.7 Boundary Conditions

Now, let $\Omega = [0, L_x] \times [0, L_y]$ our domain, for the chemoattractant we require the inhomogeneous Robin boundary condition:

$$D\frac{\partial \varphi}{\partial n} + a\varphi = b, \text{ on } \partial\Omega, \tag{18}$$

where b signals the similarity with the inhomogeneous Neumann boundary condition and regulates the exchange with the external environment.

The system of equations (7) and (8) can be now rewritten as:

$$\begin{cases} \partial_t \varphi = D\Delta\varphi + \xi \sum_{j=1}^{N_{tot,c}} \chi_{B(Y_j,R_T)} - \beta_\varphi \varphi, \\ \ddot{X}_i = \frac{\gamma}{W} \int_{B(X_i,\bar{R})} \chi(\varphi(x,t)) \nabla\varphi(x,t) w_i(x) dx + \sum_{j:Y_j \in B(X_i,R_2)\setminus\{X_i\}} K(Y_j - X_i) \\ + \sum_{j:X_j \in B(X_i,R_4)\setminus\{X_i\}} K(X_j - X_i) - \mu V_i. \end{cases} \tag{19}$$

3.2.8 Stochastic Model

The deterministic model (19) can be reformulated in a stochastic version for Eq. (8) in order to take into account the intermittent motion of ICs composed of a walking phase and of a zigzag phase, as shown in [51].

As a possible approach to the problem, here we describe such characteristic walk by Brownian motion as in [31]. This revealed to be effective in reproducing the randomness of cell trajectories.

The stochastic equation for ICs motion is:

$$\ddot{X}_i = \gamma F_1(\chi(\varphi)\nabla\varphi) + F_2(X, Y) + F_3(X) - \mu(\dot{X}_i - \sigma\psi). \tag{20}$$

Equation (20) contains the stochastic contribution, where $\psi(t)$ is a Gaussian white noise, and σ is the standard deviation of ICs trajectories. With this formulation, ICs are not only subjected to mechanical forces such as adhesion or repulsion, but also to random factors that might be related to unknown cell mechanisms.

The stochastic equation above can be decoupled as follows:

$$\dot{V}_i = \frac{\gamma}{\mathcal{W}} \int_{B(X_i,\bar{R})} \chi(\varphi(X,t))\nabla\varphi(X,t) w_i(X) dX + \sum_{j:Y_j \in B(X_i,R_2)\setminus\{X_i\}} K(Y_j - X_i) \tag{21}$$

$$+ \sum_{j:X_j \in B(X_i,R_4)\setminus\{X_i\}} K(X_j - X_i) - \mu V_i,$$

$$\dot{X}_i = V_i + \sigma\psi(t). \tag{22}$$

for $i = 1, \ldots, N_{tot,I}$.

The procedure to estimate parameter σ based on experimental data are provided in Sect. 5.2.3, but more details are in [8].

3.3 *Future Directions: Mean-Field Limits and Nonlocal Models [36]*

The hybrid model above was developed to build a bridge between granular CoC experimental data and macroscopic models, in order to gain further insights on the dynamics and short range interactions between cells. This goes towards the direction of the mean-field limit to unveil the statistical properties of the model. In this framework, here we shortly present a generalization of the kinetic model given by Cucker-Smale dynamical system developed in [36], where the mean-field limit of a general class of deterministic hybrid macro-micro models towards to a nonlocal Vlasov-type equation was derived rigorously in Wasserstein's type topologies and the formal derivation of a new nonlocal hydrodynamic macroscale model was obtained.

Let us describe the general class of particle systems we will handle in this section.

Consider on $\mathbf{R}^{2dN} \ni ((x_i(t))_{i=1,\dots,N}, (v_i(t))_{i=1,\dots,N}) := (X(t), V(t))$ the following vector field

$$\begin{cases} \dot{x_i}(t) = v_i \\ \dot{v_i}(t) = F_i(t, X(t), V(t)) \end{cases} i = 1, \dots, N, \ (X(0), V(0)) = (X^{in}, V^{in}) : \tag{23}$$

where

$$F_i(t, X, V) = \frac{1}{N}\sum_{j=1}^{N} \gamma(v_i - v_j, x_i - x_j) + \eta \nabla_x \varphi(x_i, t) + F_{ext}(x_i, v_i), \tag{24}$$

with γ the collective interaction function, F_{ext} an external force and φ satisfies the equation

$$\partial_s \varphi(x, s) = D\Delta_x \varphi - \kappa\varphi + f(x, X(s)), \ s \in [0, t], \ \varphi(x, 0) = \varpi^{in} \tag{25}$$

for some $\kappa, D, \eta \geq 0$ and the function f of the form

$$f(x, X) = \frac{1}{N}\sum_{j=1}^{N} \chi(x - x_i), \ \chi \in C_c^1. \tag{26}$$

The function $\gamma : \mathbf{R}^d \times \mathbf{R}^b \to \mathbf{R} \times \mathbf{R}^d$ is supposed to be Lipschitz continuous.

For any fixed function φ^{in} and any t, N we define the mapping $\Phi_N = \Phi$ by

$$\begin{cases} \Phi_N^t : \qquad \mathbf{R}^{2dN} \longrightarrow \mathbf{R}^{2dN} \\ Z^{in} = (X^{in}, V^{in}) \longrightarrow Z(t) = (X(t), V(t)) \text{ solution of (23)-(24)-(25)-(26).} \end{cases} \tag{27}$$

Note that Φ_N is not a flow.

We would like to derive a kinetic Vlasov-like model corresponding to system (23)-(24)-(25)-(26), that is the one particle (non-linear) PDE satisfied by the first marginal of the push-forward $\Phi\#\rho^{in}$ where $\rho^{in} \in \mathcal{P}(\mathbf{R}^{2dN})$, the space of probability measures on $\mathbf{R}^{2dN}$ and Φ_N is the mapping defined by (27). We recall that the pushforward of a measure μ by a measurable function Φ is $\Phi\#\mu$ defined by $\int \varphi d(\Phi\#\mu) := \int (\varphi \circ \Phi) d\mu$ for every test function φ.

The nonlocal Vlasov model (in the phase space (x, v)) can be written this way:

$$\partial_t \rho + v \cdot \nabla_x \rho = \nabla_v(\nu(t, x, v)\rho), \ \rho^0 = \rho^{in} \tag{28}$$

where

$$\nu(t, x, v) = \gamma * \rho(x, v) + \eta \nabla_x \psi(x) + F_{ext}(x, v) \tag{29}$$

and ψ satisfies

$$\partial_s \psi(x) = D\Delta_x \psi - \kappa\psi + g(x, \rho), \psi^0 = \varphi^{in}, \tag{30}$$

with

$$g(x, \rho) = \int_{\mathbf{R}^{2d}} \chi(x - y)\rho(y, \xi)dyd\xi. \tag{31}$$

In [36] it is proved that, if ρ^{in} is a compactly supported probability on $\mathbf{R}^{2dN}$, Φ_N the mapping generated by the particles system (23)-(24)-(25)-(26) as defined by (27), and ρ the solution to the nonlocal Vlasov equation (28)-(29)-(30)-(31) with initial condition ρ^{in}, then the first marginal

$$(\Phi_N\#(\rho^{in})^{\otimes N})_{N;1} := \int_{\mathbf{R}^{2d(N-1)}} \Phi_N\#(\rho^{in})^{\otimes N}(x, v)dx_2 \dots dx_N dv_2 \dots dv_N$$

converges in the topology of the Wasserstein distance to ρ, when $N \to +\infty$. One interesting consequence of this result is the connection with a new class of nonlocal Euler systems. Let $\mu^{in}, u^{in}, \varphi^{in}$ be some functions such that the nonlocal Euler system

$$\begin{cases} \partial_t \mu + \nabla(u\mu) = 0 \\ \partial_t(\mu u) + \nabla(\mu(u)^{\otimes 2}) = \mu \int \gamma(\cdot - y, u(\cdot) - u(y))\mu(y)dy + \eta\mu\nabla\psi + \mu F \\ \partial_s \psi = D\Delta\psi - \kappa\psi + \chi * \mu, \; s \in [0, t], \\ (\mu^0, u^0, \psi^0) = (\mu^{in}, u^{in}, \varphi^{in}) \in H^s, \; s > \frac{d}{2} + 1. \end{cases} \tag{32}$$

has a unique solution $\mu, u \in C([0, t]; H^s) \cap C^1([0, T]; H^{s-1})$, $\psi \in C([0, t]; H^s) \cap C^1([0, T]; H^{s-2}) \cap L^2(0, T; H^{s+1})$ and let

$$\rho^{in} = \mu^{in}(x)\delta(v - u^{in}(x)).$$

Then, for any $t \in [0, T]$, the measure $\rho_{mk} := \mu(x)\delta(v - u(x))$ is a monokinetic solution to the nonlocal Vlasov equation (28)-(29)-(30)-(31) and so again the first marginal $(\Phi_N\#(\rho^{in})^{\otimes N})_{N;1}$ converges in the topology of the Wasserstein distance to ρ_{mk}, when $N \to +\infty$.

Notice that the Euler system (32) can be compared to the so-called Preziosi model of vasculogenesis, where the nonlocal term in our Euler system is replaced by a phenomenological pressure gradient local in the density, see the chapter [42] also available at the link http://dimat2.polito.it/~preziosi/pubs/vasqsenz.pdf. This new class of models in very promising in terms of a multiscale approach to OoC, since it carries the microscopic features of cells interactions up to the macroscopic level, which is usable from a computational point of view.

4 Numerical Approximation

We define equispaced $x_i := i\Delta x$, $y_j := j\Delta y$ and $t_n := n\Delta t$ with $\Delta x, \Delta y, \Delta t > 0$ and $i = 0, \ldots, N_x + 1$, $j = 0, \ldots, N_y + 1$; for the channel $[0, L]$ we discretize it as $x_i = i\Delta x$, with $i = 0, \ldots, N$. For a more structured presentation, we introduce the operators

$$\delta_x^2 u_{i,j}^n := u_{i+1,j}^n - 2u_{i,j}^n + u_{i-1,j}^n,\ \delta_y^2 u_{i,j}^n := u_{i,j+1}^n - 2u_{i,j}^n + u_{i,j-1}^n,$$
$$\delta_x^0 u_{i,j}^n := u_{i+1,j}^n - u_{i-1,j}^n,\ \delta_y^0 u_{i,j}^n := u_{i,j+1}^n - u_{i,j-1}^n,$$
$$\delta_x^1 u_{i,j}^n := u_{i+1,j}^n - u_{i,j}^n,\ \delta_y^1 u_{i,j}^n := u_{i,j+1}^n - u_{i,j}^n.$$

4.1 Numerical Schemes for the Approximation of the Models (1)–(4)

In this section we briefly report the numerical approximation of the adopted 2D-doubly-parabolic (1) and 1D-hyperbolic-parabolic (4) models. Further details can be found in [3].

In order to present the numerical scheme, we write a simpler model with respect to (1) but sharing the main features of it:

$$\begin{cases} \partial_t u = D_u \Delta u - \nabla \cdot F + g(x, y, t, u), \\ \partial_t \varphi = D_\varphi \Delta \varphi + au - b\varphi, \end{cases} \tag{33}$$

with u as the density of individuals, φ as the density of chemoattractant, and with $F = u\,f$. From now on, the two components of the drift term $f = \chi(\varphi)\nabla\varphi$ will be indicated as:

$$\mathbf{f(x, y, t)} := \begin{pmatrix} f^x(x, y, t) \\ f^y(x, y, t) \end{pmatrix}.$$

For the mono-dimensional channel we consider simpler model sharing the same characteristics of (4), which reads as:

$$\begin{cases} \partial_t u_c + \partial_x v_c = g(x, t, u_c), \\ \partial_t v_c + \lambda_c^2 \partial_x u_c = F_c - v_c, \\ \partial_t \varphi_c = D_{\varphi_c} \partial_{xx} \varphi_c + a_c u_c - b_c \varphi_c, \end{cases} \tag{34}$$

with $F_c = u_c f_c$. The systems above have to be complemented with smooth initial conditions for the unknowns u, φ and also v for system (34); initial data will be specified in Sect. 5.1. On the boundary, we consider for all the quantities homogeneous Neumann conditions, i.e. no-flux boundary conditions. We also mention that hyperbolic model (34) requires an analytical monotonicity criteria to avoid unphysical solutions, see [35]. With regard to our former model (4), for the immune cell density M the monotonicity condition reads as:

$$\frac{k_1}{(k_2 + \varphi_c)^2} |\partial_x \varphi_c| \leq \sqrt{D_M}, \tag{35}$$

and it needs to be verified in the computational domain to ensure non-negative solutions.

We remark that the no-flux conditions boundary conditions used in our simulations are needed to have the mass-conservation of all the quantities. However, since in the laboratory experiment there is an inflow of cells from the outer boundaries, they are not realistic and more general ones will be considered in the future.

Notice that it is necessary to use implicit schemes to consider the presence of stiff source terms. For this reason, for the approximation of the time derivatives we use Crank-Nicolson (CN) method on the diffusion and source term, which is a second order implicit method, and an explicit central method for the convection term. Moreover, since CN is only A-stable but not L-stable, we also need to choose a Δt small enough to avoid spurious oscillations of the solution at intermediate states.

Because of the explicit term, we have numerical restrictions on the mesh grid and time step. Furthermore, we introduce artificial viscosity to avoid oscillations due to not suitable mesh grid size in dominant convection regime, which is often the case in chemotaxis models. Finally, the implicit-explicit numerical method used to compute the solutions for the density u in (33) inside the 2D domain Ω_l is:

$$u_{i,j}^{n+1} = u_{i,j}^n + D_u \frac{\Delta t}{2} \left[\frac{\delta_x^2 (u_{i,j}^n + u_{i,j}^{n+1})}{\Delta x^2} + \frac{\delta_y^2 (u_{i,j}^n + u_{i,j}^{n+1})}{\Delta y^2} \right]$$

$$- \frac{\Delta t}{4} \left[\frac{\delta_x^0 F_{i,j}^{x,n}}{\Delta x} + \frac{\delta_y^0 F_{i,j}^{y,n}}{\Delta y} \right]$$

$$+\frac{\Delta t}{2}\left(g_{i,j}^{n}+g_{i,j}^{n+1}\right)\underbrace{-\Delta t\left[\frac{\delta_x^2\theta_{i,j}^n}{2\Delta x}+\frac{\delta_y^2\theta_{i,j}^n}{2\Delta y}\right]}_{\text{artificial viscosity}},$$

$$i=1,\ldots,N_x,\; j=1,\ldots,N_y, \tag{36}$$

with

$$\theta_{i,j}^n := \chi(\varphi_{i,k}^n)u_{i,j}^n|\nabla\varphi_{i,j}^n|.$$

As can be seen, the function θ used for the artificial viscosity is almost identical to f with the exception of using the absolute value of $\nabla\varphi$. By using this, we increase artificial viscosity only where the gradient of the chemoattractant increases. This reduces the restriction on the meshgrid due to the condition induced by the cell Péclet number [41]. The numerical transmission condition on the left of node $1L$ ($i=N_x+1,\ j=j_a,\ldots,j_b$) is:

$$\begin{aligned}
u_{N_x+1,j}^{n+1} &= u_{N_x+1,j}^{n} - D_u\tfrac{\Delta t}{\Delta x^2}\delta_x^1(u_{N_x,j}^n+u_{N_x,j}^{n+1})\\
&+D_u\tfrac{\Delta t}{2\Delta y^2}\delta_y^2\left(u_{N_x+1,j}^n+u_{N_x+1,j}^{n+1}\right)\\
&+\tfrac{\Delta t}{\Delta x}\left(F_{N_x+1,j}^{x,n}+F_{N_x,j}^{x,n}\right)-\tfrac{\Delta t}{\Delta y}\left(F_{N_x+1,j+1}^{y,n}-F_{N_x+1,j-1}^{y,n}\right)\\
&+\tfrac{\Delta t}{2}\left(g_{N_x+1,j}^n+g_{N_x+1,j}^{n+1}\right)-\Delta t\left(\frac{\delta_x^1\theta_{N_x,j}^n}{\Delta x}+\frac{\delta_y^2\theta_{N_x+1,j}^n}{2\Delta y}\right)\\
&\text{(additional term for transmission condition)}\\
&\underbrace{+K\frac{\Delta t}{\Delta x}\left(u_0^{c,n}-u_{N_x+1,j}^n+u_0^{c,n+1}-u_{N_x+1,j}^{n+1}\right)}.
\end{aligned} \tag{37}$$

The role of permeability coefficient K in the positivity of (37) is discussed in Sect. 4.1.1.

For the external boundaries (the edges of the chamber Ω_l except at the junctions $j = j_a, \ldots, j_b$), we use:

$$
\begin{cases}
u_{i,0}^{n+1} = u_{i,0}^{n} + D_u \frac{\Delta t}{2\Delta x^2}\delta_x^2\left(u_{i,0}^{n} + u_{i,0}^{n+1}\right) + D_u \frac{\Delta t}{\Delta y^2}\delta_y^1\left(u_{i,0}^{n} + u_{i,0}^{n+1}\right) \\
\quad -\frac{\Delta t}{2\Delta x}\delta_x^0 F_{i,0}^{x,n} - \frac{\Delta t}{\Delta y}\left(F_{i,0}^{y,n} + F_{i,1}^{y,n}\right) \\
\quad +\frac{\Delta t}{2}\left(g_{i,0}^{n} + g_{i,0}^{n+1}\right) - \Delta t\left(\frac{\delta_x^2\theta_{i,0}^n}{2\Delta x} + \frac{\delta_y^1\theta_{i,0}^n}{\Delta y}\right), \quad i = 1, \ldots, N_x, \\
u_{i,N_y+1}^{n+1} = u_{i,N_y+1}^{n} + D_u \frac{\Delta t}{2\Delta x^2}\delta_x^2\left(u_{i,N_y+1}^{n} + u_{i,N_y+1}^{n+1}\right) \\
\quad -D_u \frac{\Delta t}{\Delta y^2}\delta_y^1\left(u_{i,N_y}^{n} + u_{i,N_y}^{n+1}\right) \\
\quad -\frac{\Delta t}{2\Delta x}\delta_x^0 F_{i,N_y+1}^{x,n} + \frac{\Delta t}{\Delta y}\left(F_{i,N_y}^{y,n} + F_{i,N_y+1}^{y,n}\right) \\
\quad +\frac{\Delta t}{2}\left(g_{i,N_y+1}^{n} + g_{i,N_y+1}^{n+1}\right) - \Delta t\left(\frac{\delta_x^2\theta_{i,N_y+1}^n}{2\Delta x} + \frac{\delta_y^1\theta_{i,N_y}^n}{\Delta y}\right), \quad i=1, \ldots, N_x, \\
u_{0,j}^{n+1} = u_{0,j}^{n} + D_u \frac{\Delta t}{\Delta x^2}\delta_x^1\left(u_{0,j}^{n} + u_{0,j}^{n+1}\right) + D_u \frac{\Delta t}{2\Delta y^2}\delta_y^2\left(u_{0,j}^{n} + u_{0,j}^{n+1}\right) \\
\quad -\frac{\Delta t}{\Delta x}\left(F_{0,j}^{x,n} + F_{1,j}^{x,n}\right) - \frac{\Delta t}{2\Delta y}\delta_y^0 F_{0,j}^{y,n} \\
\quad +\frac{\Delta t}{2}\left(g_{0,j}^{n} + g_{0,j}^{n+1}\right) - \Delta t\left(\frac{\delta_x^1\theta_{0,j}^n}{\Delta x} + \frac{\delta_y^2\theta_{0,j}^n}{2\Delta y}\right), \quad j = 1, \ldots, N_y, \\
u_{N_x+1,j}^{n+1} = u_{N_x+1,j}^{n} - D_u \frac{\Delta t}{\Delta x^2}\delta_x^1\left(u_{N_x,j}^{n} + u_{N_x,j}^{n+1}\right) \\
\quad +D\frac{\Delta t}{2\Delta y^2}\delta_y^2\left(u_{N_x+1,j}^{n} + u_{N_x+1,j}^{n+1}\right) \\
\quad -\frac{\Delta t}{2\Delta x}\delta_x^0 F_{i,N_y+1}^{x,n} + \frac{\Delta t}{\Delta x}\left(F_{i,N_y}^{y,n} + F_{i,N_y+1}^{y,n}\right) \\
\quad +\frac{\Delta t}{2}\left(g_{N_x+1,j}^{n} + g_{N_x+1,j}^{n+1}\right) - \Delta t\left(\frac{\delta_x^1\theta_{N_x,j}^n}{\Delta x} + \frac{\delta_y^2\theta_{N_x+1,j}^n}{2\Delta y}\right), \quad j=1, \ldots, N_y, \\
\quad j \neq j_a, \ldots, j_b.
\end{cases}
\tag{38}
$$

For the corners, we use the following boundary conditions:

$$
\left\{
\begin{aligned}
u_{0,0}^{n+1} &= u_{0,0}^{n} + D_u \tfrac{\Delta t}{\Delta x^2}\delta_x^1\left(u_{0,0}^{n} + u_{0,0}^{n+1}\right) + D_u \tfrac{\Delta t}{\Delta y^2}\delta_y^1\left(u_{0,0}^{n} + u_{0,0}^{n+1}\right)\\
&\quad -\tfrac{\Delta t}{\Delta x}\left(F_{0,0}^{x,n} + F_{1,0}^{x,n}\right) - \tfrac{\Delta t}{\Delta y}\left(F_{0,0}^{y,n} + F_{0,1}^{y,n}\right)\\
&\quad +\tfrac{\Delta t}{2}\left(g_{0,0}^{n} + g_{0,0}^{n+1}\right)\\
&\quad -\Delta t\left(\frac{\delta_x^1\theta_{0,0}^n}{\Delta x} + \frac{\delta_y^0\theta_{0,0}^n}{\Delta y}\right),\\
u_{N_x+1,0}^{n+1} &= u_{N_x+1,0}^{n} - D_u \tfrac{\Delta t}{\Delta x^2}\delta_x^1\left(u_{N_x,0}^{n} + u_{N_x,0}^{n+1}\right)\\
&\quad +D_u \tfrac{\Delta t}{\Delta y^2}\delta_y^1\left(u_{N_x+1,0}^{n} + u_{N_x+1,0}^{n+1}\right)\\
&\quad -\tfrac{\Delta t}{\Delta x}\left(F_{N_x+1,0}^{x,n} + F_{N_x,0}^{x,n}\right) - \tfrac{\Delta t}{\Delta y}\left(F_{N_x+1,0}^{y,n} + F_{N_x+1,1}^{y,n}\right)\\
&\quad +\tfrac{\Delta t}{2}\left(g_{N_x+1,0}^{n} + g_{N_x+1,0}^{n+1}\right)\\
&\quad -\Delta t\left(\frac{\delta_x^1\theta_{N_x,0}^n}{\Delta x} + \frac{\delta_y^1\theta_{N_x+1,0}^n}{\Delta y}\right),\\
u_{0,N_y+1}^{n+1} &= u_{0,N_y+1}^{n} + D_u \tfrac{\Delta t}{\Delta x^2}\delta_x^1\left(u_{0,N_y+1}^{n} + u_{0,N_y+1}^{n+1}\right)\\
&\quad -D_u \tfrac{\Delta t}{\Delta y^2}\delta_y^1\left(u_{0,N_y}^{n} + u_{0,N_y}^{n+1}\right)\\
&\quad -\tfrac{\Delta t}{\Delta x}\left(F_{0,N_y+1}^{x,n} + F_{1,N_y+1}^{x,n}\right) - \tfrac{\Delta t}{\Delta y}\left(F_{0,N_y+1}^{y,n} + F_{0,N_y}^{y,n}\right)\\
&\quad +\tfrac{\Delta t}{2}\left(g_{0,N_y+1}^{n} + g_{0,N_y+1}^{n+1}\right)\\
&\quad -\Delta t\left(\frac{\delta_x^1\theta_{0,N_y+1}^n}{\Delta x} + \frac{\delta_y^1\theta_{0,N_y}^n}{\Delta y}\right),\\
u_{N_x+1,N_y+1}^{n+1} &= u_{N_x+1,N_y+1}^{n} - D_u \tfrac{\Delta t}{\Delta x^2}\delta_x^1\left(u_{N_x,N_y+1}^{n} + u_{N_x,N_y+1}^{n+1}\right)\\
&\quad -D_u \tfrac{\Delta t}{\Delta y^2}\delta_y^1\left(u_{N_x+1,N_y}^{n} + u_{N_x+1,N_y}^{n+1}\right)\\
&\quad -\tfrac{\Delta t}{\Delta x}\left(F_{N_x+1,N_y+1}^{x,n} + F_{N_x,N_y+1}^{x,n}\right)\\
&\quad -\tfrac{\Delta t}{\Delta y}\left(F_{N_x+1,N_y+1}^{y,n} + F_{N_x+1,N_y}^{y,n}\right)\\
&\quad +\tfrac{\Delta t}{2}\left(g_{N_x+1,N_y+1}^{n} + g_{N_x+1,N_y+1}^{n+1}\right)\\
&\quad -\Delta t\left(\frac{\delta_x^1\theta_{N_x,N_y+1}^n}{\Delta x} + \frac{\delta_y^1\theta_{N_x+1,N_y}^n}{\Delta y}\right).
\end{aligned}
\right.
\tag{39}
$$

Similarly, for the chemoattractant φ, we have the implicit-explicit scheme in the interior points of the 2D domain:

$$
\begin{aligned}
\varphi_{i,j}^{n+1} = \varphi_{i,j}^{n} + D_{\varphi}\frac{\Delta t}{2}\left[\frac{\delta_x^2\left(\varphi_{i,j}^{n}+\varphi_{i,j}^{n+1}\right)}{\Delta x^2} + \frac{\delta_y^2\left(\varphi_{i,j}^{n}+\varphi_{i,j}^{n+1}\right)}{\Delta y^2}\right] \\
\frac{\Delta t}{2}(a(u_{i,j}^{n} + u_{i,j}^{n+1}) - \frac{\Delta t}{2}(b(\varphi_{i,j}^{n} + \varphi_{i,j}^{n+1}).
\end{aligned}
\tag{40}
$$

At the boundaries and the corners of the numerical schemes for φ, we use, respectively, conditions:

$$
\begin{cases}
\varphi_{0,0}^{n+1} &= \varphi_{0,0}^{n} + D_{\varphi}\frac{\Delta t}{\Delta x^2}\delta_x^1(\varphi_{0,0}^{n} + \varphi_{0,0}^{n+1}) + 2D_{\varphi}\frac{\Delta t}{\Delta y^2}\delta_y^1(\varphi_{0,0}^{n} + \varphi_{0,0}^{n+1}) \\
&\quad + a\Delta t(u_{0,0}^{n} + u_{0,0}^{n+1}) - b\Delta t(\varphi_{0,0}^{n} + \varphi_{0,0}^{n+1}), \\
\varphi_{N_x+1,0}^{n+1} &= \varphi_{N_x+1,0}^{n} - D_{\varphi}\frac{\Delta t}{\Delta x^2}\delta_x^1(\varphi_{N_x+1,0}^{n} + \varphi_{N_x+1,0}^{n+1}) \\
&\quad + D_{\varphi}\frac{\Delta t}{\Delta y^2}\delta_y^1(\varphi_{N_x+1,0}^{n} + \varphi_{N_x+1,0}^{n}) \\
&\quad + a\Delta t(u_{N_x+1,0}^{n} + u_{N_x+1,0}^{n+1}) - b\Delta t(\varphi_{N_x+1,0}^{n} + \varphi_{N_x+1,0}^{n+1}), \\
\varphi_{N_x+1,N_y+1}^{n+1} &= \varphi_{Nx+1,N_y+1}^{n} - D_{\varphi}\frac{\Delta t}{\Delta x^2}\delta_x^1(\varphi_{N_x,N_y+1}^{n} + \varphi_{N_x,N_y+1}^{n+1}) \\
&\quad - D_{\varphi}\frac{\Delta t}{\Delta y^2}\delta_x^1(\varphi_{N_x+1,N_y+1}^{n} + \varphi_{N_x+1,N_y}^{n}) \\
&\quad + a\Delta t u_{N_x+1,N_y+1}^{n} - b\Delta t\varphi_{N_x+1,N_y+1}^{n}, \\
\varphi_{0,N_y+1}^{n+1} &= \varphi_{0,0}^{n} + D_{\varphi}\frac{\Delta t}{\Delta x^2}\delta_x^1(\varphi_{1,N_y+1}^{n} + \varphi_{1,N_y+1}^{n+1}) \\
&\quad - D_{\varphi}\frac{\Delta t}{\Delta y^2}\delta_y^1(\varphi_{0,N_y}^{n} + \varphi_{0,N_y}^{n+1}) + a\Delta t u_{0,N_y+1}^{n} - b\Delta t\varphi_{0,N_y+1}^{n}.
\end{cases}
\tag{41}
$$

For the borders, we use:

$$
\begin{cases}
\varphi_{i,0}^{n+1} &= \varphi_{i,0}^{n} + D_{\varphi}\frac{\Delta t}{\Delta x^2}\delta_x^2(\varphi_{i,0}^{n} + \varphi_{i,0}^{n+1}) + 2D_{\varphi}\frac{\Delta t}{\Delta y^2}\delta_y^1(\varphi_{i,0}^{n} + \varphi_{i,0}^{n+1}) \\
&\quad + a\Delta t u_{i,0}^{n} - b\Delta t\varphi_{i,0}^{n}, \\
\varphi_{i,N_y+1}^{n+1} &= \varphi_{i,N_y+1}^{n} + D_{\varphi}\frac{\Delta t}{\Delta x^2}\delta_x^2(\varphi_{i,N_y+1}^{n} + \varphi_{i,N_y+1}^{n+1}) \\
&\quad - 2D_{\varphi}\frac{\Delta t}{\Delta y^2}\delta_y^1(\varphi_{i,N_y}^{n} + \varphi_{i,N_y}^{n+1}) + a\Delta t u_{i,N_y+1}^{n} - b\Delta t\varphi_{i,N_y+1}^{n}, \\
\varphi_{0,j}^{n+1} &= \varphi_{0,j}^{n} + D_{\varphi}\frac{\Delta t}{\Delta y^2}\delta_y^2(\varphi_{0,j}^{n} + \varphi_{0,j}^{n+1}) + 2D_{\varphi}\frac{\Delta t}{\Delta x^2}\delta_x^1(\varphi_{0,j}^{n} + \varphi_{0,j}^{n+1}) \\
&\quad + a\Delta t u_{0,j}^{n} - b\Delta t\varphi_{0,j}^{n}, \\
\varphi_{N_x+1,j}^{n+1} &= \varphi_{N_x+1,j}^{n} + D_{\varphi}\frac{\Delta t}{\Delta y^2}\delta_y^2(\varphi_{N_x+1,j}^{n} + \varphi_{N_x+1,j}^{n+1}) \\
&\quad - 2D_{\varphi}\frac{\Delta t}{\Delta x^2}\delta_x^1(\varphi_{N_x,j}^{n} + \varphi_{N_x,j}^{n+1}) + a\Delta t u_{N_x+1,j}^{n} - b\Delta t\varphi_{N_x+1,j}^{n}.
\end{cases}
\tag{42}
$$

Note that for $i = N_x + 1$ the last formula in (42) is applied for $j = 1, \dots, N_y,\ j \neq j_a, \dots, j_b$.

Remark 1 If we consider a two-dimensional domain Ω_r connected to the right endpoint of the one-dimensional corridor C, the complete numerical scheme for the left domain Ω_l described above can be considered.

The main difference is that the transmission conditions at the interface between the box and the channel (the left for the box Ω_l and the right for the corridor) are reversed to the left for the corridor and the right for the box Ω_r. In the numerical scheme, the change only affects the channel C, where we have transmission conditions also for $u_{N+1}^{c,n}$ (resp. $v_{N+1}^{c,n}$). The same boundary condition can be used without transmission conditions, with only the additional term derived from the KK-condition and it must be added as well for $u_0^{c,n}$ (resp. $v_0^{c,n}$).

For the computation of solutions of (4) on the one-dimensional channel C an implicit version of the second-order AHO scheme, i.e. AHO^2 scheme, formerly introduced on a line in [37] and extended on networks in [7] is applied, in order to ensure the stability of numerical solutions in the channels:

$$\begin{cases} u_i^{c,n+1} = u_i^{c,n} + \lambda \frac{\Delta t}{2\Delta x} \delta_x^2 u_i^{c,n} - \left(\frac{\Delta t}{2\Delta x} - \frac{\Delta t}{4\lambda}\right) \delta_x^0 v_i^{c,n} + \frac{\Delta t}{4\lambda} \delta_x^0 F_{c,i}^n \\ \qquad + \frac{\Delta t}{4} \left(g_{i-1}^n + 2g_i^n + g_{i+1}^n\right), \\ v_i^{c,n+1} = v_i^{c,n} - \lambda^2 \frac{\Delta t}{2\Delta x} \delta_x^0 u_i^{c,n} + \left(\frac{\lambda \Delta t}{2\Delta x} - \frac{\Delta t}{4}\right) \delta_x^2 v_i^{c,n} + \frac{\Delta t}{4} \delta_x^2 F_{c,i}^n \\ \qquad + \lambda \frac{\Delta t}{4} \left(g_{i-1}^n - g_{i+1}^n\right), \end{cases} \tag{43}$$

with mass-preserving boundary conditions (including the additional source term g) at the external boundaries. The scheme (43) is endowed with transmission conditions for the flux and for the density of cells, respectively:

$$v_0^{c,n+1} = -K(b-a) u_0^{c,n+1} + K \Delta y \sum_{j=j_a}^{j_b} \left(u_{N_x+1,j}^n + u_{N_x+1,j}^{n+1}\right), \tag{44}$$

and

$$\begin{aligned} u_0^{c,n+1} &= u_0^{c,n} + \lambda \frac{\Delta t}{\Delta x} \delta_x^1 u_0^{c,n+1} - \left(\frac{\Delta t}{\Delta x} - \frac{\Delta t}{2\lambda}\right) \left(v_0^{c,n+1} + v_1^{c,n+1}\right) \\ &\quad - K \frac{\Delta t}{\Delta x} \Delta y \sum_{j=j_a}^{j_b} \left(u_0^{c,n+1} + u_0^{c,n} - u_{N_x+1,j}^{n+1} - u_{N_x+1,j}^n\right) \\ &\quad - \frac{\Delta t}{2\lambda} \left(F_{c,0}^{n+1} + F_{c,1}^{n+1}\right) + \frac{\Delta t}{2} \left(g_0^{n+1} + g_1^{n+1}\right). \end{aligned} \tag{45}$$

4.1.1 Stability at Interfaces

Note that, in order to ensure the positivity of the quantities in the above formulas deriving from the KK conditions—i.e., (37) for the 2D domain and (45) for the 1D domain- we also need to take care of the ratio between the KK coefficient K and the space discretization steps. In particular, one needs to ensure that $K\frac{\Delta t}{\Delta x}$ and, respectively, $K\frac{\Delta t}{\Delta x}\Delta y$ is not too big in order to damp possible high oscillations produced by the term in parentheses.

Moreover, as previously discussed, we need to check that the numerical monotonicity condition is satisfied:

$$\frac{k_1}{\left(k_2+\varphi_{i,j}^n\right)^2}|\partial_{x,i,j}^n\varphi_{i,j}^n| \leq \sqrt{D_M} \tag{46}$$

in the computational domain in order to ensure non-negative solutions.

Now, we consider the interface between the 2D and 1D domains. If we assume $g=0$ and $F=\chi(\varphi)\nabla\varphi$, the first equation in the 2D parabolic system (33), rewrites as $\partial_t u = D\Delta u - \nabla\cdot(u\cdot f)$, then for the new version of 2D transmission condition the monotonicity is preserved when:

$$1-D_u\frac{\Delta t}{\Delta x^2}-D_u\frac{\Delta t}{\Delta y^2}+\frac{\Delta t}{\Delta x}f_{N_x+1,j}^{x,n}-\frac{\Delta t}{\Delta x}|f_{N_x+1,j}^{x,n}|-\frac{\Delta t}{\Delta y}|f_{N_y+1,j}^{y,n}|-\frac{\Delta t}{\Delta x}K>0, \tag{47}$$

which gives us the stability condition for the left side of the interface.

The monotonicity is ensured if we assume the following inequality for the explicit KK-transmission (44) and (45):

$$\begin{aligned}&(1+\rho)-2\lambda\frac{\Delta t}{\Delta x}\rho+\frac{\Delta t}{2}(\rho-1)-\frac{\Delta t}{\Delta x}K(b-a)(1+\rho)-\frac{\Delta t}{2\lambda}F_{c,0}^n>0\\&\Leftrightarrow \Delta t\leq\frac{1+\rho}{\frac{2\lambda}{\Delta x}\rho-\frac{\rho-1}{2}+K\frac{(b-a)(1+\rho)}{\Delta x}+\frac{F_{c,0}^n}{2\lambda}},\end{aligned} \tag{48}$$

with $\rho := \frac{\lambda-(b-a)K}{\lambda+(b-a)K}$. For the implicit AHO^2 scheme (43), the condition above reads as:

$$\Delta t \leq \frac{1+\rho}{K\frac{(b-a)(1+\rho)}{\Delta x}+\frac{F_{c,0}^n}{2\lambda}}.$$

In Fig. 1, the time step restriction (48) is depicted for a qualitative understanding of the effect of the Kedem–Katchalsky constant K and of the channel width σ on the time step Δt. As expected, the time step Δt must be chosen smaller when either K or σ increases. Furthermore for $K=0$ we recover the time step restriction of the AHO^2-scheme $\Delta t \leq \frac{4\Delta x}{\Delta x+4\lambda} = 2\cdot 10^{-3}$. Since the values of K are typically of

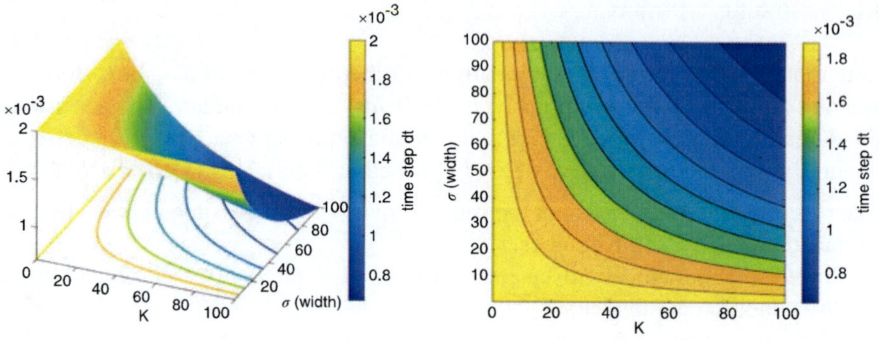

Fig. 1 Time step restriction (48) Δt for the hyperbolic transmission condition with $\Delta x = 0.01$, $\Delta y = 0.1$ and $\lambda = 5$ for different K and channel widths $(b - a)$ for the transmission between the 2D parabolic and the 1D hyperbolic interface

similar magnitude to the diffusion coefficients, the additional stability restrictions caused by the hyperbolic part of the transmission conditions are minimal.

Comparing time step restrictions with each other, it is evident that the restriction for the 2D parabolic transmission condition dominates the full model.

For the sake of completeness, we underline that at each time step a non-linear equation system must be solved, for which Newton–Krylov subspace methods [23] can be used, which take advantage of the mostly sparse structure of the Jacobian matrix.

4.2 *Numerical Schemes for the Approximation of the Model (7)–(8)*

In this section we present numerical simulations based on finite difference schemes employed for equations (7) coupled with (8), with the aim of showing the effect of variation of some key parameters of the model. We consider a square domain $\Omega = [100, 500] \times [100, 500]$, which represents an area, located near the bottom of the chamber where the TCs are positioned, having horizontal and vertical size $L_x = L_y = 400\,\mu\text{m}$. We introduce a discretization on L_x in $N - 1$ subintervals of length $\Delta x = \dfrac{L_x}{N-1}$ and a discretization on L_y in $M - 1$ subintervals of length $\Delta y = \dfrac{L_y}{M-1}$. In all of performed tests, the time and spatial steps are chosen as $\Delta t = 10\,\text{s}$, i.e. representing $1/6$ of the video footage timeframe (of 2 min) and $\Delta x = \Delta y = 4\,\mu\text{m}$, i.e. the size of ICs radius.

Then we introduce a Cartesian grid Ω_Δ consisting of grid points (x_n, y_m), where $x_n = n\Delta x$, for $n = 0, \ldots, N-1$ and $y_m = m\Delta y$, for $m = 0, \ldots, M-1$. The same can be done for the time interval $[0, T]$, in this case if Δt is the time step, t_k will be the k-th temporal step, i.e. $t_k = k\Delta t$, for $k = 0, \ldots, N_{\Delta t}$.

With the notation $u^k_{n,m}$ we denote the approximation of a function $u(x, y, t)$ at the grid point (x_n, y_m, t_k). Besides, since experimentally it was observed that ICs leave the domain Ω, to manage the entrance and exit of cells and avoid numerical instabilities, we added a ghost grid to Ω_d, where the cells lie after having left the main domain. To construct the grid, we considered the extended x interval $[-L_{x^*}, L_x + L_{x^*}]$, with $L_{x^*} > 0$, with nodes $x_n = n\Delta x$, for $n = -N^*, \ldots, 0, \ldots, N^*$, and the extended y interval $[-L_{y^*}, L_y + L_{y^*}]$, with $L_{y^*} > 0$, with nodes $y_m = m\Delta y$, for $m = -M^*, \ldots, 0, \ldots, M^*$. The equations were solved only on Ω_d.

4.2.1 Discretization of the PDE (Eq. (7))

The parabolic Eq. (7) is composed of the diffusion term, the source term, and the stiff degradation term $-\eta f$.

In order to eliminate this last quantity we perform the classical exponential transformation: $f(\mathbf{x}, t) = e^{-\eta t} u(\mathbf{x}, t)$, which leads to the diffusion equation with source for $u(\mathbf{x}, t)$:

$$\partial_t u = D\Delta u + e^{\eta t} \xi \sum_{j=1}^{N_{tot,c}} \chi_{\mathbf{B}(\mathbf{Y}_j, R_T)}. \tag{49}$$

For this equation we apply a central difference scheme in space, i.e. the 5-point stencil for the Laplacian, and the parabolic Crank–Nicolson scheme in time.

The numerical scheme can be written as:

$$\frac{u^{k+1}_{n,m} - u^k_{n,m}}{\Delta t} = \frac{D}{2}\left(D^2_x u^{k+1} + D^2_y u^{k+1}\right) + \frac{D}{2}\left(D^2_x u^k + D^2_y u^k\right)$$
$$+ \frac{1}{2} e^{\eta(k+1)\Delta t} \xi \sum_{j=1}^{N_{tot,c}} \chi_{\mathbf{B}(\mathbf{Y}^{k+1}_j, R_T)} + \frac{1}{2} e^{\eta k \Delta t} \xi \sum_{j=1}^{N_{tot,c}} \chi_{\mathbf{B}(\mathbf{Y}^k_j, R_T)},$$

where $D^2_x u$ (and analogously $D^2_y u$) is defined as the central difference:

$$D^2_x u^k = \frac{u^k_{n-1,m} - 2u^k_{n,m} + u^k_{n+1,m}}{\Delta x^2}.$$

4.2.2 Boundary Conditions

Using the exponential transformation above, the associated boundary conditions are:

$$D\frac{\partial u}{\partial \mathbf{n}} + au = e^{\eta t}b,$$

that we rewrite as $\frac{\partial u}{\partial \mathbf{n}} + pu = q(t)$, with $p = \frac{a}{D}$ and $q(t) = e^{\eta t}\frac{b}{D}$. Besides, to distinguish the values of p and q on the different sides of Ω, we number them as follows: p_1 and q_1 are assumed on $y = 0$, p_2 and q_2 on $x = L_x$, p_3 and q_3 on $y = L_y$ and p_4 and q_4 on $x = 0$.

For the discretization of the boundary conditions we use a central finite difference scheme. Following [45], on $y = 0$ and $y = L_y$, we have:

$$\frac{\partial u}{\partial y} + p_s u - q_s = \frac{u^k_{n,m+1} - u^k_{n,m-1}}{2\Delta y} + r^k_s u_{n,m} - h^k_s, \tag{50}$$

and we proceed analogously for the vertical conditions, with the signs of r_s and h_s depending on the incoming/outgoing flow. For instance on $y = 0$, we have $r_1 = -p_1$ and $h_1 = -q_1$, on $y = L_y$, we have $r_3 = p_3$ and $h_3 = q_3$, on $x = L_x$, we have $r_4 = -p_4$ and $h_4 = -q_4$ and on $x = 0$ we have $r_2 = p_2$ and $h_2 = q_2$ in order to have an incoming chemical flux through the boundary.

4.2.3 Discretization of the ODE (8)

As in [8], the equation of motion (8) is reduced to a first order system with

$$\begin{aligned}\dot{\mathbf{V}}_i = &\frac{\gamma}{\mathcal{W}}\int_{\mathbf{B}(\mathbf{X}_i,\bar{R})} \chi(f(\mathbf{x},t))\nabla f(\mathbf{x},t)w_i(\mathbf{x})d\mathbf{x} + \sum_{j:\mathbf{Y}_j\in\mathbf{B}(\mathbf{X}_i,\mathbf{R}_2)\setminus\{\mathbf{X}_i\}} \mathbf{K}(\mathbf{Y}_j - \mathbf{X}_i)\\ &+ \sum_{j:\mathbf{X}_j\in\mathbf{B}(\mathbf{X}_i,\mathbf{R}_4)\setminus\{\mathbf{X}_i\}} \mathbf{K}(\mathbf{X}_j - \mathbf{X}_i)\\ &+ \frac{\beta}{N_i}\sum_{j:\mathbf{X}_j\in B(\mathbf{X}_i,R_3)\setminus\{\mathbf{X}_i\}} \frac{(\mathbf{V}_j - \mathbf{V}_i)}{\left(1 + \frac{\|\mathbf{X}_j-\mathbf{X}_i\|^2}{R_3^2}\right)^{\sigma}} - \mu\mathbf{V}_i,\end{aligned} \tag{51}$$

$$\dot{\mathbf{X}}_i = \mathbf{V}_i, \tag{52}$$

for $i = 1,\ldots,N_{tot,I}$. Eq. (51) is discretized with a one step IMEX method, putting in implicit the term containing $\mathbf{V}_i$ and in explicit the other addends, and Eq. (52) is solved with forward Euler method. The two-dimensional integral in (51) can be computed by a 2D quadrature formula, which due to the truncated Gaussian weight function $w_i(\mathbf{x})$ given in (10), is reduced to a sum of the discretized integrand functions on the grid points belonging to the ball $\mathbf{B}(\mathbf{X}_i, \bar{R})$. The two-dimensional integral in $\mathcal{W}$ defined in (9) is approximated by $\widetilde{\mathcal{W}} := \sum_{n,m \text{s.t.} (x_n, x_m) \in \mathbf{B}(\mathbf{X}_i^k, \bar{R})} (w_i)_{n,m}^{(k)}$. The gradients of Eq. (51) are approximated with first order differences $\nabla_{n,m} f^k \approx \left(\frac{f_{n+1,m}^k - f_{n,m}^k}{\Delta x}, \frac{f_{n,m+1}^k - f_{n,m}^k}{\Delta y} \right)$, then Eqs. (51)–(52) are discretized as follows:

$$\frac{\mathbf{V}_i^{k+1} - \mathbf{V}_i^k}{\Delta t} = \frac{\gamma}{\widetilde{\mathcal{W}}} \sum_{n,m \text{s.t.} (x_n, x_m) \in \mathbf{B}(\mathbf{X}_i^k, \bar{R})} \chi(f^k)(\nabla_{n,m} f^k)(w_i)_{n,m}^{(k)} \tag{53}$$

$$+ \sum_{j:\mathbf{X}_j^k \in \mathbf{B}(\mathbf{X}_i^k, \mathbf{R}_2) \setminus \{\mathbf{X}_i^k\}} \mathbf{K}(\mathbf{X}_j^k - \mathbf{X}_i^k) + \sum_{j:\mathbf{X}_j^k \in \mathbf{B}(\mathbf{Y}_i^k, \mathbf{R}_4) \setminus \{\mathbf{Y}_i^k\}} \mathbf{K}(\mathbf{Y}_j^k - \mathbf{X}_i^k) \tag{54}$$

$$+ \frac{\beta}{N_i} \sum_{j:\mathbf{X}_j^k \in B(\mathbf{X}_i^k, R_3) \setminus \{\mathbf{X}_i^k\}} \frac{(\mathbf{V}_j^k - \mathbf{V}_i^{k+1})}{\left(1 + \frac{\|\mathbf{X}_j^k - \mathbf{X}_i^k\|^2}{R_3^2}\right)^{\sigma}} - \mu \mathbf{V}_i^{k+1}, \tag{55}$$

$$\frac{\mathbf{X}_i^{k+1} - \mathbf{X}_i^k}{\Delta t} = \mathbf{V}_i^{k+1}. \tag{56}$$

4.3 *Discretization of the SDE (20)*

The stochastic equation of the motion (20) can be decoupled as follows:

$$\dot{\mathbf{V}}_i = \frac{\gamma}{\mathcal{W}} \int_{\mathbf{B}(\mathbf{X}_i, \bar{R})} \chi(f(\mathbf{x}, t)) \nabla f(\mathbf{x}, t) w_i(\mathbf{x}) d\mathbf{x} + \sum_{j:\mathbf{Y}_j \in \mathbf{B}(\mathbf{X}_i, \mathbf{R}_2) \setminus \{\mathbf{X}_i\}} \mathbf{K}(\mathbf{Y}_j - \mathbf{X}_i) \tag{57}$$

$$+ \sum_{j:\mathbf{X}_j \in \mathbf{B}(\mathbf{X}_i, \mathbf{R}_4) \setminus \{\mathbf{X}_i\}} \mathbf{K}(\mathbf{X}_j - \mathbf{X}_i) - \mu \mathbf{V}_i,$$

$$\dot{\mathbf{X}}_i = \mathbf{V}_i + \sigma \psi(t). \tag{58}$$

for $i = 1, \ldots, N_{tot,I}$. The discretization of Eq. (57) coincides with the one for Eq. (55), while Eq. (58) requires the application of the Euler-Maruyama method [18]. Equation (58) can be written in the differential form and use $dW(t) = \psi(t)dt$ where $dW(t)$ denotes the differential form of the Brownian motion:

$$d\mathbf{X}_i(t) = \mathbf{V}_i dt + \sigma dW(t), \tag{59}$$

where $\mathbf{X}_i(t)$ is a one dimensional Wiener process with drift $\mathbf{V}_i$ and diffusion σ. This equation is discretized with the Euler-Maruyama scheme, which is the stochastic version of the deterministic Euler scheme. The increments of the Wiener process are defined as:

$$\Delta W = W^{k+1} - W^k,$$

with $0 \leq k \leq N_{\Delta t} - 1$. The increment ΔW is a random variable with zero mean and variance equal to Δt, with $\Delta W \sim \mathcal{N}(0, \Delta t)$, and with this increment we can construct approximations by drawing normally distributed numbers from a random generator. We approximate the process (59) at the discrete time points t_k, $0 \leq k \leq N_{\Delta t} - 1$ by

$$\mathbf{X}_i^{k+1} = \mathbf{X}_i^k + \mathbf{V}_i^{k+1}\Delta t + \sigma \Delta W, \tag{60}$$

where $\Delta W = \sqrt{\Delta t} Z^k$, with Z^k being standard normal variables with mean 0 and variance 1 for all k.

5 Simulation Results

Then, the simulation algorithm built on the mathematical model can be used to reproduce dynamics and interactions observed experimentally. In a forthcoming paper, using the available data taken from video footage of laboratory experiment, we also aim at estimating accurately model parameters in order to use the simulation algorithm as a forecasting tool to predict the behavior of immune cells and to better understand the cross-talk between cancer and immune cells.

All the numerical simulations here presented are performed in MATLAB ©. Moreover, in Table 1 all the parameters used in the numerical simulations are reported.

5.1 Simulation Results Obtained by Macroscopic Model

The computational time for a simulation on the complete geometry until time $t = 50{,}000$ took about 400 s on an Intel(R) Core(TM) i7-3630 QM CPU 2.4 GHz.

The computational domain is schematized in Fig. 2 with the two chambers and 5 corridors $C_m := [0, L]$, $m = 1, \ldots, 5$ with the same width 12 μm and equispaced from each other.

The discretization grid has time step size $\Delta t = 100$ s and space size $\Delta x = 2.5$ μm, $\Delta y = 25$ μm.

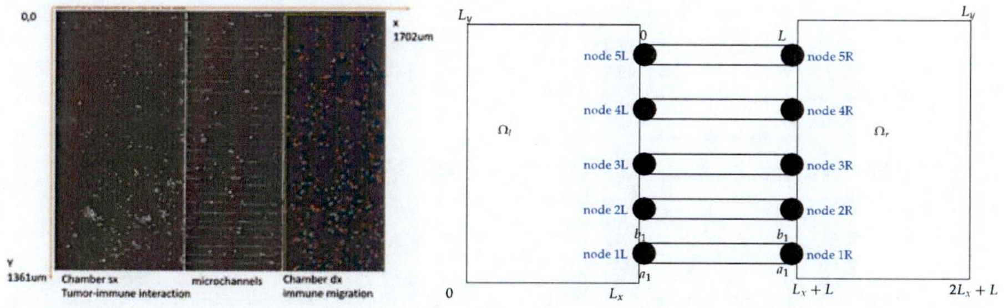

Fig. 2 Microfluidic chip geometry representing two chambers connected by multiple channels. Left panel: pre-processing of a picture taken from the video footage published in the supplementary material of Vacchelli et al. [47]. Right panel: schematization of the chip geometry used in numerical simulations of macroscopic model (1)–(4). Picture taken from Braun et al. [3]. CC BY 4.0

For the examples below, we assume the following initial condition (time $t = 0$) for the tumor cells distribution on the chip for $(x, y) \in \Omega_l$:

$$T(x, y, 0) = 5e^{-10^{-4}(x^2+(y-500)^2)} + 5e^{-10^{-4}(x^2+(y-5)^2)} + 5e^{-10^{-4}(x^2+(y-1000)^2)}, \tag{61}$$

whereas, in the corridors and the right chamber, no tumor cells are present.

For the immune cells distribution on the chip for $(x, y) \in \Omega_r$, we assign:

$$M(x, y, 0) = 5, \text{ for } x, y \in \Omega_r, \tag{62}$$

to mean that IC are disposed in the right chamber, whereas no immune cells are initially present neither in the left chamber nor in the corridors.

For the chemoattractants, we set a constantly null initial density for ω and φ in both the chambers and also in the channels.

5.1.1 Time Evolution of Macroscopic Densities

For the following numerical simulation, we perform numerical simulations being inspired by the laboratory experiment described in Sect. 2, see also [47]. It is worth noting that in the simulations here presented we let the density of tumor cells slightly diffuse in the chip environment, due to the administration of chemotherapy treatment. The results are depicted in the following Figs. 3 and 4.

In Fig. 3 the initial configuration (time $t = 0$) is depicted, while in Fig. 4 the density of the TCs and ICs for times $t = 50{,}000$ s (about 14 h) and $t = 90{,}000$ s (25 h) is represented. Note that at time $t = 0$ tumor cells occupy the left chamber only and immune cells are present in the right chamber only.

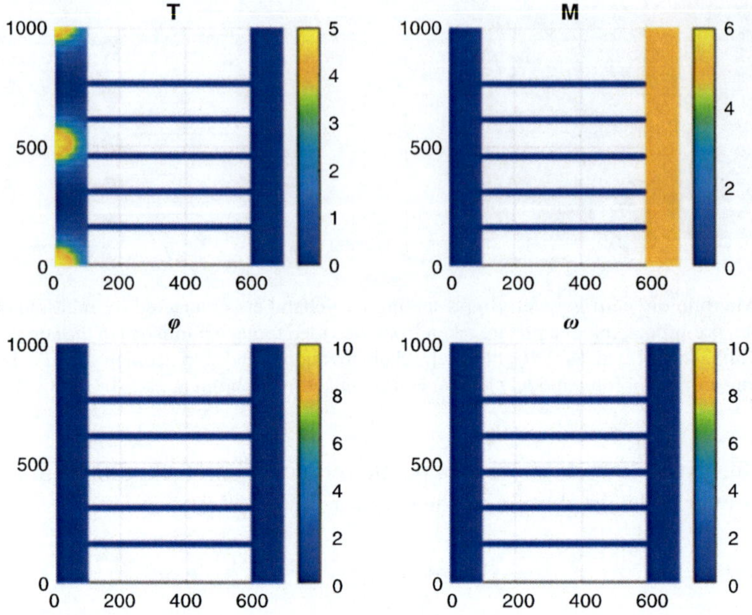

Fig. 3 Initial distribution for the model (1)–(4) at time $t = 0$

Since tumor cells are previously treated by a chemotherapy drug, they slowly diffuse around and stay confined in the left chamber Ω_l during all the simulation time; in the meanwhile they produce chemoattractant φ attracting immune cells. Immune cells M, instead, diffuse around in Ω_r, cross the channels, and after a certain time they enter the left chamber Ω_l while creating chemoattractant ω. This is due to the fact that the chemoattractant φ produced by TCs travels through channels and induces a migration of the ICs M towards the left chamber occupied by TCs. At time $t = 90{,}000$ s, we can see that the density of tumor cells is decreasing under the action of immune cells producing cytokines ω.

In the following Fig. 5, we represent the density of tumor cells and immune cells as particles, by randomly placing them according to their density. The higher the density at a given point, the more cells will be distributed randomly around that area. If the density is lower than a chosen threshold in a certain point, no cells will be represented around it.

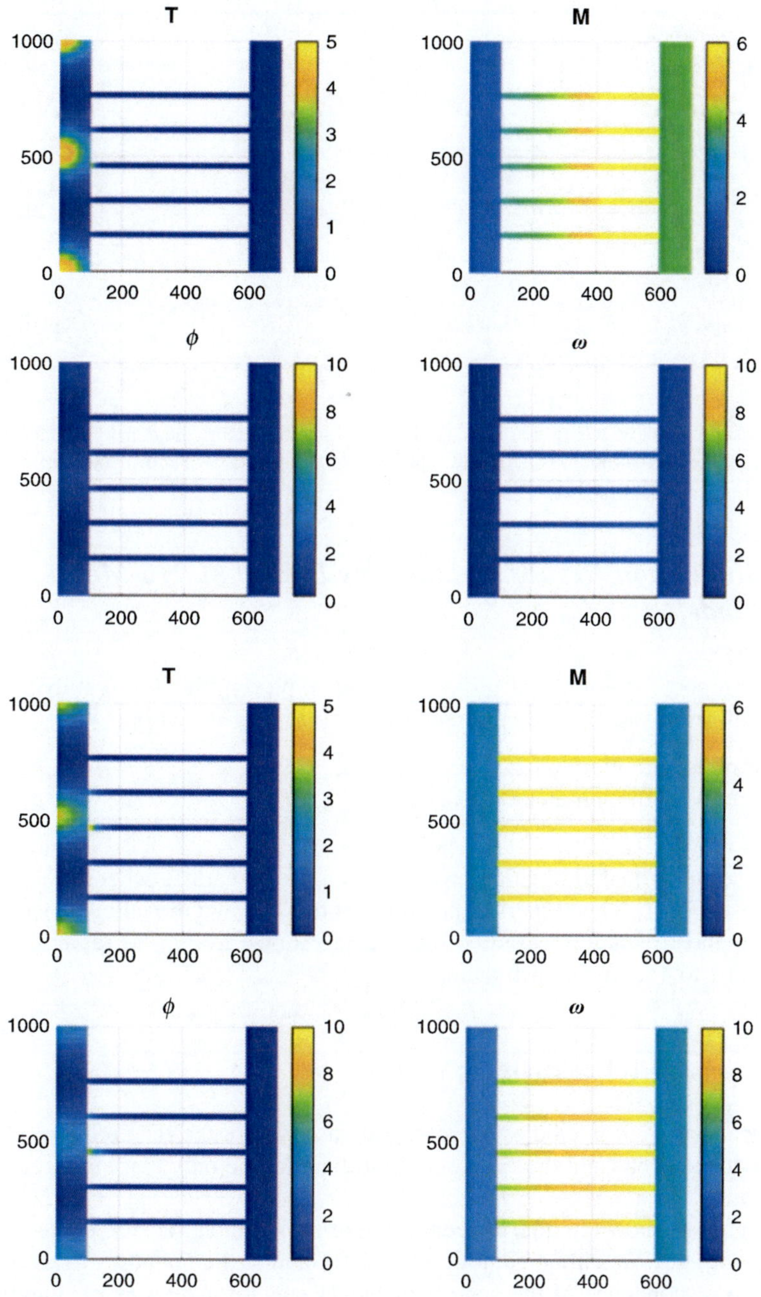

Fig. 4 Evolution of the model (1)–(4) at time $t = 50{,}000$ s (**top**) and at time $t = 90{,}000$ s (**bottom**)

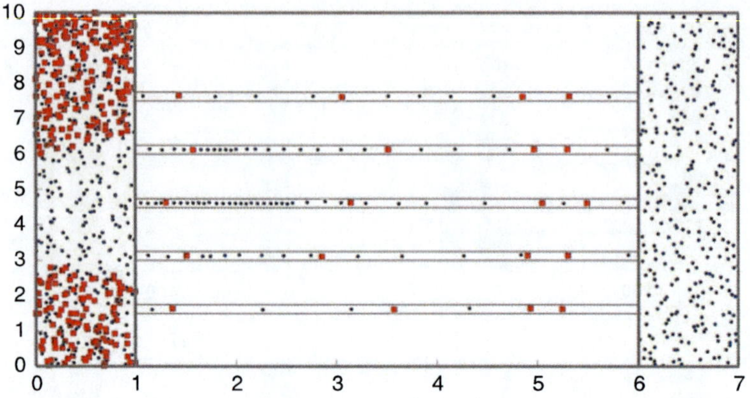

Fig. 5 Visualization of ICs (blue dots) and TCs (red squares) using the density of each quantity obtained by the macro model (1)–(4) and representing them as cells. Picture taken from Braun et al. [3]. CC BY 4.0

5.2 Simulation Results Obtained by Hybrid Macro-Micro Model

Here we report the numerical results obtained exploring the different dynamics in three significant situations represented by the following scenarios:

1. deterministic motion;
2. deterministic motion including cell death;
3. stochastic motion.

For the numerical simulations we consider the domain Ω where four TCs are present, see Fig. 6. The numerical positions of the TCs and the initial concentration of the chemoattractant released by the TCs are shown in Fig. 7. We numbered the TCs from 1 to 4 to distinguish them.

5.2.1 Scenario 1: Deterministic Motion

This scenario is considered as a prototypal case, where the reciprocal mechanical forces between cells and the chemotactic stimuli are the only guiding forces of ICs migration.

In Fig. 8 we show the final concentration of the chemoattractant, plotted as a 2D surface with a focus on the contour lines (left), and at the boundary of the domain and in correspondence of the centers of the TCs as a function of the longitudinal (center) and transverse distance (right). Cutting the surface this way, the maximum concentrations related to each TC are highlighted.

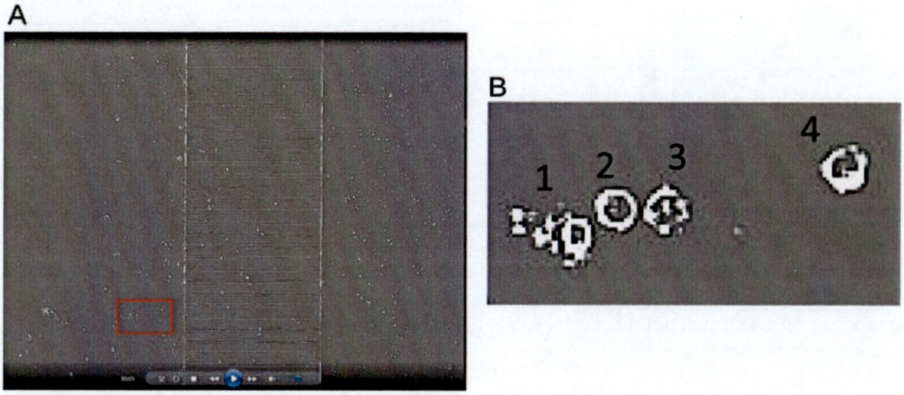

Fig. 6 Timelapse pre-processed image of microfluidic chip environment at time t = 24 h. (**a**) Full chip, in red the domain Ω used in the hybrid micro-macro model; (**b**) Focus on the area Ω. Pre-processing of a picture taken from the video footage published in the supplementary material of Vacchelli et al. [47]. CC BY 4.0

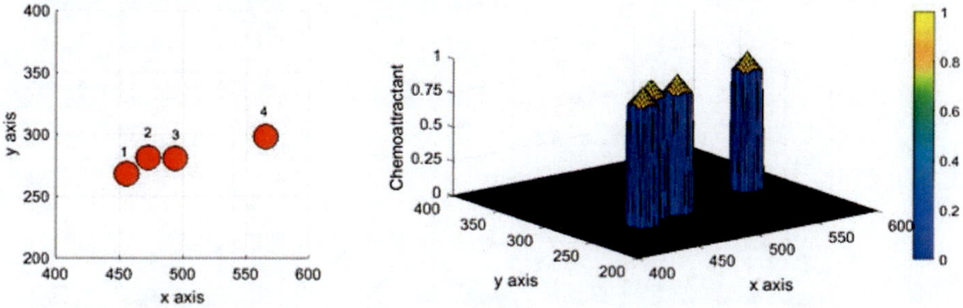

Fig. 7 TCs locations (**left**) and TCs chemoattractant (**right**) at time $t = 0$. The grid dimensions are in μm. Pictures taken from Bretti et al. [8]. CC BY 4.0

In Fig. 9 the time evolution of the migratory activity of ICs at six different times is depicted. At the initial time, only one IC enters the domain, while at time T_1 the number of ICs increases, and they are directed towards the tumor. As the time grows, most of the ICs approach the TCs and stay nearby, accumulating around them; while the others, guided by the inflow of chemical signal, move towards the left-bottom boundaries of the domain.

5.2.2 Scenario 2: Deterministic Motion Including Cell Death

In the current scenario we take into account the possible effects of TCs death over ICs motion. In this preliminary study we did not add to the model a specific term to describe the death of TCs due to ICs, but we directly *turned off* some TCs. In particular, in the following test, we turn off the third TC during the time interval

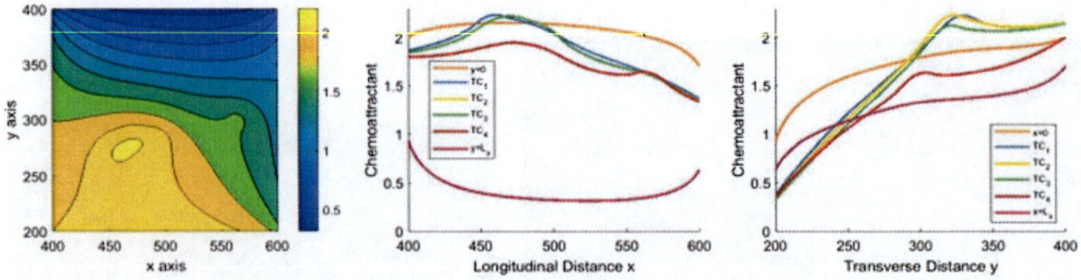

Fig. 8 Final concentration of chemoattractant φ. (**Left**) Concentration plotted in 2D, with contour lines. (**Center**) Concentration profiles as a function of the longitudinal distance. (**Right**) Concentration profiles as a function of the transverse distance. The grid dimensions are in µm. Pictures taken from Bretti et al. [8]. CC BY 4.0

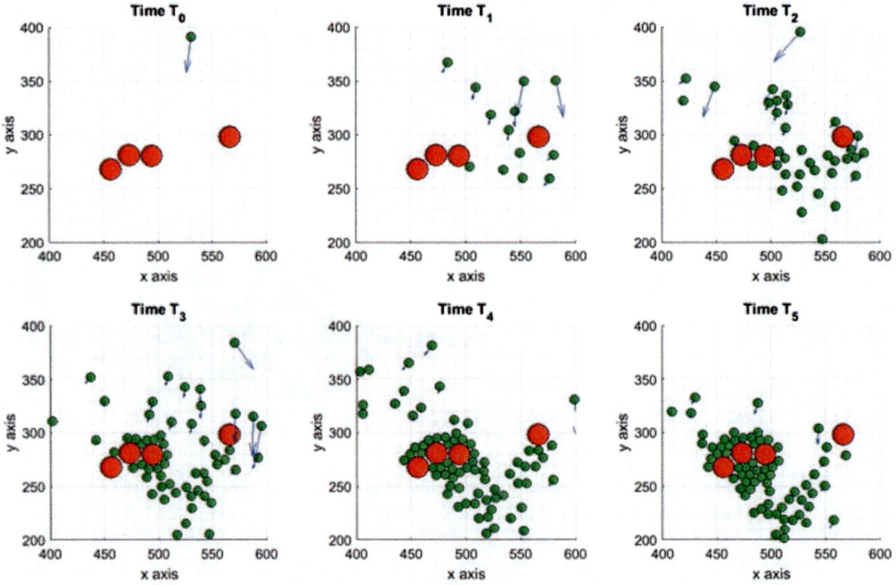

Fig. 9 Scenario 1: ICs meet all TCs and move towards the left and the right sides of Ω. ICs dynamics depicted at six consecutive times. The grid dimensions are in µm. Picture taken from Bretti et al. [8]. CC BY 4.0

$[T_2, T_3]$, and then we turn off also the fourth TC in the time interval $[T_3, T_4]$. This is a simplification of the real process, but is here applied for illustrative purposes. However, in the next future we will include in our modeling the death of cancer cells as a consequence of killing activity of ICs, adding macroscopic equations taking into account the production and the killing activity of cytokines secreted by ICs similarly to the macroscopic modeling in Sect. 3.1.

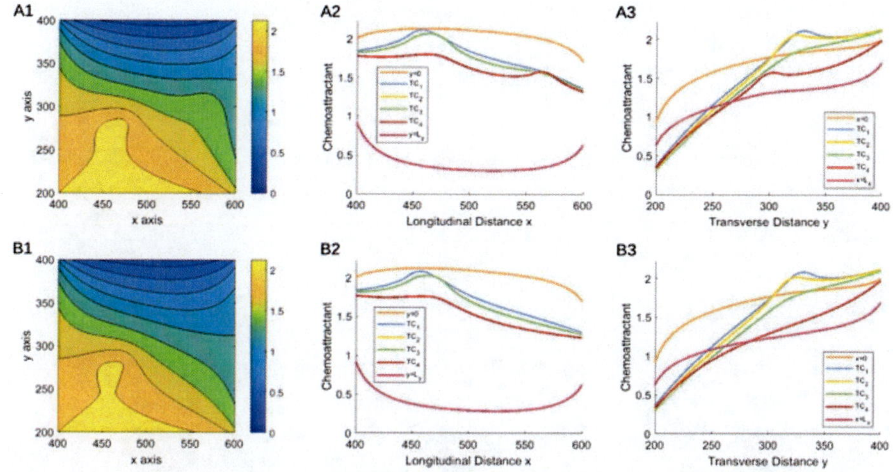

Fig. 10 Chemoattractant concentration in the time interval $[T_2, T_3]$ (**Top**) and at the final time (**Bottom**). (**a1,a2**) Concentration plotted in 2D, concentration profiles as a function of the longitudinal distance (**a2**, **b2**) and transverse (**a3**, **b3**) distance. Picture taken from Bretti et al. [8]. CC BY 4.0

Since dead TCs still remain inside the domain, the repulsion forces with ICs are still effective. A decrease in the concentration of chemoattractant is observed due to the diminution of chemical sources, as shown by Fig. 8 (initial concentration) and Fig. 10 (evolution of the concentration after cell death).

Figure 11 shows the ICs dynamics. At time T_2 both the third and the fourth cells are approached by ICs, while at time T_3 ICs start moving away from the third cell, which was previously killed and not releasing chemicals anymore. Between time T_3 and T_4 also the fourth cell is turned off and from time T_4 onwards it is not approached anymore. The effect of cell death of some tumor cells (the cells 3 and 4 depicted in Fig. 7) on ICs dynamics is depicted in Fig. 11. Please note that in the present case a different behavior respect to the one depicted in Fig. 9 is observed, since here the accumulation around dead cancer cells does not occur, as expected.

5.2.3 Scenario 3: Stochastic Motion

This third case contemplates the addition of the stochastic component (58), which modifies the deterministic IC trajectories.

Indeed, although the deterministic model allows us to control the direction of motion of the ICs biased by the chemical signal produced by TCs, the complex migratory activity of ICs is better reproduced by a stochastic model, describing the

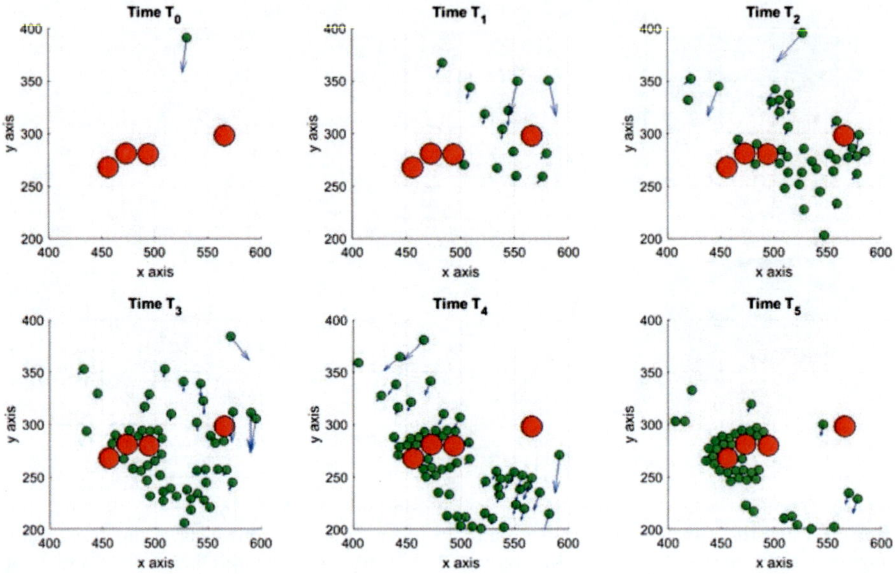

Fig. 11 Scenario 2: Cell 3 dies during the interval $[T_2, T_3]$, while cell 4 dies during the interval $[T_3, T_4]$. ICs dynamics depicted at six consecutive times. The grid dimensions are in μm. Picture taken from Bretti et al. [8]. CC BY 4.0

randomness in cell behavior. For this reason, the features of this last scenario seem to be more realistic.

Estimates of Standard Deviations Here we compute the standard deviation σ in Eq. (20) by using data. To this end, we consider the trajectory P_i of the i-th IC, for $i = 1, \ldots, N_{tot,I}$ and its corresponding smoothing S_i, obtained with a moving average (the Matlab © *smooth* function), and we compute the variance of each IC trajectory $\sigma_i^2 = \frac{1}{T_i} \sum_{k=1}^{T_i} (P_i^k - S_i^k)^2$, where T_i is the number of frames the i-th IC spends on the domain Ω and we obtain the sequence $\Sigma = \{\sigma_i^2\}_{i=1,\ldots,N_{tot,I}}$. Successively, we average the values of Σ and divide by 120 s, which corresponds to the time interval between two consecutive frames:

$$\sigma^2 = \left(\frac{1}{N_{tot,I}} \sum_{i=1}^{N_{tot,I}} \sigma_i^2 \right) / 120, \tag{63}$$

obtaining a variance expressed in seconds. Then we finally obtain the standard deviation σ of real trajectories taking the square root of the relation above.

Since the immune cell population in the experiment is heterogeneous and a clear distinction among cell species is not possible currently, a mathematical model describing an average behavior was formulated. Then, we discarded cell trajectories whose variance is too far from the average in the computation of the standard

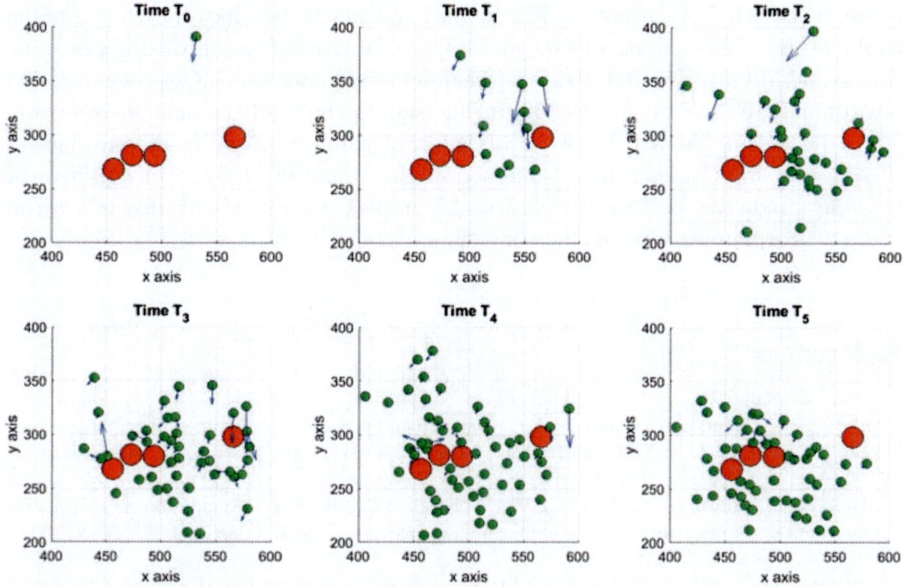

Fig. 12 Scenario 3: ICs meet all TCs and move towards the left and the right sides of Ω. ICs dynamics photographed at six consecutive times. The grid dimensions are in μm. Picture taken from Bretti et al. [8]. CC BY 4.0

deviation. In the future, having more data at our disposal we aim at working on a classification of cells based on their pathways.

ICs Dynamics Figure 12 captures ICs at six different times, while Fig. 8 shows the evolution of the chemoattractant. Differences with the deterministic dynamics shown in Fig. 9 can be highlighted. In the stochastic case, ICs are more spread.

Note that the computations in this case are performed with a finer time spacing of $\Delta t = 10$ s in order to better describe the complex mechanisms at smaller time scales. The plots are always depicted every 2 min since it is the timeframe of video recordings.

6 Conclusions

This work has shown that the algorithm developed and calibrated, although in a simplified setting and under hypoteses that have not taken into account the entire complexity of processes involved in ICD, have shown a good agreement with experimental data. Simulations have reproduced some processes experimentally recorded but not included in the calibration step (TC cell death). Future developments of this approach will go in the definition of equations to take into account heterogeneous behaviors among ICs on one hand and TCs on the other thanks to the exploitation of this mixed macro-micro approach, consideration of different time frames, addition of terms (release of chemical gradients from ICs), switch to 3D environment, both for in vitro experimental setting and mathematical environment. The expected output will be in the direction of gaining insight in the mechanism of action of migration, cell-cell interaction, death and the long term objective of reaching predictive performances of such complex processes with this in vitro-in silico platform going towards the construction of OoC digital twins.

References

1. Bayly, P.V., Taber, L.A., Carlsson, A.E.: Damped and persistent oscillations in a simple model of cell crawling. J. R. Soc. Interface **9**, 1241–1253 (2011). https://doi.org/10.1098/rsif.2011.0627
2. Boulter, E., Grall, D., Cagnol, S., Van Obberghen-Schilling, E.: Regulation of cell-matrix adhesion dynamics and Rac-1 by integrin linked kinase. FASEB J. **20**(9), 1489–1491 (2006). https://doi.org/10.1096/fj.05-4579fje
3. Braun, E.C., Bretti, G., Natalini, R.: Mass-preserving approximation of a chemotaxis multi-domain transmission model for microfluidic chips. Mathematics **9**, 688 (2021). https://doi.org/10.3390/math9060688
4. Braun, E.C., Bretti, G., Natalini, R.: Parameter estimation techniques for a chemotaxis model inspired by Cancer-on-Chip (COC) experiments. Int. J. Non-Linear Mech. **140**, 103895 (2022). https://doi.org/10.1016/j.ijnonlinmec.2021.103895
5. Bretti, G., De Gaetano, A.: An agent-based interpretation of leukocyte chemotaxis in cancer-on-chip experiments. Mathematics **10**, 1338 (2022). https://doi.org/10.3390/math10081338
6. Bretti, G., Natalini, R.: Numerical approximation of nonhomogeneous boundary conditions on networks for a hyperbolic system of chemotaxis modeling the Physarum dynamics. J. Comput. Methods Sci. Eng. **18**, 85–115 (2018)
7. Bretti, G., Natalini, R., Ribot, M.: A hyperbolic model of chemotaxis on a network: a numerical study. Math. Model. Numer. Anal. **48**, 231–258 (2014)
8. Bretti, G., De Ninno, A., Natalini, R., Peri, D., Roselli, N.: Estimation algorithm for a hybrid PDE–ODE model inspired by immunocompetent cancer-on-chip experiment. Axioms **10**, 243 (2021). https://doi.org/10.3390/axioms10040243
9. Checcoli, A., Pol, J.G., Naldi, A., et al.: Dynamical Boolean modeling of immunogenic cell death. Front Physiol. **11**, 590479 (2020). https://doi.org/10.3389/fphys.2020.590479
10. Cohen, J.E.: Mathematics is biology's next microscope, only better; biology is mathematics' next physics, only better. PLoS Biol. **2**(12), e439 (2004). https://doi.org/10.1371/journal.pbio.0020439
11. Curk, T., Marenduzzo, D., Dobnikar, J.: Chemotactic sensing towards ambient and secreted attractant drives collective behaviour of E. coli. PLoS ONE **8**, e74878 (2013). https://doi.org/10.1371/journal.pone.0074878
12. De Ninno, A., Bertani, F.R., Gerardino, A., Schiavoni, G., Musella, M., Galassi, C., Mattei, F., Sistigu, A., Businaro, L.: Microfluidic co-culture models for dissecting the immune response in in vitro tumor microenvironments. J. Vis. Exp. **170**, e61895 (2021). https://doi.org/10.1161/10.3791/61895
13. Di Costanzo, E., Natalini, R., Preziosi, L.: A hybrid mathematical model for self-organizing cell migration in the zebrafish lateral line. J. Math. Biol. **71**, 171–214 (2014). https://doi.org/10.1007/s00285-014-0812-9

14. Filipovic, N., Nikolic, M., Sustersic, T.: Biomaterials for Organ and Tissue Regeneration. New Technologies and Future Prospects Woodhead Publishing Series in Biomaterials, pp. 753–790. Woodhead Publishing, Sawston (2020)
15. Fournier, M.F., Sauser, R., Ambrosi, D., Meister, J.-J., Verkhovsky, A.: Force transmission in migrating cells. J. Cell Biol. **188**, 287–297 (2010). https://doi.org/10.1083/jcb.200906139
16. Gori, M., Simonelli, M.C., Giannitelli, S.M., Businaro, L., Trombetta, M., Rainer, A.: Investigating nonalcoholic fatty liver disease in a liver-on-a-chip microfluidic device. PLoS One **11**(7), e0159729 (2016)
17. Greenberg, J.M., Alt, W.: Stability results for a diffusion equation with functional drift approximating a chemotaxis model. Trans. Am. Math. Soc. **300**, 235–258 (1987)
18. Higham, D.J.: An algorithmic introduction to numerical simulation of stochastic differential equations. SIAM Rev. **43**(3), 525–546 (2001). https://doi.org/10.1137/S0036144500378302
19. Keller, E.F., Segel, L.A.: Initiation of slime mold aggregation viewed as an instability. J. Theor. Biol. **26**, 399–415 (1970)
20. Khetani, S.R., Bhatia, S.N.: Microscale culture of human liver cells for drug development. Nat. Biotechnol. **26**(1), 120–126 (2008)
21. Kim, H.J., Huh, D., Hamilton, G., Ingber, D.E.: Human gut-on-a-chip inhabited by microbial flora that experiences intestinal peristalsis-like motions and flow. Lab Chip **12**(12), 2165–2174 (2012)
22. Kirschner, D., Panetta, J.C.: Modeling immunotherapy of the tumor-immune interaction. J Math Biol. **37**(3), 235–252 (1998). https://doi.org/10.1007/s002850050127
23. Knoll, D.A., Keyes, D.E.: Jacobian-Free Newton-Krylov methods: a survey of approaches and applications. J. Comput. Phys. **193**, 357–397 (2004)
24. Kroemer, G., Galluzzi, L., Kepp, O., Zitvogel, L.: Immunogenic cell death in cancer therapy. Annu. Rev. Immunol. **31**, 51–72 (2013). https://doi.org/10.1146/annurev-immunol-032712-100008
25. Lapidis, I.R., Schiller, R.: Model for the chemotactic response of a bacterial population. Biophys. J. **16**, 779–789 (1976)
26. Lee, J.H., Kim, S.K., Khawar, I.A., et al.: Microfluidic co-culture of pancreatic tumor spheroids with stellate cells as a novel 3D model for investigation of stroma-mediated cell motility and drug resistance. J. Exp. Clin. Cancer Res. **37**(1), 4 (2018). https://doi.org/10.1186/s13046-017-0654-6
27. Lee, S.W.L., Seager, R.J., Litvak, F., et al.: Integrated in silico and 3D in vitro model of macrophage migration in response to physical and chemical factors in the tumor microenvironment. Integr. Biol. (Camb) **12**(4), 90–108 (2020). https://doi.org/10.1093/intbio/zyaa007
28. Lewin, T.D., Avignon, B., Tovaglieri, A., Cabon, L., Gjorevski, N., Hutchinson, L.G.: An in silico model of T cell infiltration dynamics based on an advanced in vitro system to enhance preclinical decision making in cancer immunotherapy. Front. Pharmacol **13**, 837261 (2022). https://doi.org/10.3389/fphar.2022.837261
29. Low, L.A., Mummery, C., Berridge, B.R., Austin, C.P., Tagle, D.A.: Organs-on-chips: into the next decade. Nat. Rev. Drug Discov. **20**, 1–17 (2020)
30. Mastrangeli, M., Millet, S., ORCHID partners, T., van den Eijnden-van Raaij, J.: Organ-on-chip in development: towards a roadmap for organs-on-chip. ALTEX - Altern. Anim. Exp. **36**(4), 650–668 (2019). https://doi.org/10.14573/altex.1908271
31. Miller, M.J., Wei, S.H., Parker, I., Cahalan, M.D.: Two-photon imaging of lymphocyte motility and antigen response in intact lymph node. Science **296**, 1869–1873 (2002). https://doi.org/10.1126/science.1070051
32. Miller, C.P., Shin, W., Ahn, E.H., Kim, H.J., Kim, D.H.: Engineering microphysiological immune system responses on chips. Trends Biotechnol. **38**(8), 857–872 (2020). https://doi.org/10.1016/j.tibtech.2020.01.003
33. Mittal, R., Woo, F.W., Castro, C.S., Cohen, M.A., Karanxha, J., Mittal, J., Chhibber, T., Jhaveri, V.M.: Organ-on-chip models: implications in drug discovery and clinical applications. J. Cell Physiol. **234**, 8352–8380 (2019)

34. Murray, J.D.: *Mathematical Biology II Spatial Models and Biomedical Applications*. Springer, Berlin (2003)
35. Natalini, R.: Convergence to equilibrium for the relaxation approximations of conservation laws. Commun. Pure Appl. Math. **49**, 795–823 (1996)
36. Natalini, R., Paul, T.: The Mean-Field limit for hybrid models of collective motions with chemotaxis. Preprint (2022) https://arxiv.org/abs/2107.14645v4
37. Natalini, R., Ribot, M.: An asymptotic high order mass-preserving scheme for a hyperbolic model of chemotaxis. SIAM J. Numer. Anal. **50**, 883–905 (2012)
38. Nguyen, M., et al.: Dissecting effects of anti-cancer drugs and cancer- associated fibroblasts by on-chip reconstitution of immunocompetent tumor microenvironments. Cell Rep. **25**, 3884–3893 (2018). https://doi.org/10.1016/j.celrep.2018.12.015
39. Palacio-Castañeda, V., Dumas, S., Albrecht, P., Wijgers, T.J., Descroix, S., Verdurmen, W.P.R.: A hybrid in silico and tumor-on-a-chip approach to model targeted protein behavior in 3D microenvironments. Cancers (Basel) **13**(10), 2461 (2021). https://doi.org/d10.3390/cancers13102461
40. Parlato, S., De Ninno, A., Molfetta, R., Toschi, E., Salerno, D., Mencattini, A., Romagnoli, G., Fragale, A., Roccazzello, L., Buoncervello, M., et al.: 3D Microfluidic model for evaluating immunotherapy efficacy by tracking dendritic cell behaviour toward tumor cells. Sci. Rep. **7**, 1–16 (2017). http://dx.doi.org/10.1039/D0LC00799D
41. Patankar, S.V.: Numerical Heat Transfer and Fluid Flow. McGraw-Hill, New York (1996). ISBN 0-89116-522-3
42. Preziosi, L., Astanin, S.: Modelling the formation of capillaries. In: Quarteroni, A., et al. (eds.) Complex Systems in Biomedicine, pp. 109–145. Springer, Milano (2006)
43. Quarteroni, A.: Numerical Models for Differential Problems. Springer, Berlin (2017)
44. Scianna, M., Tosin, A., Preziosi, L.: From discrete to continuous models of cell colonies: a measure-theoretic approach. arXiv (2011). arXiv:1108.1212
45. Strikwerda, J.C.: Finite Difference Schemes and Partial Differential Equations, 2nd edn. Society for Industrial & Applied Mathematics, Philadelphia (2004). https://doi.org/10.1137/1.9780898717938
46. Sung, J.H., Wang, Y., Shuler, M.L.: Strategies for using mathematical modeling approaches to design and interpret multiorgan microphysiological systems (MPS). APL Bioeng. **3**, 021501 (2019). https://doi.org/10.1063/1.5097675
47. Vacchelli, E., Ma, Y., Baracco, E.E., Sistigu, A., Enot, D.P., Pietrocola, F., Yang, H., Adjemian, S., Chaba, K., Semeraro, M., et al.: Chemotherapy-induced antitumor immunity requires formyl peptide receptor 1. Science **350**, 972–978 (2015). https://doi.org/10.1126/science.aad0779
48. Wikswo, J.P.: The relevance and potential roles of microphysiological systems in biology and medicine. Exp. Biol. Med. (Maywood) **239**(9), 1061–1072 (2014). https://doi.org/10.1177/1535370214542068
49. Wiśniewski, J.R., Hein, M.Y., Cox, J., Mann, M.: A "proteomic ruler" for protein copy number and concentration estimation without spike-in standards. Mol. Cell Proteomics **13**(12), 3497–506 (2014). https://doi.org/10.1074/mcp.M113.037309
50. Wu, Q., Liu, J., Wang, X., et al.: Organ-on-a-chip: recent breakthroughs and future prospects. Biomed. Eng. Online **19**(1), 9 (2020)
51. Yang, T.D., Park, J.-S., Choi, Y., Choi, W., Ko, T.-W., Lee, K.J.: Zigzag turning preference of freely crawling cells. PLoS ONE **6**, e20255 (2011). https://doi.org/10.1371/journal.pone.0020255

A Particle Model to Reproduce Collective Migration and Aggregation of Cells with Different Phenotypes

Annachiara Colombi and Marco Scianna

Abstract We here propose a theoretical framework where aggregates of cells are described as systems of pointwise agents. Each element is then set to move according to Newtonian laws, being both the speed and the direction of movement determined by its phenotype, which is here assumed to fall within the epithelial-mesenchymal spectrum. In particular, the latter results from the balance of a given set of behavioural stimuli, each of them defined by a direction and a weight, that quantifies its relative importance. A constraint on the sum of the weights then avoids implausible simultaneous maximization/minimization of all movement traits. Further ingredients, such as cell duplication and simple chemical dynamics, are included as well. Numerical realizations are then provided in a perspective of analysis of critical model ingredients and parameters. The proposed approach is finally applied to reproduce selected aspects of a wound healing scenario. Some hints for its improvement are given in the last part of the chapter.

1 Introduction

The evolution of a biological system, in general, results from the interconnection of many different but integrated and superimposed processes, that span a range of spatio-temporal levels. In particular, macroscopic dynamics of cell aggregates and tissues are determined by microscopic behavior and interactions of the single component individuals, which are also able to sense and actively respond to selected signals coming from the surrounding environment in a non-strictly mechanical manner. The study of such an intricate flux of information highlights the need for

A. Colombi
Institute for Mediterranean Agricultural and Forestry Systems (ISAFOM) - National Research Council (CNR), Portici, Italy

M. Scianna (✉)
Department of Mathematical Sciences, Politecnico di Torino, Torino, Italy
e-mail: marco.scianna@polito.it

G. Bretti et al. (eds.), *Mathematical Models and Computer Simulations for Biomedical Applications*, SEMA SIMAI Springer Series 33,
https://doi.org/10.1007/978-3-031-35715-2_3

multidisciplinary researchers to work in multiscale environments, that are capable of ensemble representations of biological systems while still linking their evolution to the actual cell-based behavior.

The phenomenon of migration of more or less extended cell populations represents a clear example of the necessity of such an approach. It in fact results from the synergic integration and superimposition of at least two modes of movement, i.e., *single-cell/individual* and *multicellular/collective*.

The former, often termed mesenchymal-like migration, mainly relies on the independent response of single individuals, that indeed act as isolated atoms, to environmental cues of biochemical and biomechanical origin. As reviewed in [19], it is for instance the case of chemokines and growth factors that may be either freely diffusing (thereby giving rise to the phenomenon of chemotaxis), or tethered to matrix macromolecules (thereby giving rise to the phenomenon of haptotaxis), or of gradients in the substrate rigidity that are instead at the basis of durotactic mechanisms. Such external inputs typically trigger signaling cascades resulting in a series of mechanisms critical to achieve an efficient directional locomotion, such as formation of functional zones within the individual body, cytoskeletal protrusion and polarization, membrane ruffles and extension, and/or overexpression of integrins (transmembrane proteins used by cells to adhere to matrix components and exert traction force). Single-cell migration is known to contribute to many physio-pathological scenarios, such as development, immune surveillance and cancer metastasis, see again [19]. It in fact allows individual agents to position themselves in tissues or secondary growths, as they do during morphogenesis and cancer, or to transiently pass through selected environments, as in the case of immune cells.

During collective migration, cells instead remain connected and communicate one with the others: this gives rise to moving cohorts with varying degrees of organization [19]. In these cases, different migratory functions and behavior are shared between the different individuals to reach a coordinated translocation, further fueled by phenotypic heterogeneity within the same population. For instance, few specialized cells typically behave as a patterning guidance for the rest of the aggregate [29]. Such leader individuals, also termed *pathfinder* or *tip* agents, display some mesenchymal-like characteristics, such as polarized morphologies, ability to detect extracellular guidance cues and, where needed, to proteolytically remodel extracellular matrix (ECM) for path generation [29]. The remaining group of cells (also named *follower* or *stalk* individuals) is instead observed to have a sort of epithelial-like behavior. They in fact tend to form tightly packaged assemblies, such as rosettes or tubular networks, and to have poor cytoskeletal dynamics and inability to autonomously undergo sustained directional movement. Such phenotypic differentiation among individuals within the same aggregate may be established by both temporary activations and long-lasting differentiation processes, eventually triggered by environmental signals and/or by epigenetic mutations [29].

All these migratory functions are coordinated between neighboring cells via their junctions proteins, including cadherins of different families and other immunoglobulin superfamily members, which can both activate intracellular chemical pathways and directly or indirectly connect to the actin and/or intermediate filament cytoskeleton, thereby providing mechanically robust but dynamic coupling. Collective migration of cell groups is central in the formation or regeneration of complex tissues. For instance, as described in [19], bidimensional monolayers of cells move either constitutively across an intact basement membrane, such as the gut intestinal epithelium, or on demand, such as epidermal keratinocytes during wound closure. Furthermore, sprouting ducts and glands often comprise distinct cell types that move together and form structured trees or networks. In a less controlled form, coordinated migration of cells also occurs during cancer invasion, see [20] and reference therein. Malignant individuals can in fact cooperate to survive and efficiently invade the host. In this respect, it is possible to distinguish cancer cells with high motility rates from those with enhanced proliferative potential [26]. In particular, the former group of agents open migratory paths for less motile but more death-resistant clones. On the opposite, highly mitotic particles fuel invasion of more motile individuals by exerting oncotic pressure and consuming critical substrate.

The migratory behavior of a cell system is here tackled with a particle-based approach, where each agent is individually represented by a material point, whose velocity varies according to first-order Newtonian dynamics. In particular, we realistically distinguish between individual speed and direction of motion: the latter accounts both for migratory contributions that affect all individuals, regardless of their type, i.e., repulsion and randomness, and for phenotypic-dependent velocity terms. In this respect, each cell is assigned state variables that identify its position along the epithelial-mesenchymal spectrum and therefore establish the intensity of the corresponding migratory modes. By the use of proper weights and constraints, we then avoid simultaneous minimization/maximization of individual response to all migratory stimuli of the same nature. The role played by mitosis and chemical variables in affecting individual and collective dynamics is taken into account as well.

Structure of the Chapter The remaining part of the chapter is organized as follows. In Sect. 2, we introduce the main model ingredients and present sample numerical realizations that shows how they work and interact to reproduce cell dynamics in different situations. In Sect. 3, we apply our approach to the representative case of a would healing assay, both replicating its main aspects and analyzing the effect of variation of critical model parameters. In Sect. 4, we finally give some conclusive remarks and hints for possible developments of the proposed theoretical environments.

2 Mathematical Framework and Representative Simulations

We focus on biological systems whose dynamics, for the sake of simplicity, are restricted on the two-dimensional space. The component cells are mathematically represented by material points with unitary mass: they are set to move and evolve on a planar domain $\Omega \subseteq \mathbb{R}^2$. In this respect, the position of the i-th particle (with $i = 1, \ldots, n(t)$, n being the total number of individuals that can vary in time) is identified by the vector $\mathbf{x}_i(t) \in \Omega$. The time variable t is finally assumed to fall within the interval $T = [0, t_F]$, with $t_F < \infty$.

Let us now describe the proposed rules for cell movement and duplication/death processes.

2.1 Cell Proliferation

Cell proliferation is modelled by including a limited set of mechanisms related to the mitotic process. In particular, we take only into account (1) of a minimum period of time between successive duplications of the same individual and (2) of a contact-inhibition of growth in the case of too high local cell density. In this respect, the i-th agent is allowed to undergo mitosis at a given time $t \in T$ if the following two conditions simultaneously occur:

$$t - t_i^{\text{dupl}} > T_i^{\text{dupl}}; \tag{1}$$

$$d_i^{\min}(t) := \min_{\substack{j=1,\ldots,n(t):\\ j \neq i}} |\mathbf{x}_i(t) - \mathbf{x}_j(t)| > d_i^{\text{cell}}. \tag{2}$$

In (1), t_i^{dupl} indicates the time instant of the previous duplication of individual i, whereas T_i^{dupl} is a measure of the characteristic duration of its mitotic cycle. Equation (2) instead assures that there is enough free space around the cell of interest, as $d_i^{\min}(t)$ evaluates its present minimal distance from the other agents and d_i^{cell} is hereafter taken sufficiently larger than its perinuclear region (whose compression typically results in an inhibition of most mechamisms relative to cell proliferation).

When individual i effectively duplicates a newborn particle, labelled by the identity number $n(t)+1$, is added to the system close to the parent's actual position, i.e.,

$$\mathbf{x}_{n(t)+1}(t) = \mathbf{x}_i(t) + \frac{d_i^{\text{cell}}}{2} \left(\cos(\theta_i(t)), \sin(\theta_i(t))\right)^T, \tag{3}$$

where $\theta_i(t)$ is a random angle uniformly distributed over $[0, 2\pi)$. The daughter cell inherits from the progenitor all its biophysical characteritics (e.g., dimensions, phenotype, and migratory behavior).

2.2 Cell Movement

The movement of the generic i-th cell (being $i = 1, \ldots, n(t)$) is determined by the following first-order model

$$\frac{d\mathbf{x}_i}{dt}(t) = v_i(t)\widehat{\mathbf{w}_i(t)}, \tag{4}$$

which can be derived from the classic second-order Newtonian approach under the assumption of overdamped force-velocity response (a consistent hypothesis for living entities, see [14, 45] and references therein). Equation (4) effectively differentiates and decouples

- cell *speed*, given by the scalar $v_i(t) \in \mathbb{R}_+$ (possibly, $v_i(t) \in [0, v_{\max}]$ to account for physiological limitations);
- cell *direction of movement*, identified by the unit vector $\widehat{\mathbf{w}_i(t)} = \dfrac{\mathbf{w}_i(t)}{|\mathbf{w}_i(t)|}$ being $\mathbf{w}_i(t) \in \mathbb{R}^2$.

The two quantities have different biophysical meanings. For instance, cell speed is related to the concept of cell *motility*, which is established by the frequency of retraction/expansion cycles of plasmamembrane (PM) motility structures, such as filopodia and pseudopodia. Such a PM ruffling is in turn highly controlled by intracellular cascades involving specific ions and molecules (e.g., calcium, Rac1, Rho, see [34, 37, 42]). The direction of movement of a cell is instead dictated by the spatial organization of its cytoskeletal filaments. They in fact align either spontaneously or in response to external inputs (due, e.g., to the presence of other individuals or to specific environmental conditions), and establish a preferred orientation of the individual body, which is necessary to have effective spatial displacement [15, 19–21]. In this respect, $\widehat{\mathbf{w}_i(t)}$ may be also interpreted as the polarization axis of the i-th cell.

Notation Remark Hereafter, we will use the notation $\widehat{\mathbf{a}}$ to identify the unit vector corresponding to a generic vector $\mathbf{a}$, i.e., $\widehat{\mathbf{a}} := \dfrac{\mathbf{a}}{|\mathbf{a}|}$ where $\mathbf{a} \in \mathbb{R}^2$.

2.2.1 Cell Repulsive Behavior and Random Movement

Each cell i, regardless its type or individual biochemical and biophysical determinants, is subjected at least to two velocity contributions:

$$\mathbf{w}_i(t) = \widehat{\mathbf{w}_i^{\text{rand}}(t)} + \widehat{\mathbf{w}_i^{\text{rep}}(t)}. \tag{5}$$

$\widehat{\mathbf{w}_i^{\text{rand}}(t)}$ is a noise term that takes into account that biological elements (not only cells but also bacteria and other organisms) crawl in a *random* fashion to explore the surrounding environment. A wide range of sophisiticated or application-related laws may be derived to determine such a model component: however, for the sake of simplicity, we opt to set

$$\widehat{\mathbf{w}_i^{\mathrm{rand}}(t)} = (\cos(\eta_i(t)), \sin(\eta_i(t)))^T, \tag{6}$$

where, for any time $t \in T$, $\eta_i(t)$ is a stochastic variable uniformly distributed over the interval $[0, 2\pi)$.

$\widehat{\mathbf{w}_i^{\mathrm{rep}}(t)}$ instead describes intercellular *repulsive* interactions. They are mainly due to the resistance to compression of individuals which have a spatially extended rigid body, in particular a voluminous and substantially stiff nucleus. This term is here given by a directional vector that allows the i-th agent to move away from close enough neighbours, i.e.,

$$\mathbf{w}_i^{\mathrm{rep}}(t) = \sum_{j \in \mathcal{N}_i^{\mathrm{rep}}(t)} (\widehat{\mathbf{x}_i(t) - \mathbf{x}_j(t)}), \tag{7}$$

where

$$\mathcal{N}_i^{\mathrm{rep}}(t) = \left\{ j = 1, \ldots, n(t),\, j \neq i \,:\, 0 \leq |\mathbf{x}_j(t) - \mathbf{x}_i(t)| \leq d_{ij}^{\mathrm{rep}} = d_i^{\mathrm{rep}} + d_j^{\mathrm{rep}} \right\}, \tag{8}$$

d_i^{rep} and d_j^{rep} being the radius of the perinuclear compartments of the two agents of interests. The central region of a cell is in fact poorly elastic and therefore dictates, at least in a first approximation, the repulsive behavior of the entire individual, see for instance [18, 20, 22, 44, 55] and references therein. In this respect, d_{ij}^{rep} represents a comfort space between the generic pair of cells i and j since it assures that both agents do not experience compression of their nuclear area, refer to [8] for more comprehensive comments.

Sample Simulation Let us now propose a representative numerical test to shed lights on how the so-far proposed cell dynamics integrate with proliferation mechanisms. In this respect, we place on the square domain $\Omega = [0, 600]^2\,\mu\mathrm{m}^2$ a node of $n(0) = 50$ individuals. Each of them is set to move according to laws (4)–(8) and to duplicate following the rules given in Eqs. (1)–(3). Initially, all agents are distributed within a 100 μm-diameter round region at the center of the domain, with randomly established position and direction of motion, see Fig. 1a. They share the same biophysical characteristics, that are also constant in time, i.e.,

$$T_i^{\mathrm{dupl}} = T^{\mathrm{dupl}}; \qquad d_i^{\mathrm{cell}} = d^{\mathrm{cell}}; \qquad v_i(t) = v; \qquad d_i^{\mathrm{rep}} = d^{\mathrm{rep}},$$

for any $i = 1, \ldots, n(t)$ and $t \in T$. In particular, the following set of reasonable parameter estimate is employed: $T^{\mathrm{dupl}} = 24\,\mathrm{h}$, $v = 10\,\mu\mathrm{m/h}$, and $d^{\mathrm{cell}} = 2d^{\mathrm{rep}} = 8\,\mu\mathrm{m}$. This value also indicates the relaxed intercellular distance (see the comments

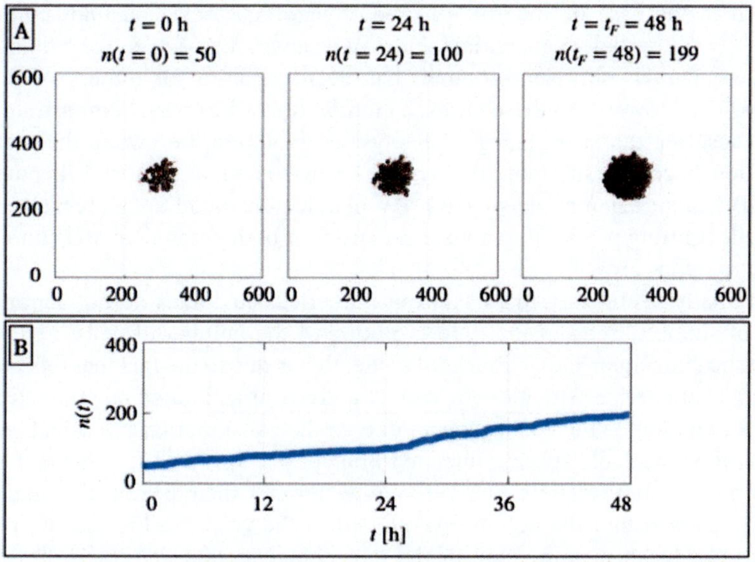

Fig. 1 Evolution of an aggregate of proliferative cells whose dynamics are regulated by random and repulsive contributions. The system is initially constituted by $n(0) = 50$ individuals. (**a**) Spatial configuration of the colony of cells at $t = 0, 24, 48$ h. The red arrows indicate the instantaneous direction of movement of each agent (i.e., $\widehat{\mathbf{w}_i(t)}$). (**b**) Time evolution of the overall number of particles n

before). Each generic agent i is then characterized by an own internal clock: in this respect, the instant of its last duplication, i.e., t_i^{dupl}, is initialized as a random variable uniformly distributed in the range $[-T^{\text{dupl}}, 0]$ and is then updated at any subsequent mitotic event. We finally assume that an individual i suddenly stops its locomotion when it falls too close to a point of the domain boundary $\partial\Omega$ (which therefore represent a physical wall), i.e., if $\text{dist}(\mathbf{x}_i(t)\,,\,\partial\Omega) \leq d^{\text{rep}}$ at a given time t (being dist the Euclidean distance function). Such a type of boundary conditions is employed for all forthcoming numerical realizations. The particle system is then observed for a period of time corresponding to 2 days, i.e., $t_{\text{F}} = 48$ h.

Figure 1 shows the evolution of the cell system. We can observe a quasi-radial extension of the aggregate with the number of individuals that is characterized by an exponential growth, that can be approximated by the following law:

$$n(t) = n(0)e^{at},$$

being $n(0) = 50$ and $a \approx 0.028\,\text{h}^{-1}$. This implies that the amount of agents almost doubles over a day, to reach a value of $n(t_{\text{F}} = 48) = 199$.

An exponentially increasing trend has been widely shown to characterize the early stages of growth of tumor masses (in terms of both volume and number of component cells). For instance, such a characteristic growth profile has been observed in the case of spheroids formed by glioblastoma cell lines cultured *in vitro* [39]. An initial exponential phase of growth, i.e., when the volume of a malignant mass is sufficiently small, has been captured for tumors of different types [6, 13]. However in these cases, a Gompertz law has been demonstrated to be able to describe the later stages of the system evolution, i.e., when the malignant population becomes substantially larger. As discussed in [7, 10, 50], purely (or reduced) Gompertzian models have been then demonstrated to be characterized by an excellent fitting potential to a wide spectrum of both preclinical and clinical data relative to tumor growth.

It is finally useful to remark that our numerical outcomes do not capture such a type of sigmoid trend of the time evolution of the number of cells, i.e., with an asymptotic convergence to a threshold value: this is due to the fact that this behavior typically characterizes situations where, at a given time, individual competition for resources initiates as the consequence of, e.g., the consumption of critical nutrients or the limited available space of the environment.

Entering in more details, the cells crawl around their position to maintain a comfort space within the colony. As plotted in the graph in Fig. 2a, the minimal interparticle distance in the case of the representative agent 1, i.e., $d_1^{\min}$ is in fact almost constantly equal to 8 μm ($= 2d^{\mathrm{rep}} = d^{\mathrm{cell}}$) with temporary drops to 4 μm in correspondence of mitotic events (see the (B) panel of the same image). In this respect, we recall that new born individuals are placed at a distance equal to

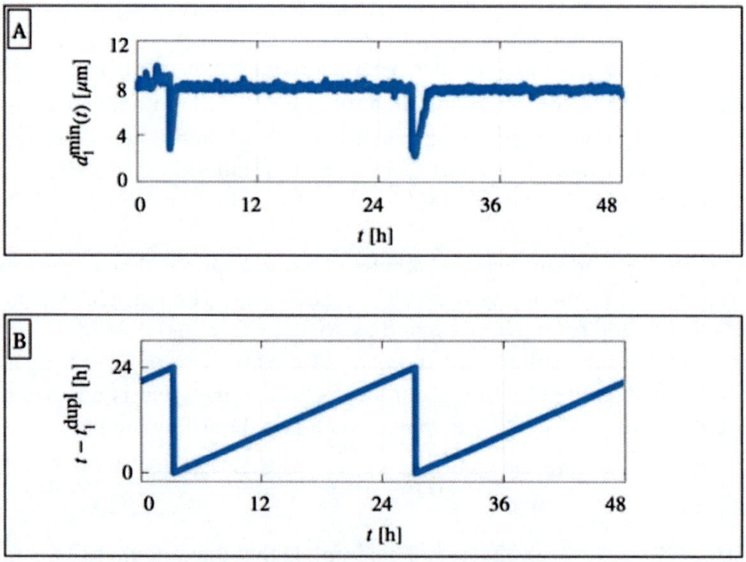

Fig. 2 (**a**) Time evolution of the minimal intercellular distance of the representative individual 1, i.e., $d_1^{\min}$, as defined in Eq. (2). (**b**) Mitotic internal clock of the representative cell 1, which is evaluated as the time lapse from its last proliferation event, see Eq. (1)

$d^{\text{cell}}/2$ from the progenitor's actual position (cf. Eq. (3)) and need a while to move sufficiently away.

Summing up, we can say that the repulsive term allows the cell aggregate to consistently arrange in a relaxed (i.e., non compressed) configuration even in the presence of a sustained proliferation.

2.2.2 Phenotypic-Related Cell Behavior

A wide range of migratory stimuli is specifically established by the individual *phenotype*. It is effectively defined at the protein scale, i.e., by gene transcription and expression levels, which are eventually affected by internal signals, environmental conditions, and stochasticity. In this respect, we hereafter assume that each agent is characterized by a phenotype falling along the *epithelial-mesenchymal (E-M) spectrum*. From a modeling perspective, this amounts to assign to the representative cell i (being $i = 1, \ldots, n(t)$, with $t \in T$) a pair of state variables $\alpha_i^{\text{epi}}, \alpha_i^{\text{mes}} \in [0, 1]$ such that $\alpha_i^{\text{epi}} + \alpha_i^{\text{mes}} = 1$. In particular, $\alpha_i^{\text{epi}} = 1$ implies that particle i has fully epithelial determinants whereas $\alpha_i^{\text{mes}} = 1$ implies that cell i has fully mesenchymal characteristics. Between such extreme values, there is a continuum range of intermediate states, that implicate the possible presence of cell variants with hybrid hallmarks.

Phenotypic-dependent contributions in individual movement are then collected in a specific term that, in the case of the i-th particle, is given by the following convex combination

$$\mathbf{w}_i^{\text{phen}}(t) = \alpha_i^{\text{epi}} \widehat{\mathbf{w}_i^{\text{epi}}(t)} + \alpha_i^{\text{mes}} \widehat{\mathbf{w}_i^{\text{mes}}(t)}, \tag{9}$$

and has to be added to the law of motion (5), i.e.,

$$\mathbf{w}_i(t) = \widehat{\mathbf{w}_i^{\text{rand}}(t)} + \widehat{\mathbf{w}_i^{\text{rep}}(t)} + \widehat{\mathbf{w}_i^{\text{phen}}(t)}. \tag{10}$$

In Eq. (9), $\mathbf{w}_i^{\text{epi}}(t)$ indeed accounts for epithelial-like velocity contributions whereas $\mathbf{w}_i^{\text{mes}}(t)$ accounts for mesenchymal-like migratory stimuli. As briefly explained in the introduction section, the former ones mainly rely on cell-cell interactions and are at the basis of collective cell movements; the latter ones are instead established by the ability of single particles to respond to environmental inputs and to individually migrate on or within substrates and tissues.

Both types of velocity terms are in turn given by the superimposition of selected migratory terms, i.e.,

$$\mathbf{w}_i^{\text{p}}(t) = \sum_{j \in \mathcal{J}_i^{\text{p}}} \beta_i^{\text{p},j} \widehat{\mathbf{w}_i^{\text{p},j}(t)}, \tag{11}$$

where p ∈ {epi, mes}. In (11), $\mathcal{J}_i^{\mathrm{p}}$ is the set of behavioral inputs that influence the dynamics of cell i as the unit vectors $\widehat{\mathbf{w}_i^{\mathrm{p},j}(t)}$ define the corresponding orientations. Finally, the coefficients $\beta_i^{\mathrm{p},j}$ are weights that describe the relative importance of each stimulus with respect to the others of the same nature. In particular, to have comparable effects on agent dynamics and to avoid simultaneous maximization/minimization of cell response to all analogous stimuli, we assume that

$$\begin{cases} \beta_i^{\mathrm{p},j} \in [0, 1]; \\ \sum_{j \in \mathcal{J}_i^{\mathrm{p}}} \beta_i^{\mathrm{p},j}(t) = 1, \end{cases} \tag{12}$$

for all $i = 1, \ldots, n(t)$ and $j \in \mathcal{J}_i^{\mathrm{p}}$, being again $t \in T$ and p ∈ {epi, mes}.

The specific phenotype of a cell has also an impact on its actual motility. For instance, epithelial individuals have been consistently seen to have decreased speed w.r.t. mesenchymal ones. This aspect is here taken into account by setting:

$$v_i(t) = \overline{v}_i(t) v_i^{\mathrm{phen}} = \overline{v}_i(t) v_{\mathrm{epi}} (1 + \alpha_i^{\mathrm{mes}}), \tag{13}$$

where $\overline{v}_i(t)$ indicates the fraction of the motility of the i-th cell that is not strictly related on its phenotype but that may depend, e.g., on the actual system temperature or on experimental manipulations [4]. In (13), v_{epi} identifies the speed of agents with fully epithelial hallmarks. Of course more sophisticated relationships between individual phenotype and motility may be employed as well.

Epithelial Migratory Modes and Related Sample Simulations The two more relevant and widely studied migratory modes of epithelial nature rely on intercellular adhesive interactions and movement alignment, respectively. The former is based on the expression and the activity of transmembrane molecules (i.e., cadherins) that connect the plasmamembranes (PMs) of pairs of cells and attract the rest of their body. In mathematical terms, in the case of the representative i-th individual, we can indeed define:

$$\mathbf{w}_i^{\mathrm{epi,\,adh}}(t) = \sum_{j \in \mathcal{N}_i^{\mathrm{adh}}(t)} (\mathbf{x}_j \widehat{(t) - \mathbf{x}_i}(t)) \tag{14}$$

with

$$\mathcal{N}_i^{\mathrm{adh}}(t) = \left\{ j = 1, \ldots, n(t),\ j \neq i\ :\ d_{ij}^{\mathrm{rep}} < |\mathbf{x}_j(t) - \mathbf{x}_i(t)| \leq d_{ij}^{\mathrm{adh}} = d_i^{\mathrm{adh}} + d_j^{\mathrm{adh}} \right\}. \tag{15}$$

d_i^{adh} and d_j^{adh} can be interpreted as the maximal extension that can be reached by the deformable adhesive structures (i.e., filopodia, pseudopodia) of each of the two interacting cells, respectively.

Cells with epithelial determinants are also able to rearrange their cytoskeletal elements (i.e., their polarization axis) in order to synchronize their movement with surrounding individuals. Such a behavior is here implemented by the following term:

$$\mathbf{w}_i^{\text{epi, align}}(t) = \text{M}(\widehat{\mathbf{w}_i(t)})_{j \in \mathcal{N}_i^{\text{align}}(t)}, \tag{16}$$

where $\text{M}(\widehat{\mathbf{w}_i(t)})_{j \in \mathcal{N}_i^{\text{align}}(t)}$ denotes the mean of the movement direction evaluated over the set of particles

$$\mathcal{N}_i^{\text{align}}(t) = \Big\{ j = 1, \ldots, n(t), \, j \neq i \, : \, d_{ij}^{\text{rep}} < |\mathbf{x}_j(t) - \mathbf{x}_i(t)| \leq d_{ij}^{\text{align}} \\ = d_i^{\text{align}} + d_j^{\text{align}} \Big\}. \tag{17}$$

where d_i^{align} and d_j^{align} measure the radius of the alignment neighborhood of each of the two agents. In particular, movement synchronization is observed to typically occur between cells that not only locally adhere but with a sufficiently large part of their membranes in contact (in order to mutually adapt their body, see [19]). In this respect, it is consistent to set $d_i^{\text{align}} < d_i^{\text{adh}}$ for any $i = 1, \ldots, n(t)$, with $t \in T$.

To show how the above-introduced velocity terms work, we now analyze the dynamics of a group of 50 cells with a fully epithelial trait, i.e., $\alpha_i^{\text{epi}} = 1$ and $\alpha_i^{\text{mes}} = 0$ for $i = 1, \ldots, 50$. We employ the same domain Ω and the same rules for the initial system configuration used in the previous Sect. 2.2.1. To avoid overcomplication, we neglect proliferation processes and random movements. Each individual is indeed set to move according to:

$$\mathbf{w}_i(t) = \widehat{\mathbf{w}_i^{\text{rep}}(t)} + \widehat{\mathbf{w}_i^{\text{epi}}(t)} = \widehat{\mathbf{w}_i^{\text{rep}}(t)} + \sum_{j \in \mathcal{J}_i^{\text{epi}}} \beta_i^{\text{epi},j} \widehat{\mathbf{w}_i^{\text{epi},j}(t)}. \tag{18}$$

Assuming that $\mathcal{J}_i^{\text{epi}} = \{\text{adh, align}\}$ for any particle i, Eq. (18) can be indeed specified as:

$$\mathbf{w}_i(t) = \widehat{\mathbf{w}_i^{\text{rep}}(t)} + \beta_i^{\text{epi,adh}} \sum_{j \in \mathcal{N}_i^{\text{adh}}(t)} (\widehat{\mathbf{x}_j(t) - \mathbf{x}_i(t)}) + \beta_i^{\text{epi,align}} \text{M}(\widehat{\mathbf{w}_i(t)})_{j \in \mathcal{N}_i^{\text{align}}(t)}, \tag{19}$$

with $\beta_i^{\mathrm{epi,adh}} + \beta_i^{\mathrm{epi,align}} = 1$. A further consistent hypothesis is that all cells have (and maintain over time) exactly the same determinants and behavior, that results in the following parameter setting:

$$v_i(t) = v_{\mathrm{epi}} = 10\,\mu\mathrm{m/h} \qquad (\text{s.t. } \overline{v}_i(t) = 1);$$

$$d_i^{\mathrm{rep}} = d^{\mathrm{rep}} = 4\,\mu\mathrm{m}; \qquad d_i^{\mathrm{adh}} = d^{\mathrm{adh}} = 10.5\,\mu\mathrm{m};$$

$$d_i^{\mathrm{align}} = d^{\mathrm{align}} = 7\,\mu\mathrm{m}; \qquad \beta_i^{\mathrm{epi,adh}} = \beta^{\mathrm{epi,adh}}; \qquad \beta_i^{\mathrm{epi,align}} = \beta^{\mathrm{epi,align}},$$

for any $i = 1, \ldots, n(t)$ and $t \in T$. We then observe different system evolutions over a time lapse of a day (i.e., $t_{\mathrm{F}} = 24\,\mathrm{h}$) in case of selected values of the weights $\beta^{\mathrm{epi,adh}}$ and $\beta^{\mathrm{epi,align}}$, that in principle establish the hierarchy of the two epithelial migratory contributions. In particular, to facilitate comparisons between the proposed scenarios, we introduce a pair of quantities:

$$d_{\min}(t) := \min_{i=1,\ldots,n(t)} d_i^{\min}(t); \tag{20}$$

$$s_{\mathrm{align}}(t) := \sqrt{\sum_{i=1}^{n(t)} \frac{|\widehat{\mathbf{w}_i(t)} - \mathbf{M}(\widehat{\mathbf{w}_j(t)})_{j \in \mathcal{N}_i^{\mathrm{align}}(t)}|^2}{n(t)}}. \tag{21}$$

The former (where $d_i^{\min}(t)$ has been previously defined in Eq. (2)) gives a measure of the compactness of the cell aggregate, whereas the latter is a measure of the level of synchronization of individual movement. In this respect, $s_{\mathrm{align}} \approx 1$ indicates uncorrelated cell displacement while $s_{\mathrm{align}} \approx 0$ indicates that each agents moves almost in the same direction of its neighboring mates.

Figures 3 and 4 (top graph) show that, for low enough values of $\beta^{\mathrm{epi,adh}}$ (i.e., < 0.3, which correspond to high values of $\beta^{\mathrm{epi,align}}$), the cells move substantially away from their initial position maintaining a comfort distance one from each other, as $d_{\min}$ is constantly close to $2d^{\mathrm{rep}}$ (which, as seen, is a sort of relaxed interparticle distance). In particular, a quasi-complete synchronization of movement within the colony is obtained for $\beta^{\mathrm{epi,adh}} \le 0.1$ (i.e., for $\beta^{\mathrm{epi,align}} > 0.9$, see the bottom graph in Fig. 4).

Larger values of $\beta^{\mathrm{epi,adh}}$ (i.e., ≥ 0.3, which correspond to $\beta^{\mathrm{epi,align}} \le 0.7$) instead result in an unrealistic collapse of the cell aggregate. The component individuals in fact undergo directionally-uncorrelated locomotion to clusterize almost in the middle of the domain without keeping a sufficient spacing (i.e., $d_{\min}$ quickly drops to 0), cf. Figs. 3 and 4. In such a range of values of the β^{epi}-weights, the repulsive stimulus is no longer able to balance the tendency of cells to maximize their contact interactions.

Mesenchymal Migratory Modes and Related Sample Simulations Mesenchymal-type migratory modes are typically based on the ability of cells to *individually*

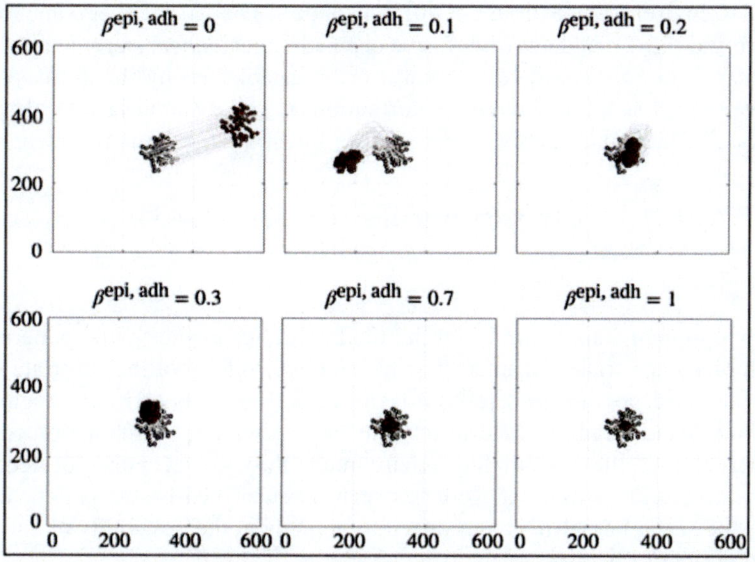

Fig. 3 Evolution in time of an aggregate of 50 fully epithelial cells, whose dynamics are regulated by Eq. (19), upon variations in the value of $\beta^{\mathrm{epi,adh}} = 1 - \beta^{\mathrm{epi,align}}$. Initial cell positions (identified by the empty black circles) and orientations are randomly established but kept the same for all numerical settings. Final cell positions, i.e., at $t_F = 24$ h, are instead indicated by the full red circles. The grey lines finally identifies individual trajectories

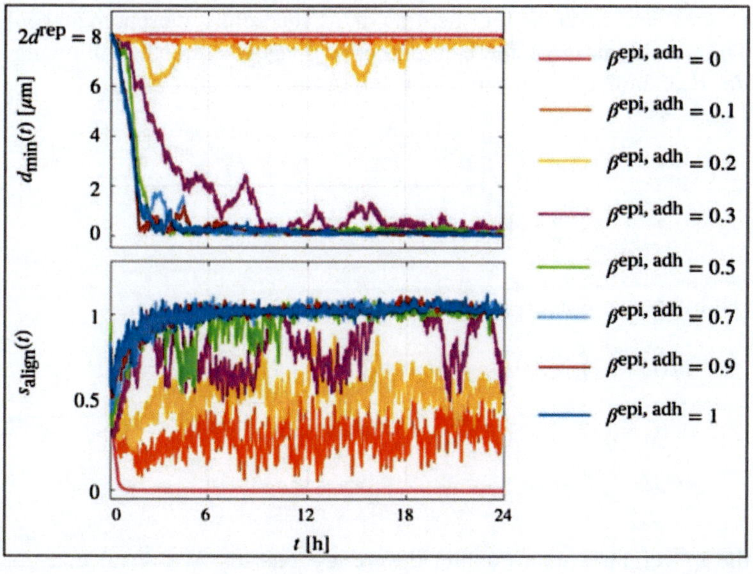

Fig. 4 Time evolution of the global minimal distance between pairs of individual within the colony $d_{\min}$ as defined in Eq. (20), and of the alignment coefficient s_{align}, as defined in Eq. (21), in the case of different values of $\beta^{\mathrm{epi,adh}}$ (and therefore of $\beta^{\mathrm{epi,align}}$)

respond to environmental stimuli of different nature. They may either come from the presence and the distribution of diffusive or fixed molecular substances (resulting in *chemotaxis* and *haptotaxis*, respectively) or be established by the stiffness of the substrate (*durotaxis*). Each of these contributions in cell dynamics can be described, in the case of the representative agent i, by an integral term of the following type:

$$\mathbf{w}_i^{\text{mes},j}(t) = \int_{\mathcal{S}_i(t)} K\big(\mathbf{y}, \mathbf{x}_i(t), \mathbf{w}_i(t)\big)\,\big(c^j(t, \mathbf{x}_i(t)) - c^j(t, \mathbf{y})\big)\,\big(\widehat{\mathbf{y} - \mathbf{x}_i(t)}\big)\,d\mathbf{y}, \tag{22}$$

where $j \in \{\text{chem, hapto, duro}\}$ and the field c^j indicates the spatial pattern either of the molecular variable of interest or of the local rigidity of the substrate, giving rise to the corresponding cell response. In Eq. (22), $S_i(t)$ is a circular region with center in $\mathbf{x}_i(t)$ and radius equal to d_i^{adh}, as any individual senses the biochemical characteristics of the surrounding environment through receptors located on its motility structures, whose extension, as seen, is quantified by the parameter d_i^{adh}. Finally, the scalar kernel K is introduced to implement the possibility of anisotropy in the probing mechanisms of the i-th particle.

We now study how the migratory behavior of a group of cells is affected by variations in their phenotypic trait. In this respect, we place a node of 50 individuals in the usual domain Ω, each of them with randomly established initial orientation and position (that, in this case, falls within the subdomain $[100, 300] \times [200, 400]$). For the sake of simplicity, we again neglect mitotic events and randomness in individual movement, assume homogeneity in cell biophysical determinants and characteristics, and account for chemotaxis as the solely mesenchymal migration mode. The combination of these hypotheses gives rises to the following law of motion for each individual $i = 1, \ldots, n = 50$:

$$\begin{aligned}\mathbf{w}_i(t) = \widehat{\mathbf{w}_i^{\text{rep}}(t)} + \alpha_i^{\text{epi}}\left[\beta_i^{\text{epi, adh}}\widehat{\mathbf{w}_i^{\text{epi,adh}}(t)} + \beta_i^{\text{epi, align}}\widehat{\mathbf{w}_i^{\text{epi,align}}(t)}\right] \\ + \alpha_i^{\text{mes}}\widehat{\mathbf{w}_i^{\text{mes,chem}}(t)},\end{aligned} \tag{23}$$

completed by the parameter setting

$$\overline{v}_i(t) = 1; \qquad v_{\text{epi}} = 10\,\mu\text{m/h}; \qquad d_i^{\text{rep}} = d^{\text{rep}} = 4\,\mu\text{m};$$

$$d_i^{\text{adh}} = d^{\text{adh}} = 10.5\,\mu\text{m}; \qquad d_i^{\text{align}} = d^{\text{align}} = 7\,\mu\text{m}; \qquad \beta_i^{\text{epi,align}} = \beta^{\text{epi,align}} = 0.9;$$

$$\beta_i^{\text{epi,adh}} = \beta^{\text{epi,adh}} = 0.1; \qquad \alpha_i^{\text{epi}} = \alpha^{\text{epi}}; \qquad \alpha_i^{\text{mes}} = \alpha^{\text{mes}},$$

where the α-weights vary to obtain different scenarios. We recall that the values of $\beta^{\text{epi,align}}$ and $\beta^{\text{epi,adh}}$ are such that the colony of cells does not collapse but its component agents maintain a sufficient spacing (see the previous part of the Section). The chemotactic velocity component in Eq. (23) has the form given in

(22), with the radius of the integration domain S_i equal to d^{adh} for all individuals. To specify the form of the kernel K, we take into account that the local amount of chemical receptors of the representative particle i (1) decreases with the distance from its center of mass (here identified by the positional vector $\mathbf{x}_i$), which is plausible in the perspective of a bidimensional approximation of its body, and (2) is larger towards its front/head (identified by the polarization vector $\widehat{\mathbf{w}_i(t)}$) than towards its rear/tail. Accordingly, we employ the following form, already used in [12]:

$$K\big(\mathbf{y}, \mathbf{x}_i(t), \mathbf{w}_i(t)\big) = K_r\Big(|\mathbf{y} - \mathbf{x}_i(t)|\Big) K_\delta\Big(\widehat{(\mathbf{y} - \mathbf{x}_i(t))} \cdot \widehat{\mathbf{w}_i(t)}\Big) \tag{24}$$

with

$$K_r(r) = \begin{cases} 1/r, & \text{if } 0 < r \le d^{\text{adh}}, \\ 0, & \text{otherwise;} \end{cases} \tag{25}$$

and

$$K_\delta(u) = \begin{cases} u^2, & \text{if } u > U, \\ 0, & \text{otherwise;} \end{cases} \tag{26}$$

with $U = 1/2$, see Fig. 5. Equation (26) implies that cell probing activity spans an angular sector which simmetrically extends for $\pi/3$ around the actual cell orientation. The integral in Eq. (22) can be therefore rewritten as

$$\begin{aligned} \mathbf{w}_i^{\text{mes,chem}}(t) = \int_0^{d^{\text{adh}}} & r K_r(r) \int_0^{2\pi} K_\delta\big(\mathbf{n}(\delta) \cdot \mathbf{w}_i(t)\big) \\ & \Big(c^j(t, \mathbf{x}_i(t)) - c^j(t, \mathbf{x}_i(t) + r\mathbf{n}(\delta))\Big)\, \mathbf{n}(\delta(t))\, d\delta\, dr, \end{aligned} \tag{27}$$

where $\mathbf{n}(\delta) = (\cos\delta, \sin\delta)$ for any $\delta \in [0, 2\pi]$.

The molecular substance that, as seen, acts as a chemotatic guidance cue for the cluster of cells is finally assumed to have a constant-in-time spatial distribution

$$c^{\text{chem}}(\mathbf{x} = (x_1, x_2) \in \Omega) = C_1(\tanh(C_2(x_1 - L)) + 1), \tag{28}$$

where $L = 600\,\mu\text{m}$. while the coefficients C_1 and C_2 are arbitrarily fixed equal to $0.5\,\mu\text{M}/\mu\text{m}^2$ and $0.02\,\mu\text{m}^{-1}$, respectively. The inclusion of a chemical element confers the proposed model a genuinely *multiscale* characteristic, as cell and subcellular levels are integrated within the same theoretical framework.

As shown in Fig. 6, cell dynamics in the case of $\alpha^{\text{mes}} = 0$ are exactly the same as those observed in the previous section for $\beta^{\text{epi,align}} = 0.9$ and $\beta^{\text{epi,adh}} = 0.1$ (cf. Fig. 3): a common direction of movement, independent from the chemical distribution, in fact emerges within the fully epithelial colony and the component

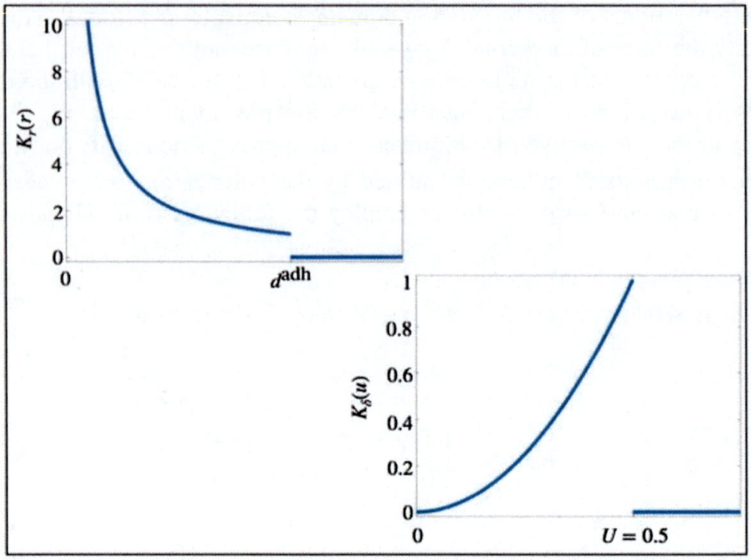

Fig. 5 Plots of the two factors that define the scalar kernel K, introduced in Eq. (22). In particular, K_r, given in Eq. (25), implements the fact that the amount chemical receptors is higher towards the center of mass of the cell (in a perspective of bidimensional approximation). K_δ, given in Eq. (26), models instead the clusterization of chemical receptors at the front region of the individual rather than at its lateral and rear parts

individuals maintain a sufficient spacing, since repulsion is able to balance such a low adhesion.

For $\alpha^{\text{mes}} \in [0.1, 0.9]$, cells are characterized by a hybrid phenotype, being therefore subjected to both epithelial and mesenchymal stimuli. In particular, higher values of α^{mes} result in more significant cell directional movement towards domain regions with greater chemical amounts, see again Fig. 6. The underlying rationale is that increments in α^{mes} imply decrements in α^{epi}, which in turn imply poor intercellular interactions. Each particle is indeed more free to independently move along the molecular gradient. Finally, for $\alpha^{\text{mes}} = 1$, the colony is composed of agents that, displaying fully mesenchymal hallmarks, migrate towards the right part of Ω as isolated entities, paying only attention to avoid superimpositions. The repulsive velocity term is in fact still active as all cells, regardless their phenotypic trait, are characterized by a spatially extended body. It is indeed useful to stress that the common direction of movement emerged in the case of a fully mesenchymal aggregate is not dictated by a collective synchronization of locomotion; rather it is the result of the fact that all individuals are stimulated, and equally respond, to the same chemotactic cue.

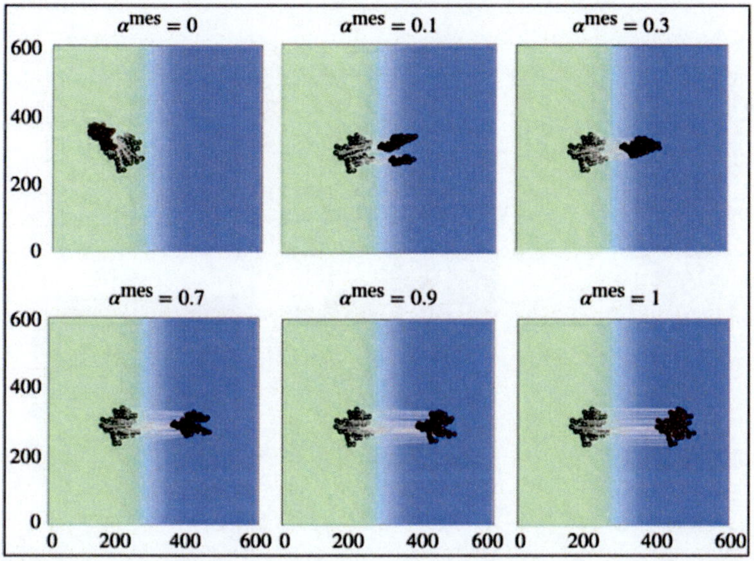

Fig. 6 Evolution in time of an aggregate of 50 cells, whose dynamics are regulated by Eq. (23), upon variations in the value of $\alpha^{mes} = 1 - \alpha^{epi}$, which establishes their (common) phenotypic trait. Initial cell positions (identified by the empty black circles) and orientations are randomly established but kept the same for all numerical settings. Final cell positions, i.e., at $t_F = 12\,h$, are instead indicated by the full red circles. The grey lines finally identify individual trajectories. The domain background shadow indicates the normalized chemical concentration, that increases passing from light to dark blues, cf. Eq. (28)

3 Model Application: Wound Healing Assay

In order to show how the modeling framework illustrated in the previous sections can be applied to realistic scenarios, we here present a comprehensive simulation of a *wound healing assay*. A wound healing experiment is considered a simple and reliable test for a quantitative evaluation of cell motility, in particular in response to molecules putatively involved in migratory processes (see Fig. 7 for representative examples and refer, for instance to [5, 17, 36, 41, 49, 56]). Procedurally, a cell population, coated with a matrix, grows to confluence in a culture plate. The monolayer is then wounded with a sharp object (for example a pipette tip), and stimulated with selected concentrations of the molecule of interest. The recolonization of the lesion by the remaining cell mass is then monitored by means of time-lapse microscopy: images are captured at regular intervals during cell migration, and then used for an accurate analysis of the population invasive capacity. Quantification is given by evaluating the rate of advance of the wound edge or the extension of the recolonized area.

Such an *in vitro* protocol recapitulates some of the main aspects of the corresponding *in vivo* phenomenon. For instance, during repair of skin or of corneal

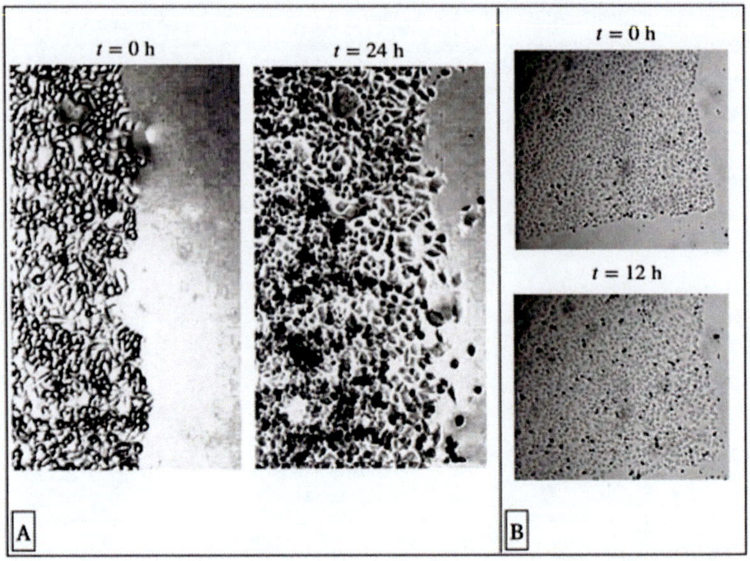

Fig. 7 Representative images taken from wound healing assays. (**a**) Culture of breast tumor-derived endothelial cells (B-TEC) stimulated by a nanomolar concentration of sodium hydrosulphide (NaHS). (**b**) Culture of bovine aortic endothelial cells (BAEC) stimulated by a micromolar concentration of hepatocyte growth factor (HGF). Unpublished pictures, courtesy of the Cellular and Molecular Angiogenesis Laboratory of the Department of Life Sciences and Systems Biology at the University of Turin, Italy

epithelium after injury, collective cell migration of keratinocytes occurs across the provisional wound bed leading to scratch closure. Keratinocytes move as a monolayer sheet that, after hours to days, eventually undergoes multilayered stratification and forms *de novo* epidermis [19, 29].

In more details, it has been widely shown that both in *in vitro* and in *in vivo* situations, the cells located at the first row of the monolayer (i.e., those close to the wound) undergo *active* movement of mesenchymal-type nature. In particular, they sense and respond to biochemical signals and are attracted across the lesion. Such environmental molecular inputs in fact generate intracellular pathways that trigger actin filament growth, cytoskeletal polarization and outward deformation of cell PM, which are at the basis of an efficient directional movement. Leader cells also use integrin transmembrane molecules to generate force on the substrate to further sustain their locomotion. They may also synthesize new basement membrane.

The rest of the cells, i.e., those located at the central part and at the rear front of the sheet, are instead characterized by epithelial-like dynamics. They are in fact *passively* dragged by cadherin-based adhesive interactions, being also able to orient their body in the same direction of the front individuals.

It is useful to underline that the force generated by leader cells is sufficient to pull and coordinate migration persistence in the case of cell sheets formed up to

10 raws behind the front edge. In the case of larger aggregates, not only leader but also some follower individuals have to develop mesenchymal-like hallmarks, i.e., to actively move, in order to have efficient translocation of the entire system and therefore wound closure, see again [19, 29].

The simulation and parameter setting employed to reproduce the biological system of interest is summarized in the following points:

- the numerical domain is $\Omega = [0, 300] \times [0, 600]\,\mu\text{m}^2$;
- initially, there are $n(0) = 600$ cells which are distributed in 8 sufficiently-spaced rows (each composed by 75 individuals) at the left part of the domain. In this respect, the remaining (right) region of Ω represents the wounded area of an experimental culture plate, see Fig. 8 (left panel);
- the i-th cell belonging to the row at the extreme right edge of the aggregate (i.e., belonging to the *front* row) is characterized by a hybrid phenotype quantified by a value of $\alpha_i^{\text{mes}} = \alpha^{\text{mes}} > 0$. Such a set of agents, indicated by orange circles in Fig. 8, is assumed to move in response to chemical signals (according to Eq. (27)), being also subjected to repulsion, adhesive interactions and alignment. They have inhibited mitotic events;

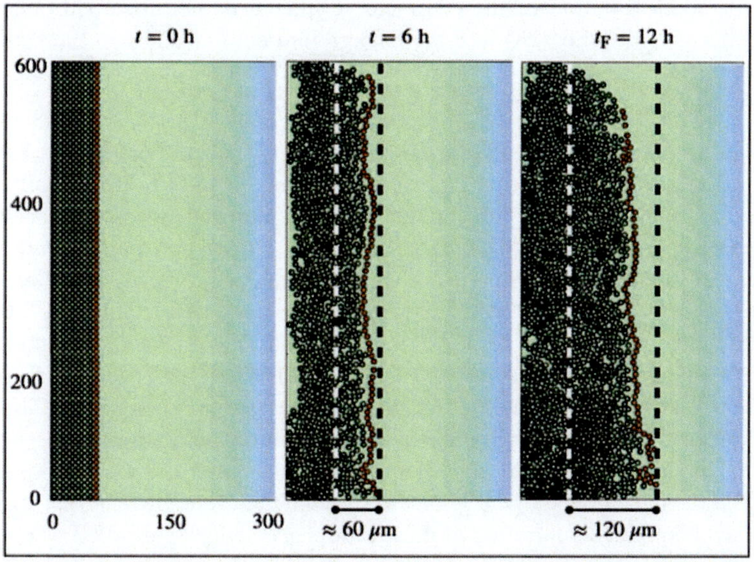

Fig. 8 Evolution in time of the wound healing process of a layer of virtual cells. Representative images taken at 0, 6, and 12 h. Orange circles identify individuals with a hybrid phenotype (i.e., with $\alpha^{\text{mes}} \neq 0$), whereas black circles indicate agents with a fully epithelial trait (i.e., with $\alpha^{\text{mes}} = 0$). The front of the cell sheet, initially located at a distance of $\approx 60\,\mu$m from the left border of the domain (see the white dashed lines in the images) advances and invades the gap at a rate of nearly $10\,\mu$m/h (as indicated by the black dashed lines in the middle and in the right panels). The domain background shadow indicates the normalized chemical concentration, that increases passing from light to dark blues, cf. Eq. (28)

- the remaining group of cells is instead assigned a fully epithelial trait (i.e., for them $\alpha^{mes} = 0$ and $\alpha^{epi} = 1$). Their dynamics include repulsion, as long as adhesive interactions and synchronization of movement. These individuals, identified by black circles in Fig. 8, are allowed to duplicate according to the rules given in Eqs. (1)–(3), with a common value of T^{dupl};
- all cells are set to have common dimensions and to share the same hierarchy of epithelial migratory stimuli. These hypotheses give rise to

$$\overline{v}_i(t) = 1; \qquad v_{epi} = 10\,\mu\text{m/h}; \qquad d_{ij}^{rep} = d^{rep} = 4\,\mu\text{m};$$

$$d_{ij}^{adh} = d^{adh} = 10.5\,\mu\text{m}; \qquad d_i^{align} = d^{align} = 7\,\mu\text{m};$$

$$\beta_i^{epi,adh} = \beta^{epi,adh} = 0.9; \qquad \beta_i^{epi,adh} = \beta^{epi,adh} = 0.1;$$

- variations in the values of the pair of parameters α^{mes} and T^{dupl} will be performed to obtain different scenarios. In this respect, we recall that the former coefficient quantifies the mesenchymal trait of the cell belonging to the front row of the aggregate, whereas the latter establishes the duplication rate of fully epithelial individuals;
- the spatial distribution of the chemical substance is then assumed constant in time and given by Eq. (28), see the background color-shadow in the simulation images;
- the system is finally observed for $t_F = 12$ h.

A substantial gap closure is obtained by setting $\alpha^{mes} = 0.1$ and $T^{dupl} = 8$ h. As shown in Fig. 8, the migration of the overall cell mass is in fact directionally biased towards the right region of the domain. In particular, the front of the cell sheet advances at approximately 10 μm/h, to reach almost the mid point of the wound at the end of the observation period (i.e., at $t_F = 12$ h). Entering in more details, the leading cells, i.e., those with also mesenchymal determinants, are observed to significantly move from their original location and to wander in their close proximity, displaying an evident capacity to invade the open space. It is useful to notice that such a front agents do not display a locomotion completely aligned to the chemical gradient, but rather show slight direction oscillations. This is due to the fact that they maintain a significant epithelial component in their dynamics, as they are characterized by $\alpha^{mes} = 0.1$ and therefore by $\alpha^{epi} = 0.9$. The individual chemotactic migration of these particles is indeed perturbed by collective intercellular interactions. Adhesiveness is instead the leading force underlying the movement of the rest of the cell mass, i.e., of the individuals behind the first row that are assigned a fully epithelial trait. Their locomotion is further reinforced by the alignment stimulus and sustained by proliferation. In fact, mitotic events push fully epithelial agents towards the free areas of the domain to avoid overcompression.

We then keep fixed $T^{dupl} = 8$ h and analyze possible variations in the healing capacity of the cell sheet upon increments in the value of α^{mes}. Such a model manipulation implies a shift towards a more significant mesenchymal behavior

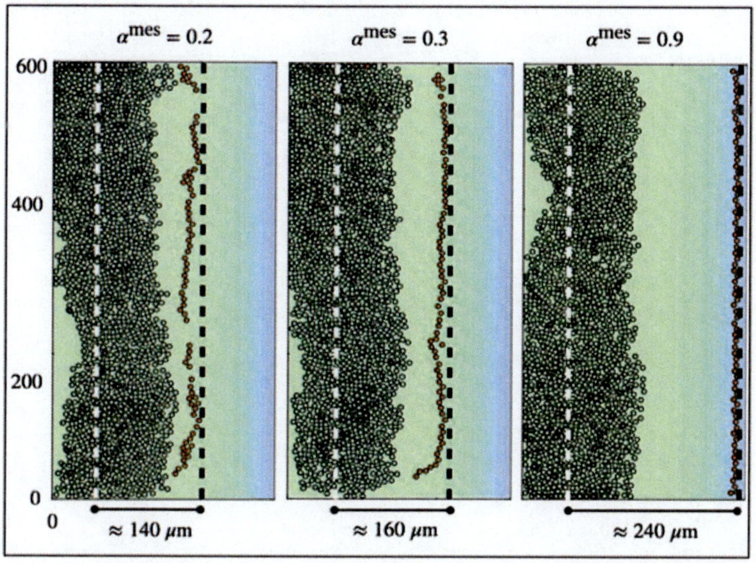

Fig. 9 Final cell configurations, i.e., at $t_F = 12\,\mathrm{h}$, in the case of different values of α^{mes}. We recall that α^{mes} quantifies the mesenchymal contribution in the movement of the cells located in the leading row of the aggregate (i.e, those identified by orange circles). The other parameters are kept the same as in the simulation shown in Fig. 8. Again, the dashed white lines indicate the initial position of the front of cell whereas the dashed black lines those at the end of the observation time. The domain background shadow indicates the normalized chemical concentration, that increases passing from light to dark blues, cf. Eq. (28)

of front individuals. As shown in Fig. 9 (which reproduces the final system configurations, i.e., at $t_F = 12\,\mathrm{h}$, in the different scenarios), in the case of α^{mes} larger than 0.1, we observe the detachment of the leading row from the rest of the aggregate. In particular, the more α^{mes} is high, the earlier the cells with a hybrid trait move away from the others and start to individually crawl along the chemical gradient, as the epithelial component in their behavior loses its relevancy. In such a range of values, intercellular adhesiveness is in fact no longer able to maintain a sufficient compactness between cells with different phenotypes. For instance, if $\alpha^{mes} = 0.9$ (i.e., if $\alpha^{epi} = 0.1$), the leading row of the cell sheet undergoes a quasi rigid translocation to reach the right border of the domain. The bulk of fully epithelial agents instead expands towards the middle of the gap mainly as a consequence of mitotic events: however, areas deprived of cells emerges at its rear part, see again Fig. 9. Increments in the mesenchymal determinants of the leading particles therefore do not imply increments in the healing capacity of the cell population: the wound is in fact not closed; rather, it is only crossed by a single-cell-width front of highly motile agents.

We finally keep fixed $\alpha^{mes} = 0.1$ and vary T^{dupl}. In this respect, we recall that increments in T^{dupl} imply decrements in the number of mitotic events for fully epithelial individuals (cf. Eq. (1)). As shown in Fig. 10, variations in T^{dupl} do not

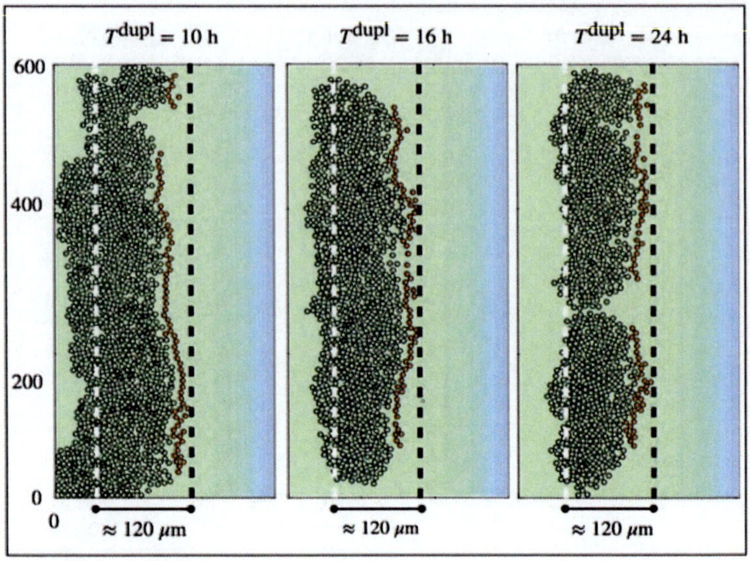

Fig. 10 Final cell configurations, i.e., at $t_F = 12\,\text{h}$, in the case of different values of T^{dupl}. We recall that T^{dupl} establishes the rate of duplication of fully epithelial individual (i.e, those identified by black circles), see Eq. (1). The other parameters are kept the same as in the simulation shown in Fig. 8. Again, the dashed white lines indicate the initial position of the front of cells, whereas the dashed black lines those at the end of the observation time. The domain background shadow indicates the normalized chemical concentration, that increases passing from light to dark blues, cf. Eq. (28)

have a significant impact on the invasive behavior of the cell sheet: the front of the aggregate in fact advances in all the proposed case of approximately $120\,\mu\text{m}$ from its initial configuration (as in the reference simulation displayed in Fig. 8). The cells with hybrid determinants are in fact subjected to adhesion and are therefore able to drag the mass of remaining particles. However, for $T^{\text{dupl}} > 8\,\text{h}$, extended gaps emerge at the rear part of the cell layer, see again Fig. 10. Mitosis is in fact not able to balance migration of fully epithelial individuals and therefore the lesion is not completely covered.

Summing up, we can conclude that the rate of invasion of the cell sheet completely depends on the mesenchymal determinants of the leading cells (i.e., on the parameter α^{mes}). The value of T^{dupl} instead establishes the possibility to have a confluent monolayer of cells during the healing process. To have a physiologic closure of the lesion proper values of both parameters are needed.

4 Conclusions

The analysis of collective cell behavior has recently become one of the main topics in theoretical biology. It is in fact at the basis of a wide spectrum of physio-pathological phenomena, ranging from morphogenesis and regeneration to tumor growth and development. In particular, multicellular aggregates efficiently move as a consequence of the cooperative and synergic interaction between individuals with different functions and migratory determinants, i.e., with different phenotypes. For instance, in several cell systems, few specialized agents with mesenchymal hallmarks are able to sense environmental signals and behave as a patterning guidance for the rest of the groupmates, which instead passively displace mainly due to adhesion. It is the case of relevant processes such as vascular sprouting, skin repair, or matrix invasion by malignant strands, see [19, 29] and references therein.

This topic has been here addressed with a microscopic particle approach, where each cell has been allowed to proliferate and move according to a first-order Newtonian approach, which distinguishes individual speed and orientation (such a type of models is widely used to describe behavior of physic-pathological aggregates of cells, as reviewed in [2, 3, 14], see also [28] for a representative work). In particular, all agents have been set to be subjected to a repulsive behavior (consequence of the spatial extension of their body) and to the possibility of random crawling.

Each cell has been then assigned two state variables, i.e., α^{mes} and α^{epi}, that identify its phenotype along the mesenchymal-epithelial spectrum. In this respect, the more α^{mes} (rsp., α^{epi}) is high the more the considered individual is characterized by mesenchymal-like (rsp., epithelial-like) dynamics. We have imposed a unitary sum of α^{mes} and α^{epi} since, as demonstrated in [24], cells may display a hybrid trait but not a simultaneous maximization of both epithelial and mesenchymal characteristics. This would give rise to a super phenotype which has not been yet observed [47].

The effective mesenchymal-like (rsp., epithelial-like) behavior has been assumed to be in turn the result of competing stimuli, each of them weighted by a coefficient $\beta^{\mathrm{mes},(\cdot)}$ (rsp., $\beta^{\mathrm{epi},(\cdot)}$) that has been set to define a sort of individual preference. The sum of the β-coefficients has been fixed equal to one in order to account for a balance between movement traits of the same nature.

Among epithelial migratory modes, we have accounted for adhesive interactions and orientation alignment. Taxis-like migrations have been instead taken as representative of mesenchymal-like dynamics. It is also useful to underline that the proposed modeling framework has included the possibility that the phenotype of a cell affects also its speed and duplication rate. In this respect, we have employed a sort of "Go-or-Grow" (GoG) hypothesis, which states an inverse correlation between cell motility and duplication potential. Such an assumption has been widely used in the theoretical literature [23, 27, 32, 57].

The coefficients α^{mes} and α^{epi}, introduced in the proposed work, play the same modeling role of the structuring variables used in the family of mathematical approaches based on the so-called *adaptive dynamics*. As reviewed in [40], these types of models are built on the idea that a cell population can be differentiated according to one or more traits (e.g., motility, type of metabolism, proliferation capacity). The evolution of the system is then mainly driven by selection and mutation. In particular, the former favors the diffusion of cell variants with the most adapt phenotype, whereas the latter allows the emergence of offsprings with slightly different determinants with respect to their progenitors. Such a modeling framework has been recently developed to reproduce selected aspects of the multicellular dynamics with a particular emphasis to solid tumor growth [9, 16, 31, 52].

The description of our approach has been supported along the text by means of representative simulations. They have been set to show how the different model ingredients integrate and contribute to obtain a wide range of cell dynamics. In particular, we have presented the evolution of colonies of individuals characterized by phenotypic homogeneity in the presence of one of the following sets of migratory stimuli:

- repulsion and randomness (in conjunction to mitosis), cf. Figs. 1 and 2;
- repulsion, adhesion, and alignment, whose relative importance has been varied, cf. Figs. 3 and 4;
- repulsion, as well as both epithelial-like and mesenchymal-like movement contributions, with different values of the corresponding α-weights, cf. Fig. 6.

More realistic numerical realizations have been finally proposed, which have reproduced the main aspects of a wound healing assay, i.e., an experimental protocol where a monolayer of cells is wounded by a pipette tip and the lesion is recolonized by the remaining cell mass, in the case of sufficient migratory capacity, eventually stimulated by chemical factors. In particular, we have accounted for a leading front of cells with hybrid phenotype, whereas the rest of individuals has been assigned fully epithelial determinants. The invasive ability of our virtual cell sheet has been then assessed upon variations either in the intensity of the mesenchymal behavior of the particles with hybrid trait or in the proliferation rate of epithelial agents.

As seen, the mesenchymal-like behavior of cells has been here set to completely depend on how each individual responds to a given chemical field (cf. Eq. (22)). Therefore, it has been necessary to implement a numerical method able to handle such a multiscale coupling from a computational point of view. In this respect, in agreement with the work [12], we have opted to combine the Heun method for the evolution in time of cell positions with a tailored Gauss–Legendre quadrature formula to solve the integral in Eq. (22), whose value determines the effective intensity and direction of the mesenchymal-type contribution to individual movement. In particular, as commented in the already-cited article [12], the Heun approach (a low order Runge-Kutta scheme) represents a reasonable compromise among accuracy and computational cost. The proposed *ad hoc* Gauss–Legendre quadrature formula is characterized by the fact that it has to be applied to each subinterval of the integration domain where the integrand function is actually regular (the direct application to the entire domain of integration is instead a feature of the standard approach). Such a modified version of the G-L method has in principle the potential to handle discontinuous distributions of environmental cues, provided that the corresponding integrand functions are Riemann-integrable and that the possible singularities are detectable.

Possible Model Improvements The proposed modeling approach can be improved in several directions. First of all, more realistic laws to describe the relationship between the phenotypic trait of a cell and some of its biophysical determinants may be derived (possibly from empirical evidence) and employed. For instance, the values of the α-coefficients may be set to affect the frequency of duplication events (here neglected for agents without fully epithelial hallmarks) or the individual speed in a non-linear fashion.

Keeping fixed its basic structure, our theoretical framework can be easily extended by the addition of a larger spectrum of migratory modes: this would simply amount in the addition of the proper velocity contribution in Eqs. (5) and/or (11). For example, it would be interesting to include inertial aspects. Even in the case of random crawling, cells in fact take time to reorganize and reorient their cytoskeletal elements to change direction of movement.

A further, and according to us very relevant, model improvement would imply the possibility that the α-coefficients vary in time, eventually in response to variations in the intracellular state or in selected environmental conditions. Such a model development would allow to describe the phenomenon of phenotypic *plasticity*, that is the ability of cells to switch back and forth among different phenotypes, eventually maintaining unaltered their genotype.

In the context of our interest, it would indeed possible to account for the so-called *epithelial-to-mesenchymal transition* (EMT), which is involved in a wide spectrum of biomedical scenarios. Cells have in fact observed to often lose stable cadherin-based adhesive interactions and apico-basal polarity, and to acquire mesenchymal features which allow them to delaminate from the rest of the aggregate and to start individual aggressive modes of migration. The inverse process may be reproduced as well, as cells with mesenchymal determinants can lose their migratory freedom and re-acquire epithelial hallmarks, including expression of junctional proteins [35].

In particular, we could assume that phenotypic conversions are triggered by the extracellular molecular landscape and/or are affected by randomness. For instance, depending on the context, mesenchymal hallmarks are seen to be induced by RTK signaling, including FGFR, VEGFR and EGFR and downstream signaling via PI3K, Akt, MAPK activation and Rho GTPases. Phenotypic switches may also results from epigenetic mutations as well.

The possibility for cells to have an evolving phenotype has been included in some other theoretical approaches. For instance, in individual-based/cellular automata models, each single cell is allowed to vary the label indicating its actual phenotype, as in the case of the well-celebrated Cellular Potts Model, see [45] and reference therein. Models based on a continuous cell description (also in the framework of the Theory of Mixtures) instead typically associate to each subpopulation a distinct density function: phenotypic conversions are then implemented by mass exchanging terms included in the evolution equations for cell dynamics, as done for instance in

[25, 54]. In the already-cited family of approaches based on adaptive dynamics and dealing with structured populations, random phenotypic transitions are accounted by including a diffusion term on the trait domain, see, e.g., [31, 52]. Differently from the cited works, in [11] phenotypic switches do imply also variations in the mathematical representation of cells: epithelial individuals are in fact collectively described with undifferentiated density functions, whereas mesenchymal agents are given pointwise representation (with the possibility of transition between the two descriptive instance given by a so-called bubble function).

In the present approach, we have included a substantially simplified description of the mechanisms underlying cell duplication. For instance, we have only accounted a given time interval between two mitotic events of the same individual (being T^{dupl} an estimate of the duration of the cell cycle) and a density-dependent inhibition of cell proliferation (i.e., an agent located in an overcrowded region is set to enter in a sort of G_0 phase and is no longer able to duplicate). However, during the different stages of the mitotic process, cells also grow: they in fact produce proteins and cytoplasmic organelles and duplicate the genetic material (i.e., chromosomes are replicated). Such a volume increment may be implemented by assuming that, for the representative i-th agent,

$$d_i^{\text{rep}} = d_i^{\text{rep}}(t) = h(t - t_i^{\text{dupl}}),$$

being h an increasing function and t_i^{dupl} the instant of the previous duplication of i (cf. Eq. (1)). In other words, the vital space of the generic cell i (which, in the case of particle models, is an indication of its volumetric extension) increases as a consequence of the time-lasting synthetization of the new material needed for its proliferation.

The proposed model, enriched the above extensions, has indeed the potential to capture and represent phenotypic heterogeneity among a given system of cells, as well as selected mechanisms underlying phenotypic plasticity. Its possible applications therefore span a wide spectrum of phenomena. In particular, our approach is particularly suited to deal with the following two scenarios:

- angiogenic processes, where a small number of endothelial cells forming the walls of pre-existing vessels acquire a leader/tip fate and represent migratory cues for the neighboring individuals with a follower/stalk behavior [18]. These mechanisms are triggered by a number of diffusing growth factors, e.g., vascular endothelial growth factor (VEGF), hepatocyte growth factors (HGF) and mediated by the well-known Delta-Notch signaling pathways [30, 53];
- growth and metastatization of solid tumors. Individuals exhibiting different phenotypic determinants have been in fact found in several types of disease, including breast cancer [1], colorectal cancer [38] and brain cancer [46]. Interestingly, it has been shown that malignant cells within the same mass exhibit such different behavior in spite of carrying the same genetic alterations [33]. Cancer cells have been also demonstrated to be able to switch between alternative phenotypic states either spontaneously or in response to ecological inputs. For

instance, nutrient-deprived malignant individuals activate downstream pathways that result in a shift towards a more aggressive behavior, which allow them to more effectively invade surrounding tissue, intravasate, extravasate and eventually establish secondary colonies. Phenotypic differentiation and conversions of genetically identical tumor cells have been shown to facilitate survival and adaptation of the entire disease as well, which can play "hide-and-seek" with multiple therapeutic regimes [43, 51].

Model extensions can eventually improve the description of wound healing processes, i.e., the representative scenario here proposed to highlight the potential of our approach. In particular, cell interactions with matrix-like coating substrates could be included by the introduction of a proper mesenchymal-like velocity term, that may depend on the spatial pattern of a microscopic variable reproducing density and orientation of ECM elements. This study would be particular relevant in a perspective of biomedical research in the perspective of the production and the test of bioengineered scaffolds which can provide optimal extracellular environments for regrowth and regeneration of tissues, for example skin, peripheral nerves, bones or cartilage. The addition of more realistic mitotic mechanisms, that may also account for the effect of bioelectric stimuli [48], would help to obtain a more accurate reproduction of healing phenomena as well.

Acknowledgments This research was partially supported by the Italian Ministry of Education, University and Research (MIUR) through the "Dipartimenti di Eccellenza" Programme (2018–2022)—Dipartimento di Scienze Matematiche "G. L. Lagrange", Politecnico di Torino (CUP: E11G18000350001). The authors are members of GNFM (Gruppo Nazionale per la Fisica Matematica) of INdAM (Istituto Nazionale di Alta Matematica), Italy.

References

1. Al-Hajj, M., Wicha, M.S., Benito-Hernandez, A., Morrison, S.J., Clarke, M.F.: Prospective identification of tumorigenic breast cancer cells. Proc. Natl. Acad. Sci. U.S.A. **100**, 3983–3988 (2003)
2. Alber, M.S., Kiskowski, M.A., Glazier, J.A., Jiang, Y.: On cellular automaton approaches to modeling biological cells. In: Rosenthal, J., Gilliam, D.S. (eds.) Mathematical Systems Theory in Biology, Communications, Computation, and Finance, pp. 1–39. Springer, New York (2003)
3. Anderson, A., Rejniak, K.: Single-cell-based models in biology and medicine. Mathematics and Biosciences in Interaction. Birkhäuser, Basel (2007)
4. Armstrong, P.B., Parenti, D.: Cell sorting in the presence of cytochalasin B. J. Cell Biol. **55**, 542–553 (1972)
5. Bement, W.M., Forscher, P., Mooseker, M.S.: A novel cytoskeletal structure involved in purse string wound closure and cell polarity maintenance. J. Cell Biol. **121**, 565–578 (1993)
6. Benzekry, S., Tracz, A., Mastri, M., Corbelli, R., Barbolosi, D., Ebos, J.M.L.: Modeling spontaneous metastasis following surgery: an in vivo-in silico approach. Cancer Res **76**, 535–547 (2016)
7. Benzekry, S., Lamont, C., Beheshti, A., Tracz, A., Ebos, J.M.L., Hlatky, L., Hahnfeldt, P.: Classical mathematical models for description and forecast of preclinical tumor growth. PLoS Comput. Biol. **10**, e1003800 (2022)

8. Carrillo, J.A., Colombi, A., Scianna, M.: Adhesion and volume constraints via nonlocal interactions determine cell organisation and migration profiles. J. Theor. Biol. **445**, 75–91 (2018)
9. Chiari, G., Delitala, M.E., Morselli, D., Scianna, M.: A hybrid modeling environment to describe aggregates of cells heterogeneous for genotype and behavior with possible phenotypic transitions. Int. J. Nonlin. Mech. **144**, 104063 (2022)
10. Collin, A., Bruhier, H., Kolosnjaj, J., Golzio, M., Rols, M.-P., Poignard, C.: Spatial mechanistic modeling for prediction of 3D multicellular spheroids behavior upon exposure to high intensity pulsed electric fields. AIMS Bioeng. **9**, 102–122 (2022)
11. Colombi, A., Scianna, M., Preziosi, L.: Coherent modelling switch between pointwise and distributed representations of cell aggregates. J. Math. Biol. **74**, 783–808 (2017)
12. Colombi, A., Falletta, S., Scianna, M., Scuderi, L.: An integro-differential non-local model for cell migration and its efficient numerical solution. Math. Comput. Simulat. **180**, 179–204 (2021)
13. Demicheli, R., Foroni, R., Ingrosso, A.: An exponential-Gompertzian description of LoVo cell tumor growth from in vivo and in vitro data. Cancer Res. **49**, 6543–6546 (1989)
14. Drasdo, D.: On selected individual-based approaches to the dynamics in multicellular systems. In: Alt, W., Chaplain, M., Griebel, M., Lenz, J. (eds.) Polymer and Cell Dynamics. Mathematics and Biosciences in Interaction, pp. 169–203. Birkhäuser, Basel (2003)
15. Etienne-Manneville, S.: Cdc42 – the centre of polarity. J. Cell Sci. **117**, 1291–1300 (2004)
16. Fiandaca, G., Bernardi, S., Scianna, M., Delitala, M.E.: A phenotype-structured model to reproduce the avascular growth of a tumor and its interaction with the surrounding environment. J. Theor. Biol. **535**, 110980 (2022)
17. Fong, E., Tzlilb, S., Tirrell, D.A.: Boundary crossing in epithelial wound healing. Proc. Natl. Acad. Sci. U.S.A. **107**, 19302–19307 (2010)
18. Friedl, P., Brocker, E.B.: The biology of cell locomotion within three-dimensional extracellularmatrix. Cell Mol. Life Sci. **57**, 41–64 (2000)
19. Friedl, P., Gilmour D.: Collective cell migration in morphogenesis, regeneration and cancer. Nat. Rev. Mol. Cell Bio. **10**, 445–457 (2009)
20. Friedl, P., Wolf, K.: Tumour-cell invasion and migration: diversity and escape mechanisms. Nat. Rev. Cancer **3**, 362–374 (2003)
21. Friedl, P., Wolf, K.: Plasticity of cell migration: a multiscale tuning model. J. Cell Biol. **188**, 11–19 (2010)
22. Friedl, P., Wolf, K., Lammerding, J.: Nuclear mechanics during cell migration. Curr. Opin. Cell Biol. **23**, 55–64 (2011)
23. Gallaher, J.A., Brown, J.S., Anderson, A.R.A.: The impact of proliferation-migration tradeoffs on phenotypic evolution in cancer. Sci. Rep. **9**, 2425 (2019)
24. Gao, C.F., Xie, Q., Su, Y.L., Koeman, J., Khoo, S.K., Gustafson, M., Knudsen, B.S., Hay, R., Shinomiya, N., Vande Woude, G.F.: Proliferation and invasion: plasticity in tumor cells. Proc. Natl. Acad. Sci. U.S.A. **102**, 10528–10533 (2005)
25. Giverso, C., Lorenzi, T., Preziosi, L.: Effective interface conditions for continuum mechanical models describing the invasion of multiple cell populations through thin membranes. Appl. Math. Lett. **125**, 107708 (2022)
26. Hanahan, D., Weinberg, R.A.: Hallmarks of cancer: the next generation. Cell **144**, 646–674 (2011)
27. Hatzikirou, H., Basanta, D., Simon, M., Schaller, K., Deutsch, A.: 'Go or grow': the key to the emergence of invasion in tumour progression? Math. Med. Biol. **29**, 49–65 (2012)
28. Joie, J., Lei, Y., Durrieu, M.-C., Colin, T., Poignard, C., Saut, O.: Migration and orientation of endothelial cells on micropatterned polymers: a simple model based on classical mechanics. DCDS-B **20**, (2015).
29. Khalil, A.A., Friedl, P.: Determinants of leader cells in collective cell migration. Integr. Biol. Camb. **2**, 568–574 (2010)
30. Liu, S., Cong, Y., Wang, D., Sun, Y., Deng, L., Liu, Y., Martin-Trevino, R., Shang, L., McDermott, S.P., Landis, M.D., Hong, S., Adams, A., D'Angelo, R., Ginestier, C., Charafe-

Jauffret, E., Clouthier, S.G., Birnbaum, D., Wong, S.T., Zhan, M., Chang, J.C., Wicha, M.S.: Breast cancer stem cells transition between epithelial and mesenchymal states reflective of their normal counterparts. Stem Cell Rep. **2**, 78–91 (2013)
31. Lorenzi, T., Venkataraman, C., Lorz, A., Chaplain, M.A.J.: The role of spatial variations of abiotic factors in mediating intratumour phenotypic heterogeneity. J. Theor. Biol. **451**, 101–110 (2018)
32. Martínez-González, A., Calvo, G.F., Pérez Romasanta, L.A., Pérez-García, V.M.: Hypoxic cell waves around necrotic cores in glioblastoma: a biomathematical model and its therapeutic implications. Bull. Math. Biol. **74**, 2875–2896 (2012)
33. Marusyk, A., Almendro, V., Polyak, K.: Intra-tumour heterogeneity: a looking glass for cancer? Nat. Rev. Cancer **12**, 323–334 (2012)
34. Munaron, L., Fiorio Pla, A.: Endothelial calcium machinery and angiogenesis: understanding physiology to interfere with pathology. Curr. Med. Chem. **16**, 4691–4703 (2009)
35. Nieto, M.A., Huang, R.Y.J., Jackson, R.A., Thiery, J.P.: EMT: 2016. Cell **166**, 21–45 (2016)
36. Nikolić, D.L., Boettiger, A.N., Bar-Sagi, D., Carbeck, J.D., Shvartsman, S.Y.: Role of boundary conditions in an experimental model of epithelial wound healing. Am. J. Physiol. Cell Physiol. **291**, C68–C75 (2006)
37. Nobes, C.D., Hall, A.: Rho GTPases control polarity, protrusion, and adhesion during cell movement. J. Cell Biol. **144**, 1235–1244 (1999)
38. OBrien, C.A., Pollett, A., Gallinger, S., Dick, J.E.: A human colon cancer cell capable of initiating tumour growth in immunodeficient mice, Nature **445**, 106–110 (2007)
39. Oraiopoulou, M.E., Tzamali, E., Tzedakis, G., Vakis, A., Papamatheakis, J., Sakkalis, V.: In vitro/in silico study on the role of doubling time heterogeneity among primary glioblastoma cell lines. In: BioMed Research International 2017 (2017)
40. Perthame, B.: Transport Equations in Biology. Springer Science & Business Media, New York (2006)
41. Poujade, M., Grasland-Mongrain, E., Hertzog, A., Jouanneau, J., Chavrier, P., Ladoux, B., Buguin, A., Silberzan, P.: Collective migration of an epithelial monolayer in response to a model wound. Proc. Natl. Acad. Sci. U.S.A. **104**, 15988–15993 (2007)
42. Ridley, A.J., Paterson, H.F., Johnston, C.L., Diekmann, D., Hall, A.: The small GTP-binding protein rac regulates growth factor-induced membrane ruffling. Cell **70**, 401–410 (1992)
43. Roesch, A.: Tumor heterogeneity and plasticity as elusive drivers for resistance to MAPK pathway inhibition in melanoma. Oncogene **34**, 2951–2957 (2015)
44. Sabeh, F., Shimizu-Hirota, R., Weiss, S.J.: Protease-dependent versus -independent cancer cell invasion programs: three-dimensional amoeboid movement revisited. J. Cell. Biol. **185,** 11–19 (2009)
45. Scianna, M., Preziosi, L.: Multiscale developments of the Cellular Potts Model. Multiscale Model. Sim. **10**, 342–382 (1995)
46. Singh, S.K., Hawkins, C., Clarke, I.D., Squire, J.A., Bayani, J., Hide, T., Henkelman, R.M., Cusimano, M.D., Dirks, P.B.: Identification of human brain tumour initiating cells. Nature **432**, 396–401 (2004)
47. Stearns, S.C.: Trade-offs in life-history evolution. Funct. Ecol. **3**, 259–268 (1989)
48. Sundelacruz, S., Li, C., Choi, Y.J., Levin, M., Kaplan, D.: Bioelectric modulation of wound healing in a 3D in vitro model of tissue-engineered bone. Biomaterials **34**, 6695–6705 (2013)
49. Suzuki, K., Saito, J., Yanai, R., Yamada, N., Chikama, T., Seki, K., Nishida, T.: Cell–matrix and cell–cell interactions during corneal epithelial wound healing. Prog. Retin. Eye Res. **22**, 113–133 (2003)
50. Vaghi, C., Rodallec, A., Fanciullino, R., Ciccolini, J., Mochel, J.P., Mastri, M., Poignard, C., Ebos, J.M.L., Benzekry, S.: Population modeling of tumor growth curves and the reduced Gompertz model improve prediction of the age of experimental tumors. PLoS Comput. Biol. **16**, e1007178 (2020)
51. Varga, J., De Oliveira, T., Greten, F.R.: The architect who never sleeps: tumor-induced plasticity. FEBS Lett. **588**, 2422–2427 (2014)

52. Villa, C., Chaplain, M.A., Lorenzi, T.: Modeling the emergence of phenotypic heterogeneity in vascularized tumors. SIAM J. Appl. Math. **81**, 434–453 (2021)
53. Williams, C.K., Li, J.L., Murga, M., Harris, A.L., Tosato, G.: Up-regulation of the Notch ligand Delta-like 4 inhibits VEGF-induced endothelial cell function. Blood **107**, 931–939 (2006)
54. Wise, S.M., Lowengrub, J.S., Frieboes, H.B., Cristini, V.: Three-dimensional multispecies nonlinear tumor growth–I. Model and numerical method. J. Theor. Biol. **253**, 524–543 (2008)
55. Wolf, K., Te Lindert, M., Krause, M., Alexander, S., Te Riet, J., Willis, A.L., Hoffman, R.M., Figdor, C.G., Weiss, S.J., Friedl, P.: Physical limits of cell migration: control by ECM space and nuclear deformation and tuning by proteolysis and traction force. J. Cell Biol. **201**, 1069–1084 (2013)
56. Zahm, J.M., Kaplan, H., Hérard, A.L., Doriot, F., Pierrot, D., Somelette, P., Puchelle, E.: Cell migration and proliferation during the in vitro wound repair of the respiratory epithelium. Cell Motil. Cytoskel. **37**, 33–43 (1997)
57. Zhigun, A., Surulescu, C., Hunt, A.: A strongly degenerate diffusion-haptotaxis model of tumour invasion under the go-or-grow dichotomy hypothesis. Math. Method. Appl. Sci. **41**, 2403–2428 (2018)

Modelling HIF-PHD Dynamics and Related Downstream Pathways

Patrizia Ferrante and Luigi Preziosi

Abstract Hypoxia can represent a challenging condition for survival at different biological scales, from cells to organisms. The efficiency of the response to decreased oxygen availability is importantly related to the Hypoxia-Inducible Factors (HIFs) that regulate the transcription of hundreds of genes whose proteins are responsible for changes in metabolism, cell cycle, vascularisation. All these downstream responses are aimed, on one side, to optimise the consumption of oxygen and, on the other side, to change the microenvironment in order to potentially create the conditions to favor oxygen delivery. In this chapter we firstly develop a mathematical model that focuses on the oxygen-dependent regulation of HIFs, on the basis of available biological experiments, we secondly use the model to mathematically investigate the role of HIFs and hypoxia on inflammation and we finally discuss the biological background of the main responses regulated by HIFs and the mathematical models that focused explicitly on HIF action.

1 Introduction

Oxygen plays a key role for many biochemical reactions that are crucial for the survival of most complex organisms. Then, a decrease in oxygen availability (named hypoxia), as well as an increase (named hyperoxia), may represent a challenge for their survival. Hypoxia characterizes both specific cellular microenvironments and high altitude environments, where there exists a progressive decrease of the partial pressure of oxygen that affects the entire body. For instance, hypoxia characterizes the core of solid tumors as a result of an increased proliferation rate and an

P. Ferrante
Candiolo Cancer Institute FPO-IRCCS, Candiolo, Italy
e-mail: patrizia.ferrante@unito.it

L. Preziosi (✉)
Department of Mathematical Sciences, Politecnico di Torino, Torino, Italy
e-mail: luigi.preziosi@polito.it

G. Bretti et al. (eds.), *Mathematical Models and Computer Simulations for Biomedical Applications*, SEMA SIMAI Springer Series 33,
https://doi.org/10.1007/978-3-031-35715-2_4

inefficient vasculature [72]. Regardless of the specific environment or cause, the main regulator of responses to hypoxia is represented by Hypoxia Inducible Factors (HIFs).

HIFs are a family of heterodimeric proteins that act as transcription factors after the dimerisation of their two subunits: an oxygen-dependent α subunit (HIF1α, HIF2α, HIF3α), and a constitutively expressed β one (HIF1β) [29].

HIFs are regulated by post-translational modifications that are catalyzed by two members of the 2-Oxoglutarate-dependent oxygenase superfamily that are activated only in the presence of molecular oxygen [1]. The main regulator belongs to the family of prolyl hydroxylases domain (PHD1-3) enzymes which directly affect the expression of HIFα. The other regulator is the factor inhibiting HIF (FIH) that acts on a specific region of HIFs, the CAD region, affecting CAD dependent gene expression. Apart from the enzymatic regulation of HIFs, other forms of regulation are discussed in this chapter, such as the one that involves RACK1 and that is affected by microenvironment acidity [21].

The activation of all HIFs orchestrates downstream crucial adaptive responses that govern relevant aspects of cell life, such as metabolism, duplication, motility, angiogenic response, inflammation [63].

In this chapter, without going in detail in the dimerization process of HIFs, we focus on the dynamics of HIF1α and HIF2α, since very little is known about the third isoform. We first propose a simple model describing HIF1 and HIF2 dynamics in response to hypoxia and their interplay with PHD2 and PHD3. We do not include FIH regulation whose absence is demonstrated to have a significant impact on metabolism but not on other related HIF signaling, including for instance angiogenesis [81].

After validating the model on the basis of known experimental data, we focus on modeling the interplay between inflammatory and hypoxic responses. In fact, since alarmin receptors can represent a potential target of both HIF1 and HIF2 [68], we will investigate the role of both HIFs on the inflammatory state.

Finally, in the last part of the chapter we describe the effects of HIF on downstream pathways governing metabolism, pH homeostasis, duplication, high altitude, angiogenic response, stiffening of the surrounding environment that leads to increased motility. We briefly review some existing mathematical models and point out directions that need further studies from the modeling point of view.

2 HIFs and PHDs

Entering more in detail in HIF regulating role of PHDs, it is known that PHDs activate in presence of oxygen and hydroxylate HIFα on two proline residues converting it in an hydroxylated form that has a great affinity for the ubiquitination complex linked to the tumour suppressor von Hippel-Lindau protein (pVHL). The consequence is that, under normoxic conditions, HIFα is rapidly degraded by the proteasomal pathway. Conversely, under hypoxic conditions, PHD catalytic

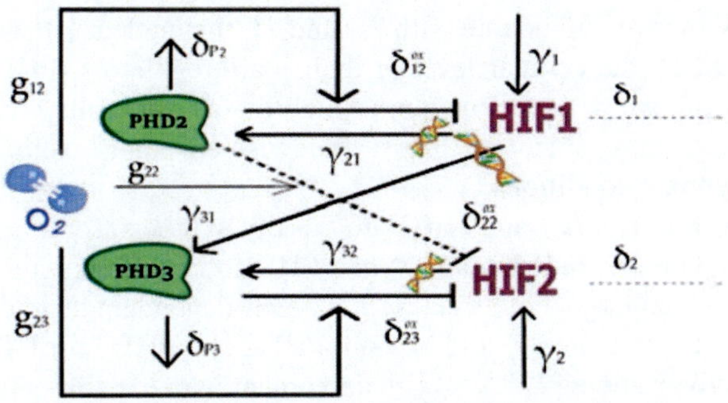

Fig. 1 HIF-PHD pathway. Normal and blockhead arrows respectively refer to promoting and inhibiting activities. When dashed it means that the related action is less relevant than that represented by a full arrows, e.g., PHD-independent degradation of HIF1 at a rate δ_1 compared to oxygen dependent degradation due to the action of PHD2 at a rate δ_{12}^{ox}. DNA schemes refer to transcriptional activities of HIFs

activity drops and, since pVHL can not recognize the un-hydroxylated HIFα, HIFα stabilizes and dimerizes with the β subunit. When this happens, they form a transcriptional complex which can bind to HIF-Response Elements (HRE) and initiate mRNA transcription.

Referring to Fig. 1, HIF1 and HIF2 are related to specific PHDs, named respectively PHD2 and PHD3, though it is found that PHD2 has also a minor regulatory activity on HIF2 [2]. PHD production is induced by HIFs with HIF1 that promotes both PHD2 and PHD3 production [2, 8, 46] and HIF2 that promotes PHD3 production only [2].

Before starting the description of the model, we need to observe that the levels of oxygen saturation defining normoxia and hypoxia are not precisely defined in the literature. Actually, it is even different in in vitro experiments and in tissues. For instance, in in vitro experiments a saturation level about 20% of oxygen is denoted as normoxia, while a level of 1% as hypoxia [4, 18, 26, 66, 71]. Finally, in [35] anoxia is defined for a value below 0.1%. On the other hand, in both normal and pathological tissues, oxygen saturation never reaches such high values even in normoxia (e.g., arterial blood O_2 is at 13.2%). So, for instance, in [45] it is observed that normoxic values for different tissues range between 3% and 7.4%. Moreover, it is observed that the median oxygenation of tumours falls approximately between 0.3% and 4.2%, with most tumours exhibiting median oxygen level below 2%. In the following we will define a hypoxic condition when $O_2 \in (0.1\%, 1\%)$, and therefore a normoxic condition above this interval and an anoxic condition below it. We observe that, in our model, oxygen is the only limiting factor for the hydroxylation mediated by PHD, implying that the other factors playing a role in this process, e.g., 2-oxoglutarate and Fe(II), are always present in sufficient quantity to allow the reaction.

In the following we will denote with $\hat{H}_1$ and $\hat{H}_2$ the nuclear HIF1 and HIF2 level and with $\hat{P}_2$ and $\hat{P}_3$ the cellular level of their main regulators (PHD2 and PHD3). On the other hand, we will take oxygen availability as externally controlled, so that it is a given function of time, typically a piecewise constant function ranging from normoxic to hypoxic conditions.

Referring to Fig. 1, HIF1 and HIF2 are basally synthesized at a rate γ_1 and γ_2, respectively, and are degraded in an oxygen/PHD-independent way at a rate δ_1 and δ_2. However, they undergo a more relevant oxygen-dependent degradation triggered by PHD2 for HIF1 at a rate δ_{12}^{ox} and by both PHD2 and PHD3 for HIF2, at rates δ_{22}^{ox} and δ_{23}^{ox}, respectively, being $\delta_{23}^{ox} > \delta_{22}^{ox}$ in agreement with experimental evidence [2].

So, the evolution of HIF1 and HIF2 is modelled by the following ODEs

$$\frac{d\hat{H}_1}{d\hat{t}} = \gamma_1 - \delta_1 \hat{H}_1 - \delta_{12}^{ox} g_{12}(\hat{O}_2) F_{12}(\hat{H}_1) \hat{P}_2 \,, \tag{1}$$

$$\frac{d\hat{H}_2}{d\hat{t}} = \gamma_2 - \delta_2 \hat{H}_2 - \delta_{22}^{ox} g_{22}(\hat{O}_2) F_{22}(\hat{H}_2) \hat{P}_2 - \delta_{23}^{ox} g_{23}(\hat{O}_2) F_{23}(\hat{H}_2) \hat{P}_3 \,, \tag{2}$$

where

$$g_{hk}(\hat{O}_2) = \frac{\hat{O}_2}{K_{hk}^{ox} + \hat{O}_2} \,, \qquad \text{and} \qquad F_{hk}(\hat{H}_h) = \frac{\hat{H}_h}{K_{hk} + \hat{H}_h} \,, \tag{3}$$

are Michaelis-Menten terms describing the degradation of HIF1 and HIF2 by PHD2 and PHD3 (as given by the F terms), that is promoted by the presence of oxygen (as given by the g terms).

In turn, HIF1 induces both PHD2 and PHD3, while HIF2 induces PHD3 only, and is regulated by both PHD3 and, to a minor extent, by PHD2 itself [2, 66]. For the moment, we will simply denote the production terms of PHD2 and PHD3 by the functions $\hat{\Gamma}_2(\hat{H}_1)$ and $\hat{\Gamma}_3(\hat{H}_1, \hat{H}_2)$ just requiring that they satisfy some biologically sound hypothesis, namely that they are increasing functions of HIF concentrations and vanish in the limit case of vanishing HIFs, i.e.,

$$\begin{aligned} &\Gamma_2(0) = \Gamma_3(0,0) = 0 \,, \quad \text{and} \\ &\frac{\partial \Gamma_2}{\partial H_1}(H_1) > 0 \,, \quad \frac{\partial \Gamma_3}{\partial H_1}(H_1, H_2) > 0 \,, \quad \frac{\partial \Gamma_3}{\partial H_2}(H_1, H_2) > 0 \,, \quad \forall H_1, H_2 \,. \end{aligned} \tag{4}$$

The simplest functions satisfying (4), that will be used in the simulations to follow, are the linear forms

$$\hat{\Gamma}_2(\hat{H}_1) = \gamma_{21} \hat{H}_1 \,, \qquad \hat{\Gamma}_3(\hat{H}_1, \hat{H}_2) = \gamma_{31} \hat{H}_1 + \gamma_{32} \hat{H}_2 \,. \tag{5}$$

So, the equations governing the evolution of PHDs are

$$\frac{d\hat{P}_2}{d\hat{t}} = \hat{\Gamma}_2(\hat{H}_1) - \delta_{P_2}\hat{P}_2\,, \tag{6}$$

$$\frac{d\hat{P}_3}{d\hat{t}} = \hat{\Gamma}_3(\hat{H}_1, \hat{H}_2) - \delta_{P_3}\hat{P}_3\,. \tag{7}$$

where δ_{P_2} and δ_{P_3} are the degradation rates of PHD2 and PHD3, respectively.

To clarify the general notation used in this chapter, we observe that hatted variables are dimensional as well as hatted functions (e.g., $\hat{\Gamma}_2$ and $\hat{\Gamma}_3$), while non-hatted functions, such as g_{hk} and F_{hk} in (3), are dimensionless. In addition, the double index hk refers to the action on h due to the presence of k. So, for instance, δ_{23}^{ox} refers to the degradation of HIF2 due to the presence of PHD3.

Overall, HIF-PHD dynamics are modelled by (1), (2), (6), and (7). However, in the following it will be useful to rewrite the system in dimensionless form introducing the following scaled variables

$$H_1 = \frac{\hat{H}_1}{K_{12}}\,, \qquad H_2 = \frac{\hat{H}_2}{K_{23}}\,, \qquad P_2 = \frac{\delta_{P_2}}{\gamma_{21}K_{12}}\hat{P}_2\,, \qquad P_3 = \frac{\delta_{P_3}}{\gamma_{32}K_{23}}\hat{P}_3\,,$$

$$O_2 = \frac{\hat{O}_2}{K_{12}^{ox}}\,, \qquad t = \delta_{P_2}\hat{t}\,.$$

In addition, the magnitude of $\hat{\Gamma}_2$, and $\hat{\Gamma}_3$ are scaled with $\gamma_{21}K_{12}$, and $\gamma_{32}K_{23}$, respectively, so that we can write

$$\Gamma_2(H_1) = H_1 \qquad \text{and} \qquad \Gamma_3(H_1, H_2) = \tilde{\gamma}H_1 + H_2 \qquad \text{with} \qquad \tilde{\gamma} = \frac{\gamma_{31}K_{12}}{\gamma_{32}K_{23}}\,. \tag{8}$$

The resulting dimensionless HIF-PHD model then writes as

$$\begin{cases} \tau\dfrac{dH_1}{dt} = 1 - \zeta H_1 - \eta g_{12}(O_2)F_{12}(H_1)P_2 \\ \dfrac{dP_2}{dt} = \Gamma_2(H_1) - P_2 \\ \tau_2\dfrac{dH_2}{dt} = 1 - \zeta_2 H_2 - \eta_2 g_{22}(O_2)F_{22}(H_2)P_2 - \eta_3 g_{23}(O_2)F_{23}(H_2)P_3 \\ \tau_P\dfrac{dP_3}{dt} = \Gamma_3(H_1, H_2) - P_3 \end{cases} \tag{9}$$

with

$$
\begin{aligned}
g_{12}(O_2) &= \frac{O_2}{1+O_2}\,, \quad g_{22}(O_2) = \frac{O_2}{\tilde{K}_{22}^{ox}+O_2}\,, \quad g_{23}(O_2) = \frac{O_2}{\tilde{K}_{23}^{ox}+O_2}\,,\\
F_{12}(H_1) &= \frac{H_1}{1+H_1}\,, \quad F_{22}(H_2) = \frac{H_2}{\tilde{K}_{22}+H_2}\,, \quad F_{23}(H_2) = \frac{H_2}{1+H_2}\,,
\end{aligned}
\tag{10}
$$

and

$$
\begin{gathered}
\tau=\frac{K_{12}\delta_{P_2}}{\gamma_1}\,, \qquad \tau_2=\frac{K_{23}\delta_{P_2}}{\gamma_2}\,, \qquad \tau_p=\frac{\delta_{P_2}}{\delta_{P_3}}\,, \qquad \zeta=\frac{K_{12}\delta_1}{\gamma_1}\,, \quad \zeta_2=\frac{K_{23}\delta_2}{\gamma_2}\,,\\
\eta=K_{12}\frac{\delta_{12}^{ox}\gamma_{21}}{\gamma_1\delta_{P_2}}\,, \quad \eta_2=K_{12}\frac{\delta_{22}^{ox}\gamma_{21}}{\gamma_2\delta_{P_2}}\,, \quad \eta_3=K_{23}\frac{\delta_{23}^{ox}\gamma_{32}}{\gamma_2\delta_{P_3}}\,, \quad \tilde{K}_{22}^{ox}=\frac{K_{22}^{ox}}{K_{12}^{ox}}\,, \quad \tilde{K}_{23}^{ox}=\frac{K_{23}^{ox}}{K_{12}^{ox}}\,,\\
\tilde{K}_{22}=\frac{K_{22}}{K_{23}}\,.
\end{gathered}
$$

Then, substituting (10) and the linear forms (8) the system writes

$$
\begin{cases}
\tau\dfrac{dH_1}{dt} = 1-\zeta H_1 - \eta\dfrac{O_2}{1+O_2}\dfrac{H_1}{1+H_1}P_2\\
\dfrac{dP_2}{dt} = H_1 - P_2\\
\tau_2\dfrac{dH_2}{dt} = 1-\zeta_2 H_2 - \eta_2\dfrac{O_2}{\tilde{K}_{22}^{ox}+O_2}\dfrac{H_2}{\tilde{K}_{22}+H_2}P_2 - \eta_3\dfrac{O_2}{\tilde{K}_{23}^{ox}+O_2}\dfrac{H_2}{1+H_2}P_3\\
\tau_P\dfrac{dP_3}{dt} = \tilde{\gamma}H_1 + H_2 - P_3
\end{cases}
\tag{11}
$$

Table 1 report the reference parameter values. A deeper discussion on their identification is given in [20].

2.1 Equilibrium States

The equilibria of the system (9) can be found first observing that

$$
P_2^{eq} = \Gamma_2(H_1^{eq}) \qquad \text{and} \qquad P_3^{eq} = \Gamma_3(H_1^{eq}, H_2^{eq})\,. \tag{12}
$$

The substitution of the former in $(9)_1$ allows to get

$$
\zeta H_1^{eq} + \eta g_{12}(O_2)F_{12}(H_1^{eq})\Gamma_2(H_1^{eq}) = 1\,, \tag{13}
$$

and therefore to implicitly identify H_1^{eq}.

Table 1 Parameters used as reference values. M-M stands for Michaelis-Menten

Parameter (unit)	HIF1	HIF2	PHD2	PHD3
Basal production (μM/s)	$\gamma_1 = 10^{-1}$	$\gamma_2 = 3 \cdot 10^{-1}$		
Basal degradation (s^{-1})	$\delta_1 = 10^{-5}$ [4, 49]	$\delta_2 = 10^{-5}$	$\delta_{P_2} = 10^{-5}$ [4, 49]	$\delta_{P_3} = 10^{-5}$

Parameter (unit)	HIF1→PHD2	HIF1→PHD3	HIF2→PHD3
PHD production (s^{-1})	$\gamma_{21} = 4 \cdot 10^{-5}$	$\gamma_{31} = 4 \cdot 10^{-5}$	$\gamma_{32} = 2 \cdot 10^{-5}$
Parameter (unit)	HIF1←PHD2	HIF2←PHD2	HIF2←PHD3
HIF degradation by PHD (s^{-1})	$\delta_{12}^{ox} = 10^{-2}$[18, 53]	$\delta_{22}^{ox} = 7 \cdot 10^{-3}$[18]	$\delta_{23}^{ox} = 10^{-4}$
M-M constant (O_2) (μM)	$K_{12}^{ox} = 100$ [18, 31, 36]	$K_{22}^{ox} = 70$ [71]	$K_{23}^{ox} = 100$
M-M constant (HIF) (μM)	$K_{12} = 14$[18, 53]	$K_{22} = 15$ [71]	$K_{23} = 150$ [36]

Reference dimensionless parameters

$\eta = 5.6$	$\eta_2 = 1.3067$	$\eta_3 = 0.1$
$\tau = 1.4 \cdot 10^{-3}$	$\tau_2 = 5 \cdot 10^{-2}$	$\tau_P = 1$
$\zeta = 1.4 \cdot 10^{-3}$	$\zeta_2 = 5 \cdot 10^{-2}$	$\sigma = 0.19$
$\tilde{K}_{22}^{ox} = 0.7$	$\tilde{K}_{23}^{ox} = 1$	$\tilde{K}_{22} = 0.1$

Because of the positivity and the increasing trend required for the functions Γ_2 and F_{12} and of the fact that $\Gamma_2(H_1 = 0) = 0$ (see Eq. (4)), Eq. (13) has only one solution. Actually, since the term g_{12} increases with the concentration of oxygen, the derivative of the l.h.s. of (13) at any fixed H_1 increases with O_2 as well, and then we can also state that both H_1^{eq} and P_2^{eq} decrease with the level of available oxygen, as expected and shown in Fig. 2.

Finally, substituting (12) in the r.h.s. of $(9)_3$, we have that it vanishes when

$$\zeta_2 H_2^{eq} + \eta_2 g_{22}(O_2) F_{22}(H_2^{eq}) \Gamma_2(H_1^{eq}) + \eta_3 g_{23}(O_2) F_{23}(H_2^{eq}) \Gamma_3(H_1^{eq}, H_2^{eq}) = 1\,, \tag{14}$$

where H_1^{eq} can be considered known from the solution of Eq. (13). This equation allows to implicitly determine H_2^{eq}. In particular, the above-mentioned properties of F_{22}, F_{23}, Γ_2 and Γ_3 (see (4)) allow to state again that there is only one equilibrium state H_2^{eq}, and that, as expected, it decreases with the level of oxygen. The same holds for P_3^{eq}.

To be specific, let us take the dimensionless forms of the linear dependencies (8). In this case we can explicitly get

$$H_1^{eq} = P_2^{eq} = \frac{1 - \zeta + \sqrt{(1+\zeta)^2 + \dfrac{4\eta O_2}{1 + O_2}}}{2\zeta + \dfrac{2\eta O_2}{1 + O_2}}\,. \tag{15}$$

The dependence of this equilibrium value in terms of the dimensionless parameters is reported in Fig. 2. In particular, Fig. 2a and b describes how H_1^{eq} (and therefore

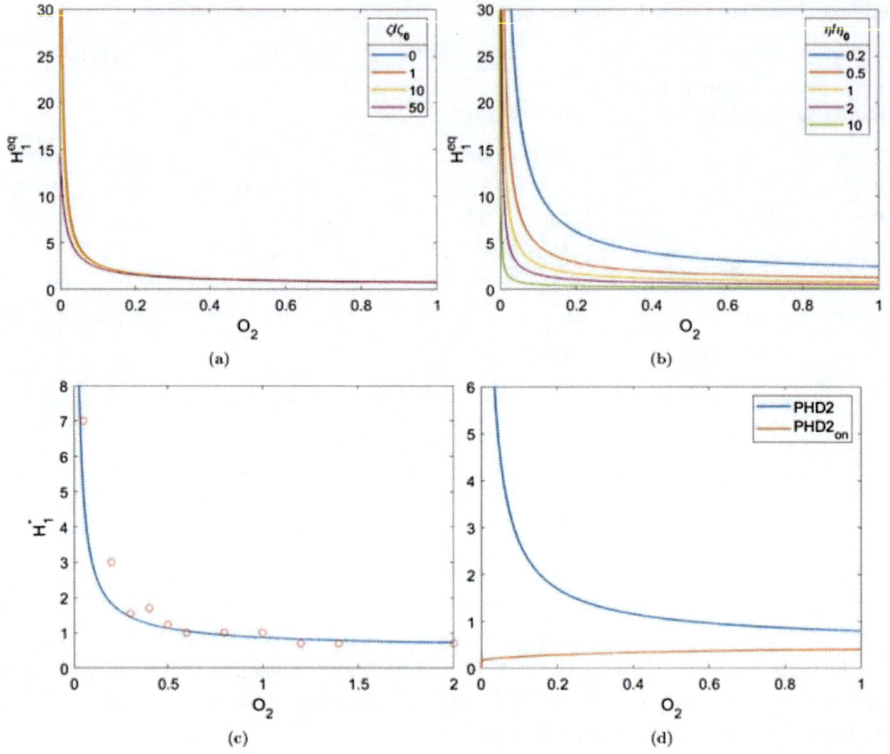

Fig. 2 Dependence of nondimensional HIF1 and PHD2 levels plotted as a function of O_2 normalized with respect to K_{12}. For reference, $O_2 = 1$ corresponds to a saturation of 10% and then to normoxic conditions in tissues. Equilibrium HIF level when (**a**) ζ and (**b**) η change, increasing from top to bottom. In (**c**) the reference curve is compared with the experimental data on HeLa cells reported in [30]. Following [30], HIF1 expression data are normalized with respect to the value obtained at 6% O_2 corresponding to $O_2 = 0.6$. (**d**) Level of PHD2 and of enzymatically active PHD2 at equilibrium, i.e., $P_2 g_{12}(O_2)$

P_2^{eq}) decrease for increasing concentrations of O_2 for different values of the parameter ζ and η, respectively.

We observe that in some articles (e.g., [71]) it is reported that the level of PHD2 increases with the oxygen level, which seems to be in contradiction with what just stated. Actually, they refer to the enzymatically active PHD2, (i.e., bound to oxygen), which is obtained by multiplying P_2 by the Michaelis-Menten term g_{12}. If this is plotted, an increasing curve is obtained as shown in Fig. 2d.

From the identification of the equilibrium states it is possible in principle to get some information on the parameters, that result independent of the normalization that characterized experimental data for calibration reasons. In fact, in experiments, data are usually normalized with respect to a value chosen as reference. The idea is to compare the equilibrium values of H_1^{eq} in normoxia and hypoxia (say H_n^{eq} and

H_h^{eq}, respectively). In fact, they have both to satisfy (15) for the two levels of O_2. Named r the ratio $r = H_h^{eq}/H_n^{eq}$, that is larger than 1, then for a fixed ζ it is possible to estimate η. For instance, in the limit of negligible ζ one has

$$\eta = \frac{g_{12}(O_n) - r g_{12}(O_h)}{[g_{12}(O_n) - r^2 g_{12}(O_h)]^2}(r-1)r\,,$$

where O_n is the normoxic value and O_h the hypoxic one.

2.2 *The Limit* $\zeta \to 0$

The experimental value reported in Table 1 shows that $\zeta \ll 1$. In fact, as will be discussed in Sect. 2.3, the role of δ_1 as well as that of δ_2 and therefore of the related dimensionless parameters ζ and ζ_2 becomes important only in anoxic conditions, when the complete lack of oxygen hampers the activation of PHDs, so that, for instance, the two terms at the denominator of (15) become comparable. Thus, in the limit $\zeta \to 0$ Eq. (15) simplifies to

$$H_1^{eq} = P_2^{eq} = \frac{1+O_2}{2\eta O_2}\left(1+\sqrt{1+\frac{4\eta O_2}{1+O_2}}\right). \tag{16}$$

Also the equation giving H_2^{eq} simplifies to

$$\eta_2 g_{22}(O_2)\frac{H_1^{eq}H_2^{eq}}{\tilde{K}_{22}+H_2^{eq}} + \eta_3 g_{23}(O_2)\frac{H_2^{eq}}{1+H_2^{eq}}(\tilde{\gamma}H_1^{eq}+H_2^{eq}) = 1\,, \tag{17}$$

which, however, can not be easily solved explicitly.

2.3 *The Anoxic Limit*

In the limit of no oxygen at all, the equilibrium is trivially given by the balance of HIF production and the PHD-independent degradation

$$H_1^{eq} = P_2^{eq} = \frac{1}{\zeta}\,, \qquad H_2^{eq} = P_3^{eq} = \frac{1}{\zeta_2}\,.$$

For very small values of oxygen corresponding to hypoxic conditions we can take g_{hk} as simply proportional to the oxygen level because the denominator is essentially constant. In addition, we can look for an equilibrium solution that is a perturbation of order O_2 of the equilibrium above, explicitly obtaining the approximations

$$H_1^{eq} \approx \frac{1}{\zeta}\left[1 - \frac{\eta}{1+\zeta}\Gamma_2\left(\frac{1}{\zeta}\right) O_2\right],$$

and

$$H_2^{eq} \approx \frac{1}{\zeta_2}\left\{1 - \left[\frac{\eta_2}{\tilde{K}_{12}^{ox}(1+\tilde{K}_{22}\zeta_2)}\Gamma_2\left(\frac{1}{\zeta}\right) + \frac{\eta_3}{\tilde{K}_{12}^{ox}(1+\zeta_2)}\Gamma_3\left(\frac{1}{\zeta}, \frac{1}{\zeta_2}\right)\right] O_2\right\},$$

that for $\zeta_2 \ll 1$ further simplifies to

$$H_2^{eq} \approx \frac{1}{\zeta_2}\left\{1 - \left[\frac{\eta_2}{\tilde{K}_{12}^{ox}}\Gamma_2\left(\frac{1}{\zeta}\right) + \frac{\eta_3}{\tilde{K}_{12}^{ox}}\Gamma_3\left(\frac{1}{\zeta}, \frac{1}{\zeta_2}\right)\right] O_2\right\},$$

We need to mention that in some experiments (see, for instance, [29]) it seems that the level of HIF1 drops, counterintuitively, dramatically in anoxic condition. This may to be linked to alteration of microenvironmental pH. Referring to Sect. 4.2 for more details, we here mention, for instance, that Honda et al. [27] demonstrated that the decrease of HIF1α under anoxia is independent of PHD-dependent degradation of HIF1. On the contrary the accumulation of HIF1α is affected by changes of the extracellular pH (pH_e). In this respect, Filatova et al. [21] found that the link between pH_e and HIFs was represented by the competition of HSP90 protein with RACK1. Under physiological conditions, RACK1 interacts with HIF leading to HIF proteasomal degradation. This pathway is altered by acidic stress that prevents HIF degradation through the expression of the stress-induced chaperone protein HSP90 that thus represents a crucial node in the interplay between microenvironment acidity and HIF activation [21].

2.4 HIF-PHD Dynamics

From the experimental point of view the dynamics of HIF1 is characterized by the presence of a well defined peak that activates in one or 2 h after a sudden drop of oxygen level [4]. HIF2 evolution also presents a peak, that is however less pronounced and appears after many hours or even a day later [26, 65]. Though these general characteristics are well known, the timing, duration, and intensity of the peak is tissue dependent, and even cell dependent [4].

In the following we will discuss the simulation of the proposed model using the value of the parameters given in Table 1. Most degradation rates are measured, but direct measurements of the production terms are not available because of difficulties inherited by the need to suitably calibrate experiments. In this case we only have some evaluated values used in the simulations in [4, 10, 39].

Having this in mind, as a first step, we focus on HIF1-PHD2 evolution which is the most studied one. In particular, we use the experimental data on HeLa cells reported by Bagnall et al. [4] and Leedale et al. [39]. In these experiments, cells were imaged every 5 min after a switch from 20.9% to 1% O_2 using time lapse confocal microscopy. In their graphs cells were artificially synchronized in the cell cycle, using the mitosis time as $t = 0$. Then, the evolution of the HIF1 level in different cells is reported. These graphs are merged in Fig. 3a normalizing the fluorescence data with respect to their maximum. In fact, Bagnall et al. [4] advice that the quantitative fluorescence output among separate experiments or among cells in the same experiment are not comparable for reasons mainly related to fluorescence

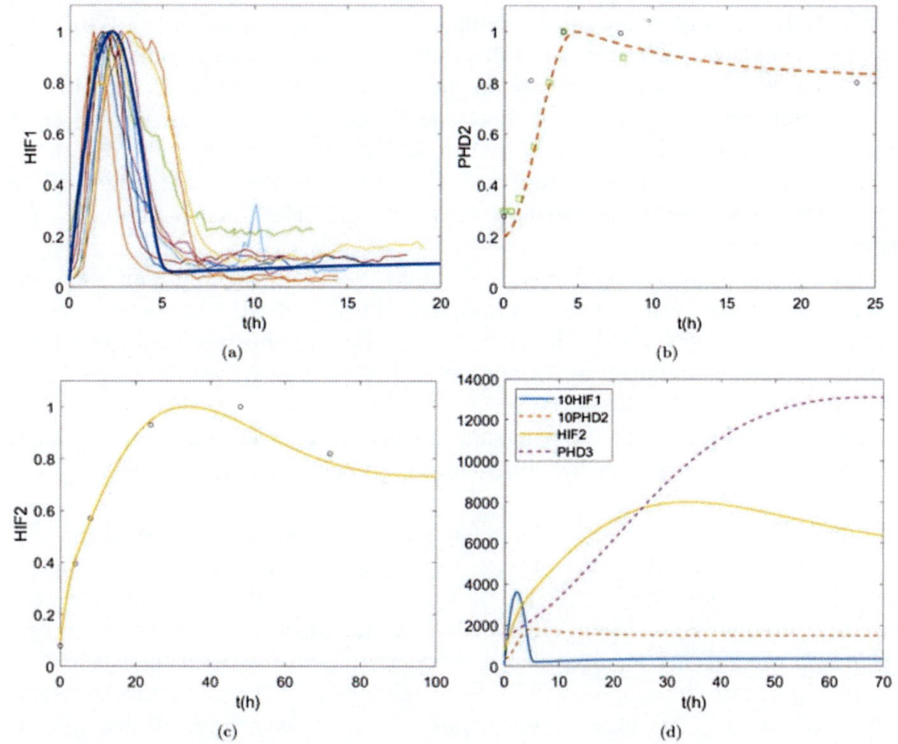

Fig. 3 Dynamics of HIF1, HIF2, PHD2, and PHD3 compared with experiments. The system is suddenly de-oxygenated (1% O_2) starting from a normoxic conditions (21% O_2). (**a**) Experimental HIF1 data on HeLa cells reported in [4] normalized with respect to their maximum values. Each curve refers to a different HeLa cell. The thick blue curve refers to the numerical result obtained using the parameters reported in Table 1. (**b**) Normalized PHD2 evolution compared with data on on HeLa cells (green circles) and on neuroblastoma cells (black circles), respectively reported in [4] and [26]. (**c**) HIF2 dynamics compared with experimental data on breast cancer cells reported in [65] normalized with respect to their maximum values. (**d**) Temporal dynamics of non normalized HIF1, HIF2, PHD2, and PHD3 for the reference values of the parameters in Table 1

calibration in the experimental protocol and to the method of transient transfection which normally leads to different basal HIF content estimation.

As already observed, from their experimental data it is evident that even within the same cell type, the response presents differences from cell to cell. So, for any cell they looked for the set of parameters yielding the best fit between the experimental data and the output of their model. In this way they obtained a distribution of parameter values. In particular, their model includes a rate of production of PHD2, i.e. our γ_{21}, for which they found fitting values ranging in the interval $0.5 - 4 \cdot 10^{-5}\mathrm{s}^{-1}$, then mediated to $1.5 \cdot 10^{-5}\mathrm{s}^{-1}$. We opt for a different strategy that exploits the fact that some parameters are evaluated in other studies using different methodologies. Having set their value, we look for the values of the remaining parameters, essentially the production parameters, that fit the experimental behavior in Fig. 3a. In this way we are able to identify $\gamma_1 = 10^{-1}\ \mu\mathrm{M/s}$ and $\gamma_{21} = 4 \cdot 10^{-5}\mathrm{s}^{-1}$ as good candidates, where we notice that the latter is still in the range found in [39].

Bagnall et al. [4] also give data on the evolution of PHD2 (see Fig. 3b), but unfortunately only for one case without specifying the coupled HIF evolution. In Fig. 3b we also report data from experiments on neuroblastoma cells performed by Holmquist et al. [26] that show a similar trend. We notice that values of the parameters previously identified lead to a trend of PHD2 that is similar to the experimental one.

Regarding γ_{31} we assume the same value as γ_{21}, since they represent the same biological function, that is the transcriptional influence of HIF1 on its target genes and $\gamma_{32} = \frac{\gamma_{21}}{2} = 2 \cdot 10^{-5}\mathrm{s}^{-1}$ also on the basis of the experimental results on MCF7 breast cancer cells obtained by Wenger et al. [65]. Similarly, we set values of the same order for $\gamma_2 = 3\gamma_1 = 3 \cdot 10^{-1}\mathrm{s}^{-1}$.

Overall, Fig. 3a, b, and c show that the choices above lead to temporal evolutions of HIFs and PHDs that compare well with experimental results reported in [4, 26, 30].

Finally, in Fig. 3d we report the reference simulation in dimensional time. In response to hypoxia, HIF1 starts its over-expression and PHD2 follows on a slower time scale to control HIF1 expression. This leads to the development of a peak in HIF over-expression. Then, HIF1 activation decreases with the formation of a small undershoot, that seems to be also present in some experimental behaviors [4, 39]. After that, the solution slowly increases to reach the new equilibrium value still given by (15), but now for the hypoxic level of oxygen. HIF2 dynamics is qualitatively similar, but with less evident peaks and longer characteristic times.

Since HIF response depends on tissue type and even presents differences from cell to cell, in Fig. 4 we report how the response changes changing the dimensionless parameters one at a time starting from the reference values given in Table 1. We do not report the results obtained as ζ changes because they are small and then the results do not change considerably in the range of physiological acceptable values (see [20] for some results in this direction, as well as for the dependence of the evolution of PHD2 from the dimensionless parameters).

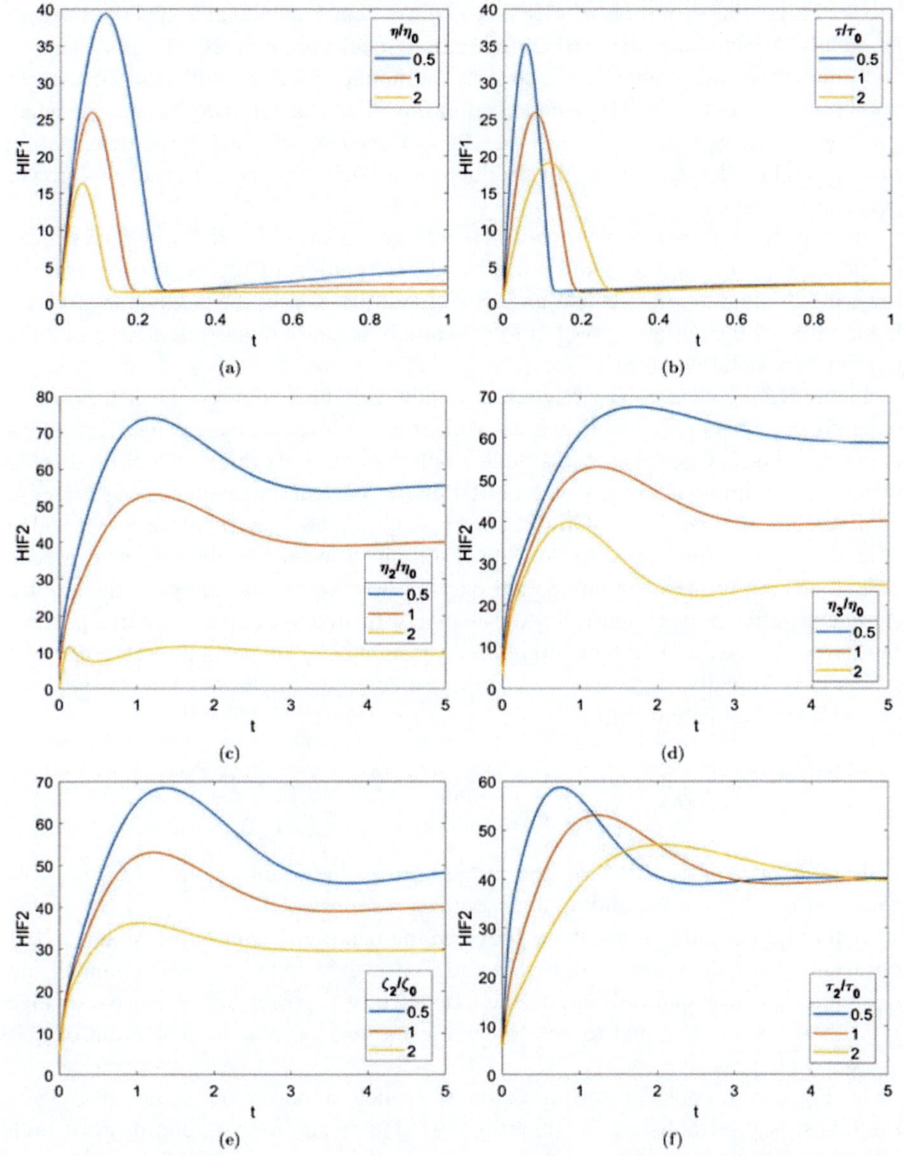

Fig. 4 HIF1 and HIF2 responses when starting from normoxic situation with 21% O_2 the system is suddenly brought to hypoxic state with 1% O_2. Starting from the reference values $\eta_0 = 5.6$, $\zeta_0 = 1.4 \cdot 10^{-3}$ and $\tau_0 = 1.4 \cdot 10^{-3}$, in **(a)** $H_1(t)$ is given when η changes and in **(b)** when τ changes. The other plots report the behavior of $H_2(t)$ when **(c)** η_2, **(d)** η_3, **(e)** ζ_2, **(f)** τ_2 change

In Fig. 4a, the evolution of HIF1 is given when η is changed. This parameter is mainly affected by the PHD2 regulation mechanism of HIF1. Increasing η from the reference value leads to less pronounced peaks with the instant of maximum expression of HIF1 achieved earlier. Instead, referring to Fig. 4b when τ is increased, which can for instance be obtained by increasing the degradation rate of PHD2 or by decreasing the production of HIF1, slower responses with lower peaks are obtained.

Increasing all dimensionless parameters in the equation for H_2 has also a similar influence on its evolution, consisting in an anticipation of the peak time and the lowering of the maximum activation value. In all these cases the behaviour of $H_1(t)$ is the same as the yellow curves in Fig. 4a and b, because this is not affected by the relative parameter changes.

In the left column of Fig. 5 we report how the dimensionless peak time t_{peak} depends on η and τ, respectively, as well as an estimate of how long HIF1 over-expression lasts, denoted by t_{over}. Such a latter quantity is evaluated looking for the instant of maximum convexity of the HIF1 graph. We can see that they both decrease as the value of η increases. On the contrary, t_{peak} and t_{over} increase as the value of τ increases. Moreover, plotting the results in a log-log scale puts in evidence first of all that the relative curves are parallel both when we change η for fixed τ (with a negative power) and when we change τ for fixed η (with a positive power). Furthermore, the fact that data align along straight lines suggests the existence of a power law. Actually, t_{peak} and t_{over} seem to share the same power law. In fact, one can well approximate them by

$$t_{peak} \approx 5.18\sqrt{\frac{\tau}{\eta}} \qquad \text{and} \qquad t_{over} \approx 11.2\sqrt{\frac{\tau}{\eta}},$$

with an error of at most 8% in the range used in the graph, mainly for very low values of η and τ corresponding to a very fast response.

In the right column of Fig. 5 we focus on the maximum over-expression of HIF1 observing that it decreases with both η and τ. However, in this case at variance from t_{peak} and t_{over} we can only argue the existence of a power law when we change τ (with an exponent equal to -0.46), while the log-log plot as a function of η is evidently concave.

In Fig. 6 we consider the situation in which a normoxic value of oxygen saturation is re-established, as done in [39]. This leads to a sudden drop of both HIF1 and HIF2 and a subsequent increase up to the normoxic equilibrium value.

Finally in Fig. 7 we highlight the difference in response when the system starts from a value of oxygen saturation used in in vitro experiments (i.e., 21%) and the one found in in vivo tissues (e.g., 4%). We can see that in vivo conditions leads to a more delayed peak occurrence of HIF1, characterized by a longer duration and by a higher value. As a consequence, also PHD2 reaches higher values. However, the main features of the dynamics of the system are conserved. In a similar way, also HIF2 and PHD3 reach higher values in in vivo conditions but HIF2 loses the peak of activation.

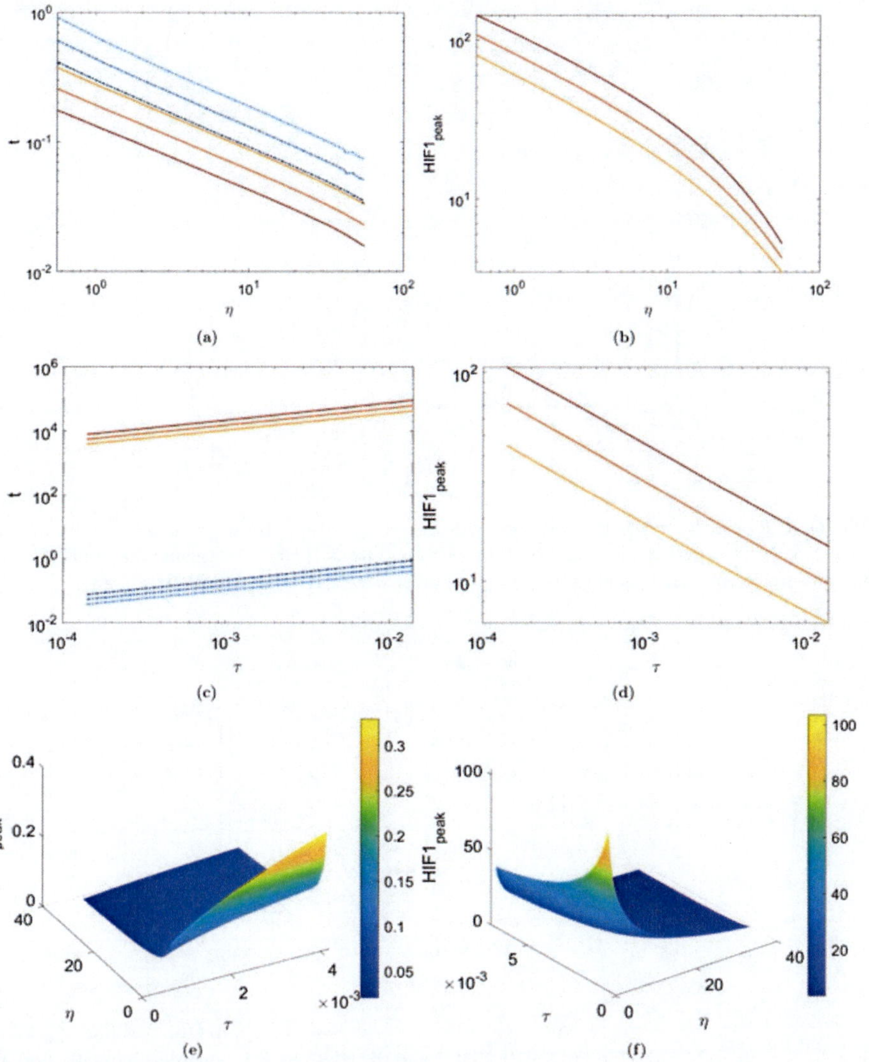

Fig. 5 On the left, dependence of the dimensionless time to reach the peak of HIF1 activation (red curves) and the temporal duration of the peak (blue curves). On the right, dependence of the maximum value of HIF1. In (**a**) and (**b**) as a function of η when $\tau/\tau_0 = 0.5, 1, 2$ with $\tau_0 = 1.4 \cdot 10^{-3}$ In (**c**) and (**d**) as a function of τ when $\eta/\eta_0 = 0.5, 1, 2$ with $\eta_0 = 5.6$. In all cases $\zeta = 1.4 \cdot 10^{-3}$. In (**e**) and (**f**) the peak time and maximum expression of HIF1 is given in terms of both η and τ

Before concluding this section we need to remind that, as already stated, HIF is also regulated by FIH. Of course, including it in the model as well as describing in more detail the dimerization process and the consequent transcription activity as done in [49, 79] complicates the model, leading to the need of experimentally

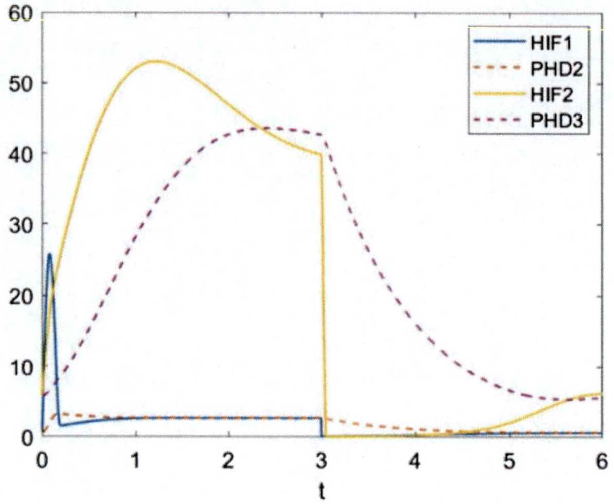

Fig. 6 Response to hypoxic condition with O_2=1% starting from normoxic conditions with O_2=21% and to a re-establishment of normoxic conditions at $t = 3$ (corresponding to 83.33 h). In dimensional terms HIF1 reaches its maximum level after 2.16 h and HIF2 after 33.88 h

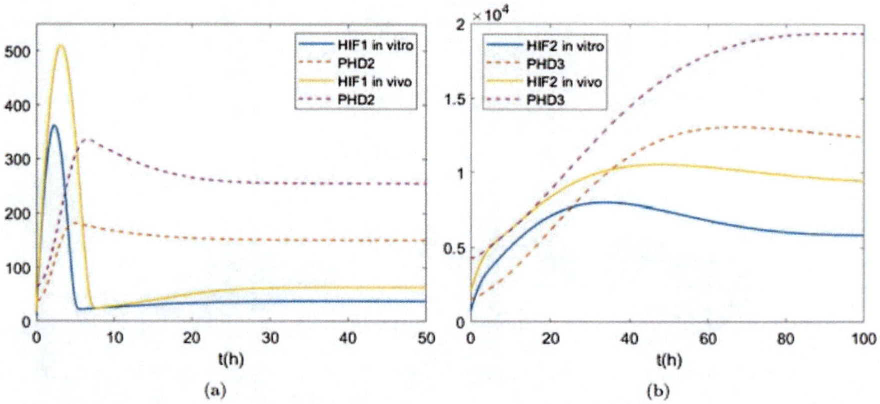

Fig. 7 Simulations mimicking (**a**) HIF1-PHD2 and (**b**) HIF2-PHD3 dynamics in in vitro and in in vivo conditions. In the former case, O_2 drops from 21% to 1%, while in the latter case from 4% to 0.5%

identifying more parameters. This is partially done in [49, 79] where however the peak dynamics is not obtained. In fact, the aim of these papers is to focus on the dependence of the equilibrium values on the availability of oxygen, rather than on their temporal evolution. For sake of completeness, we also mention that other models dealing with the regulation of HIF1 by PHD2 are proposed in [57, 59, 62].

3 Hypoxia and Inflammation

One of the effects of HIF over-expression is the promotion of tumor-associated inflammatory signaling through the involvement of the Nuclear Factor k-light-chain-enhancer of activated B cells (NFkB). NFkB is considered the master regulator of inflammatory responses since it regulates both innate (macrophages, dendritic cells) and adaptative (T helper) immune cells. It responds to various exogenous and endogenous inflammatory stimuli [19, 51, 67] such as microbial components, cytokines (TNFα, Il-1), and alarmins (HMGB1, defensins, Il-33), that are an ensemble of molecules released from damaged or dying cells [51]. Inflammation may precede or follow tumor development. For instance, the immune deregulation characterizing the inflammatory bowel disease greatly increases the risk of colorectal cancer. Also lung inflammation in chronic obstructive pulmonary disease leads to a higher lung cancer risk [23]. However, inflammation can also accompany tumor development with a crucial role played by HIF activation that allow cancer cells to respond to necrotic stimuli leading to NFkB activation.

The coexistence of hypoxia and inflammation has important physiopathological consequences also in non-cancer diseases. One example is given by obesity, in which the mismatch between hypertrophic expansion of adipocytes and blood perfusion leads to hypoxia. The response in terms of HIF activation in adipocytes is the leading cause of the chronic inflammatory state that, in turn, is considered to be one of the main driving forces for the development of insulin resistance and type 2 diabetes in obese individuals [80]. The involvement of HIFs in the low-grade chronic inflammation in obesity was reported by several studies [70], in particular, in mice, it was demonstrated that was the chronic activation of HIF2, and not of HIF1, to be the cause of inflammation through activation of NFkB pathway [41].

In the following section we will briefly present the model proposed in [20] that aimed at describing how acute and chronic inflammation scenarios can be triggered by the activation of HIF through alarmin receptors and NFkB. Then in Sect. 3.2 we will briefly describe the model proposed in [16] that aimed at investigating the relation between IL-15 and HIF mediated by the AKT-mTOR-STAT3 pathway.

3.1 HIF-Alarmin-NFkB Dynamics

As already mentioned, solid tumour environment is typically hypoxic and characterized by the presence of many alarmins released, for instance, by necrotic cells. One of the consequences of HIF activation is the over-expression of alarmin receptors [55], that, activated by the alarmins present in the environment, generate a signal cascade that ends with the activation of NFkB. This, in turn, increases the expression of alarmin receptors triggering an inflammatory state [19, 51, 67], that can become chronic in tumours [19].

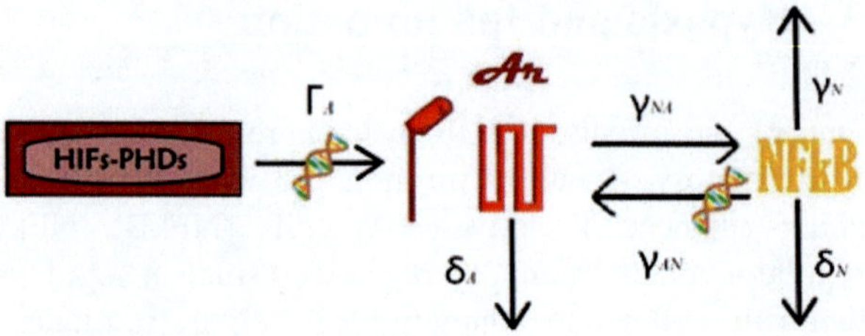

Fig. 8 Sketch of HIF-Alarmin-NFkB pathway and related parameters

In [20] we proposed a model aimed at describing how possible inflammation scenarios can be triggered by the physiological and pathological activation of HIF through alarmin receptors and NFkB. For sake of simplicity, we assumed to be under severe hypoxia in the way that alarmins are present in the microenvironment in such a way that the limiting factor for the inflammatory response is represented by alarmin receptors rather than alarmin themselves. Referring to Fig. 8 alarmin receptors, denoted by A_r, are induced by both NFkB, at a rate γ_{AN}, and by both HIF1 and HIF2, as hypothesized in [37]. Here, we take the activation rate as a linear combination of $\hat{H}_1$ and $\hat{H}_2$. On the other hand, they undergo degradation at a rate δ_A.

In turn, NFkB, denoted by N, is assumed to be produced both basally, at a rate γ_N, and via an Hill-type action of the following type, also on the basis of [69],

$$F_N(\hat{A}_r) = \frac{\hat{A}_r^2}{K_A^2 + \hat{A}_r^2}, \tag{18}$$

related to alarmin expression $\hat{A}_r$, and to be degraded at a constant rate δ_N. So, the following equations

$$\begin{cases} \dfrac{d\hat{A}_r}{d\hat{t}} = \gamma_{A1}\hat{H}_1 + \gamma_{A1}\hat{H}_2 + \gamma_{AN}\hat{N} - \delta_A\hat{A}_r \\ \dfrac{d\hat{N}}{d\hat{t}} = \gamma_N + \gamma_{NA}F_N(\hat{A}_r) - \delta_N\hat{N} \end{cases} \tag{19}$$

can be added to the HIF-PHD model presented in Sect. 2.

Actually, in [20] it is shown that the same stability results are achieved under very general and biologically sound hypothesis on the activation rate of alarmins by HIFs, e.g., that it is an increasing function of both $\hat{H}_1$ and $\hat{H}_2$.

If we introduce the dimensionless variables

$$A_r = \frac{\hat{A}_r}{K_A}, \quad N = \frac{\delta_N}{\gamma_{NA}}\hat{N},$$

we can rewrite (19) in the dimensionless form

$$\begin{cases} \tau_A \dfrac{dA_r}{dt} = p(H_1 + rH_2) + \beta N - A_r \,, \\ \tau_N \dfrac{dN}{dt} = \alpha + F_N(A_r) - N \,, \end{cases} \tag{20}$$

where

$$\tau_A = \frac{\delta_{P_2}}{\delta_A}\,, \quad \tau_N = \frac{\delta_{P_2}}{\delta_N}\,, \quad \alpha = \frac{\gamma_N}{\gamma_{NA}}\,, \quad \beta = \frac{\gamma_{AN}\gamma_{NA}}{\delta_A \delta_N K_A}\,,$$

$$p = \frac{\gamma_{A1} K_{12}}{\delta_A K_A} \quad \text{and} \quad r = \frac{\gamma_{A2} K_{23}}{\gamma_{A1} K_{12}}\,.$$

Equilibria can be identified by the curve

$$p(H_1^{eq} + rH_2^{eq}) = A_r^{eq} - \beta \left[\alpha + \frac{(A_r^{eq})^2}{1 + (A_r^{eq})^2} \right]$$

where H_1^{eq} and H_2^{eq} can be considered given by the analysis in Sect. 2. Then

$$N^{eq} = \alpha + F_N(A_r^{eq})\,. \tag{21}$$

As discussed more in detail in [20], according to the values of the parameters, a bistable situation can occur, as shown in Fig. 9. This implies the existence of two turning points, and then of a lower and an upper branch (both stable) that can be respectively related to under-expression and over-expression of inflammatory

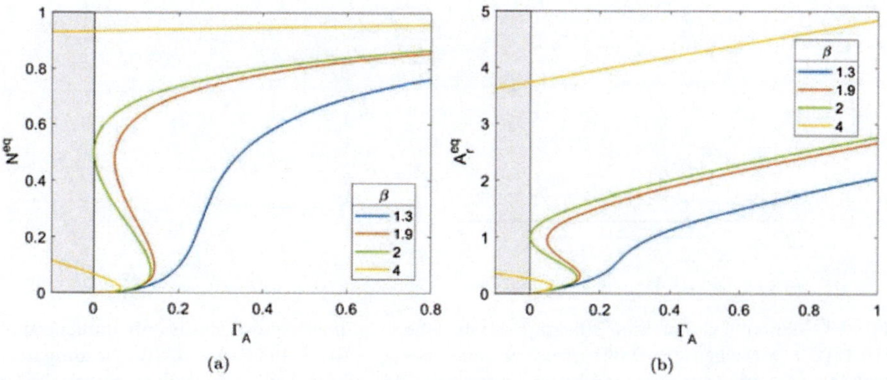

Fig. 9 Equilibrium values of NFkB N^{eq} in (**a**) and alarmin receptors A_r^{eq} in (**b**) are reported as a function of $\Gamma_A = p[H_1^{eq}(O_2) + rH_2^{eq}(O_2)]$. We recall that negative values of Γ_A are unphysical

proteins and therefore to a sudden activation of an inflammatory state that can become chronic. Specifically, denoting $\Gamma_A = p[H_1^{eq}(O_2) + r H_2^{eq}(O_2)]$ and then by $\bar{\Gamma}_1$ and $\bar{\Gamma}_2$ the values of Γ_A corresponding to the turning points, for $\Gamma_A \in (\bar{\Gamma}_1, \bar{\Gamma}_2)$ there are three equilibria, one unstable and two stable, and outside this interval there is only one stable equilibrium. Then, according to the evolution of $H_1(t)$ and $H_2(t)$, and therefore of $\Gamma_A(t)$, several scenarios are possible.

For instance, in the case in which alarmin receptors are only triggered by one HIF, e.g. HIF1 and then $\gamma_{A2} = 0$, inflammation is triggered if $\Gamma_A(t)$ overcomes $\bar{\Gamma}_2$ and lasts for a period similar to the over-expression of HIF1 if Γ_{hyp} drops below $\bar{\Gamma}_1$ as shown in Fig. 10a. So, even though hypoxia persists, the inflammatory response is temporary. This graphically corresponds to the situation in which starting from the lower branch the system jumps to the upper inflammatory branch because Γ_A has overcome the turning point. However, when the overshoot of HIF1 is over, the systems jumps back to the lower non-inflammatory branch. On the other hand, denoting by Γ_A^{hyp} the hypoxic equilibrium value of Γ_A, if $\Gamma_A^{hyp} > \bar{\Gamma}_1$, then the system chronically remains in the inflammatory states also when HIF over-expression is over, leading to a persistent inflammatory response, as shown in Fig. 10b.

If, instead, Γ_A is more sensitive to HIF2 than HIF1, it can happen that $\Gamma_A(t)$ does not reach $\bar{\Gamma}_2$ during HIF1 over-expression, but will when HIF2 over-expresses. One then has an inflammatory response that starts later, as shown in Fig. 11a where it is evident that N and A_r stay under-express for a while and the inflammatory state persists till the end of HIF2 over-expression, if the asymptotic value of $\Gamma_A(t)$ is lower than $\bar{\Gamma}_1$. Then, also in this case the inflammatory state can become chronic or not according to the fact that $\Gamma_A^{hyp} > \bar{\Gamma}_1$ or not. In the latter case the inflammatory

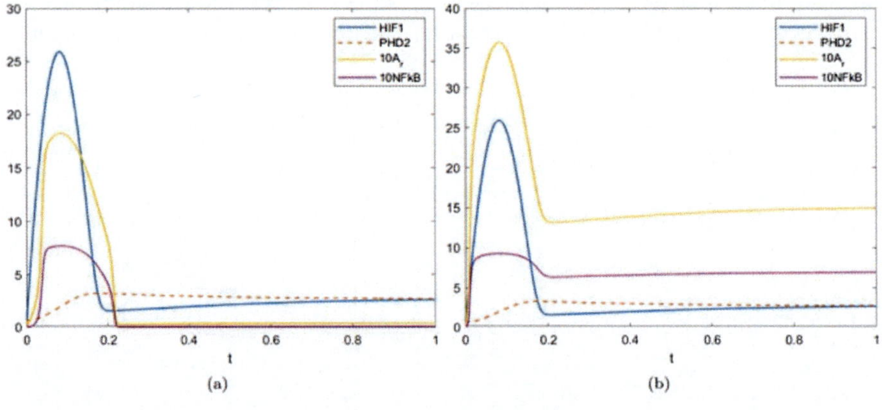

(a) (b)

Fig. 10 Temporal evolution of all variables when the inflammatory response is only influenced by HIF1, i.e. $r = 0$ when $\alpha = 0.021$, $\beta = 1.9$, and $\tau_A = \tau_N = 10^{-5}$. In (**a**) $p = 3 \cdot 10^{-4}$ leading to an early acute response represented by an activation of NFkB and then a very low asymptotic value after going back to normoxic conditions, while in (**b**) $p = 5 \cdot 10^{-3}$ leading to a chronic response, as indicated by a persistently high asymptotic value of NFkB concentration

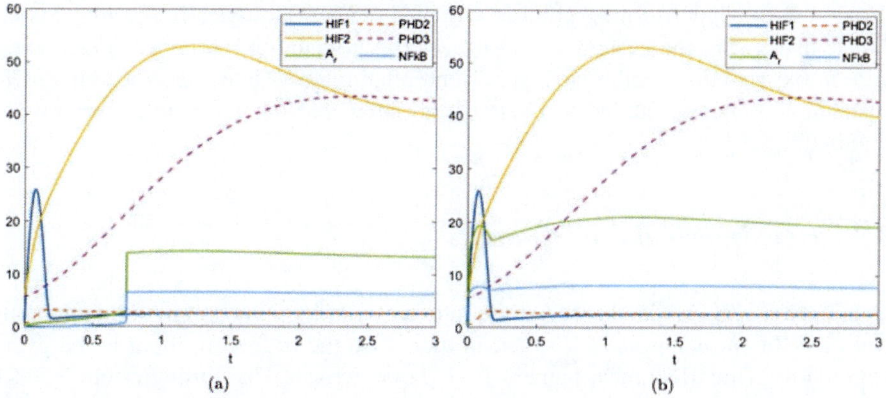

Fig. 11 Temporal evolution of all variables when the inflammatory response is induced by HIF2 only (**a**) and by both HIF1 and HIF2 activation (**b**). In particular, in (**a**) $p = 10^{-2}$ and $r = 1$; in (**b**) $p = 10^{-4}$ and $r = 30$, in both cases $\beta = 1.9$. Both scenarios lead to a persistent inflammatory response

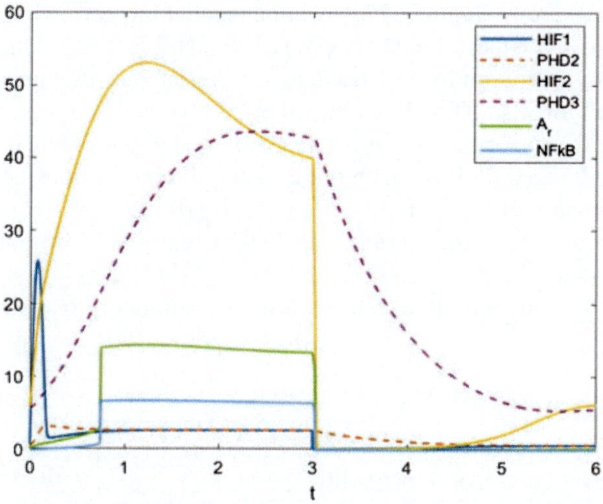

Fig. 12 Temporary hypoxic and inflammatory response that are stopped once normoxia is restored, at $t = 3$. $p = 10^{-4}$, $r = 30$ and $\beta = 1.9$

response lasts till the end of HIF2 over-expression. If also the HIF1-related peak is higher than $\bar{\Gamma}_2$, then the inflammatory state starts earlier, i.e., already when HIF1 activates and will last longer, as shown in Fig. 11b.

Finally in Fig. 12 we consider what happens when normoxia is restored. In this case, the increased level of oxygen suppresses HIF response that, then, switches off the activation of NFkB. However, we need to mention that crossing the value $\beta = 2$ might lead to dramatically different evolutions. In fact, referring to Fig. 9, for $\beta < 2$ the upper turning point is achieved for non biologically sound negative values of Γ_A. So, the system would irreversibly remain on the upper branch even after re-oxygenation, leading to a persistent inflammatory state which becomes self-sustaining. A more detailed analysis of the latter and other possible scenarios is reported in [20].

3.2 HIF-Interleukine Dynamics

Coulibaly et al. [16] also looked at the interaction of hypoxia and inflammation but from a different perspective. In fact, in their case the triggering input is the over-expression of interleukines, namely IL-15, that activate HIF through both STAT3 (Signal Transducer and Activator of Transcription 3) and NFkB, as well as AKT (also known as protein kinase B), that together with mTOR (mammalian Target of Rapamycin), enters into the regulation of cell proliferation. More precisely, referring to Fig. 13, IL-15 promotes NFkB, STAT3 and AKT and the first two promote the activation of HIF1, while the latter promotes mTOR that forms a three-nodes positive feedback promoting STAT3. As in our model the network is characterized by another positive feedback loop between HIF and NFkB. Instead, the fact that HIF has an inhibitory action on mTOR leads to the presence of two negative feedback loops, one involving also STAT3 and the other involving NFkB.

Actually, with respect to the mathematical model presented in the previous section, the model in [16] develops more in detail the interactions between HIF1α, HIF1β, the complex HIF1, HIF-1α-mRNA, and HIF-1α-aOH. Conversely, HIF2 is not considered and the interaction with PHD is extremely simplified and taken constant.

The part of the model relating HIF with the inflammatory response can then be written as

$$\begin{cases} \dfrac{dC_{AKT}}{dt} = \Gamma_{AKT}(C_{IL}) + \gamma_{AS}\dfrac{C_{STAT}^2}{K_{STAT}^2 + C_{STAT}^2} - \delta_{AKT}C_{AKT} \\ \dfrac{dC_{mTOR}}{dt} = \dfrac{\alpha_{mTOR} + \beta_{mTOR}C_{AKT}}{K_{mTOR} + \hat{H}_1} - \delta_{mTOR}C_{mTOR} \\ \dfrac{dC_{STAT}}{dt} = \Gamma_{STAT}(C_{IL}) + \gamma_{ST}C_{mTOR} - \delta_{STAT}C_{STAT} \\ \dfrac{dN}{dt} = \Gamma_N(C_{IL}) + \gamma_{NT}C_{mTOR} + \gamma_{NH}\hat{H}_1 - \delta_N N \end{cases} \tag{22}$$

Fig. 13 IL15-NFkB pathway

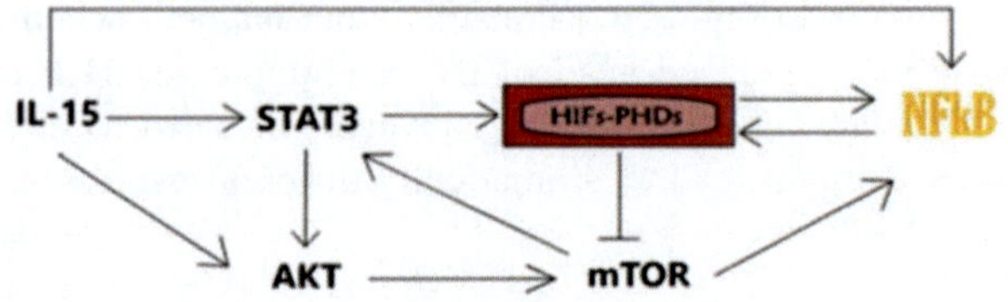

where Γ_{AKT} and Γ_{STAT} and Γ_N are linear functions of the concentration C_{IL} of IL-15, i.e. $\Gamma_\alpha = a_\alpha + b_\alpha C_{IL}$, which is considered as given. The system above is then coupled to other five equation describing all the process that from HIF1α leads to the HIF1 dimer. Actually, the terms $\hat{H}_1$ in the second and fourth equations refer more specifically to the HIF1 complex. Conversely, HIF1α mRNA is activated by STAT3 and NFkB.

The aim of the model was then to explain the mechanisms leading to the accumulation of HIF1α in natural killer cells. They then found that STAT and NFkB are fundamental regulators of IL-15 induced HIF1α enrichment, highlighting the synergistic effect of IL-15 and hypoxia.

4 Modelling Other HIF-Related Downstream Pathways

4.1 *HIF and Metabolism*

One of the crucial and most well known consequences of the activation of HIF is the switch from an aerobic to an anaerobic metabolism that leads to the fermentation of glucose to lactate. Under physiological conditions and under moderate hypoxia, the oxidation of glucose into pyruvate, through glycolysis, is followed by the mitochondrial oxidation of the pyruvate, through tricarboxylic acid (TCA) cycle and oxidative phosphorylation [15]. On the contrary, as sketched in Fig. 14, under hypoxia, HIFs (mainly HIF1) promote the expression of genes encoding enzymes that, on the one side, reduce the mitochondrial production of ATP by the activation of PDK1 (pyruvate dehydrogenase kinase 1) [34] and, on the other side, promote anaerobic ATP production through glycolysis activating PGK (phosphoglycerate

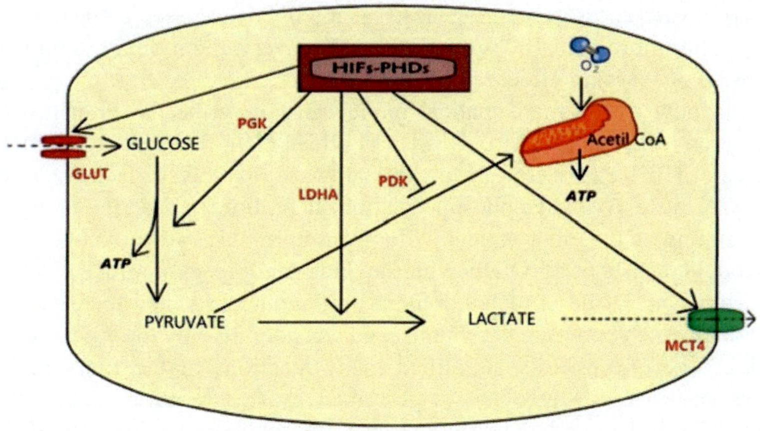

Fig. 14 HIF-related regulation of aerobic and anaerobic metabolic pathway

kinase) and LDHA (lactate dehydrogenase A) [64]. Moreover, HIFs also up-regulate genes coding for glucose receptor and the lactate/H^+ symporter MCT4, further increasing anaerobic glycolysis [15]. An additional consequence of this metabolic reprogramming is a decrease in the levels of reactive oxygen species (ROS), whose production is mainly related to mitochondrial respiratory chain [83].

Although severe hypoxia leads to anaerobic glycolysis because of the lack of the oxygen, reoxygenation does not necessarily leads to aerobic metabolism. Indeed, even in well oxygenated conditions, cancer cells may display HIF stabilisation and anaerobic metabolism [58], a peculiar choice which is known as Warburg effect or aerobic glycolysis. The Warburg effect was originally reported in ascites cancer cells of the mouse [74], in which, despite the presence of oxygen, an increased level of glucose and production of lactate was documented. These characteristics were later on demonstrated in several tumors and even in non-cancer cells, such as in activated effector T cells, which are involved in adaptative immune response [61]. This cellular choice may seem irrational because of the energetic inefficiency of the fermentation of glucose which only yields 2 molecules of ATP per 1 molecule glucose (vs. 34–38 from glucose oxidation). Actually, the reason why many proliferating cells display the Warburg effect is still not fully understood but several hypotheses have been advanced to explain it. One of these focuses on the high rate of glucose fermentation that leads to ATP production 10–100 times faster than the oxidation of glucose in the mitochondria. However, ATP demand for cancer growth does not seem to be so high to justify this constant metabolic switch without preferring other metabolic pathways [40]. Another proposal is related to the fact that the increased uptake and consumption of glucose may be used as carbon source to promote the synthesis of nucleotides, lipids and proteins that are important for cell proliferation and survival. However, most of the glucose is not retained for this scope [40]. Another hypothesis is related to the consequences that the altered metabolism has on the tumor microenvironment. In fact, the increased production of lactate may lead, both directly and indirectly, to favourable consequences on cancer progression in terms of invasiveness and angiogenesis, as it will be discussed in the next section. However, after more than 100 years from its discovery, the intimate significate and mechanisms of Warburg effect remain unclear.

Coming now to the mathematical models that describe the influence of HIF activation on cell metabolism, Bocharov et al. [10] focused in detail on the influences of HIF1 on the four metabolic pathways (glycolysis, lactic acid fermentation, TCA cycle, oxidative phosphorylation) that allow to describe the metabolic reprogramming as the environment shifts from normoxia to hypoxia (see Fig. 14). Their model consists of 16 ODEs with the underlying reaction processes localized in the cytoplasm, in the mitochondria and in the nucleus. They focused on HIF1, without distinguish between its subunits, and considering only the regulation played by PHD2. They necessarily simplified many reactions of the four biochemical pathways considered. Glycolysis is described by two reaction rates, which are influenced by HIF1, and they summarized the eight reaction of the TCA into three, taking trace only of the NADH and of the ATP produced. Finally, the oxidative phosphorylation is modeled through one reaction rate that gathers the four enzymatic complexes and that depends on the membrane potential, on NADH, and on oxygen concentration.

Another study by Hashemzadeh et al. [24] addressed the relation between oxygen availability and the production of lactate. Also in this study, only the influence of HIF1 was evaluated. The model consists in 26 ODEs that describe the changes in the concentration of enzymes and metabolites involved in the glycolytic pathway with the aim to provide a quantitative relationship between the hypoxia intensity and the intracellular lactate levels. They do not consider any regulation of HIF, but use the dependence of the equilibrium values on oxygen concentration. Moreover, of the three esoergonic, thus irreversible, reactions of the glycolysis, the model is able to predict the pivotal role of the enzyme that regulates one of these, the phosphofructokinase-1.

Pressley et al. [56] focused on HIFα stability as the availability of oxygen changes temporarily both in a regular and in a stochastic way in order to address the issue related to Warburg effect. They explored different strategies (constant, instantaneous and fast/slow responses) that cancer cells may adopt, in terms of HIF expression, to face the change in oxygenation. In their work the Warburg effect is the strategy represented by the constant activation of HIF, that is not affected by changes in oxygen availability. They do not consider any regulation of HIF but for each strategy they assume a different mathematical dependence of HIFα on oxygen. Through a cost-benefit analysis of these strategies, based on the net growth rate, they conclude that the constant stabilisation of HIFα (Warburg effect), regardless of the availability of oxygen, may represent a selective advantage for cancer cells.

4.2 *HIF and pH*

Under severe hypoxia, the production of more lactate and the inefficient wash-out of H^+ ions released into the extracellular medium lead to a decrease in extracellular pH (pH_e) and therefore to acidification of the microenvironment that is detrimental for the cell. On the contrary, the intracellular pH (pH_i) is finely maintained in an optimal range, usually reaching a higher pH_i than normal cells, with positive consequences on cell proliferation and evasion of apoptosis [15].

A mild alkaline pH_i is related to the increased expression and activity of many H^+ transporters, such as MCT4 and Na^+/H^+ exchanger isoform 1, whose genes are targets of HIF1 regulation, and Na^+-HCO_3^- transporter, that, on the contrary is hypoxia-independent. Another important HIF target that promotes pH_e acidification is represented by CAIX, a zinc metalloenzyme, that hydrates CO_2 into HCO_3^- and H^+ ions, further increasing H^+ outside of the cell, thus decreasing pH_e.

Overall, a reversed pH gradient develops between tumors and normal tissues [15] with positive effects on cancer progression. Indeed, on the one side, mild pH_i supports cell migration through a pH-dependent increase in the activity of numerous actin-binding proteins [75], that drive membrane protrusion in migrating cells, and promotes cell survival during growth factor limitation also in non-cancer cells [33].

On the other side, acidic pH_e induces VEGF expression in an HIF-HRE independent way further increasing angiogenesis [78]. It also promotes morphological and dynamic changes towards a more invasive and migratory cell phenotype [3, 52]. Moreover, the lactate itself, far from being a simple end-product of the glycolysis, was demonstrated to have a direct involvement in cancer progression. Indeed, it stimulates endothelial cells to increase their VEGF production, and activates the ERK/STAT3 signaling inducing macrophage polarization favoring proliferation and migration [48]. Finally, it was demonstrated to be directly related to the inhibition of tumour surveillance by T and Natural Killer cells [11].

As well as pH has direct influences on cancer progression, it is not surprising that it can also influence HIF stability.

Filatova et al. [21] reported a narrow pH window (6.6–6.8) for acidosis-induced upregulation of HIF1/2α which is independent of PHD-regulation and already detectable under normoxia. The authors demonstrated the involvement of the HSP90 protein in this activation: HSP90 is a chaperone that acts protecting proteins, like transcription facts, from their degradation under dangerous conditions, such as heat stress. Specifically, HSP90 together with additional co-chaperones competes with the protein RACK1 for the association with HIF, preventing its degradation and leading to HIF stabilisation. However, the authors did not investigate the effect of severe acidosis on HSP90 expression. In fact, they do not investigate further the decrease of HIF for pH_e lower than 6.6.

To the best of our knowledge, there are no mathematical models that explicitly describe the influence of HIF on pH regulation. However, there are some studies that address the remodulation of pH in response to hypoxia. For instance, Piasentin et al. [54] show a dominance of the H^+ exchanges in pH regulation in normoxic regions and of HCO_3^- exchanges in more hypoxic areas. They furthermore show the importance of CAIX upregulation in maintaining a mild alkaline pH_i.

4.3 HIF and Cell Cycle

Cell division consists in a precisely ordered sequence of biochemical events that are highly regulated by several signaling proteins (growth factors, G proteins, protein kinases). The main regulators are the cyclin and the cyclin-dependent protein kinases (CDKs) which act at specific points of the four stages (G1, S, G2, M) of the cell cycle modulating, through phosphorilation, the activity of crucial proteins, such as lamin, condensin, myosin and the retinoblastoma protein. The dysregulation of this process impacts genomic integrity favouring cancerogenesis. One of the cause of such dysregulation is represented by hypoxia and several studies demonstrated a direct involvement of HIFs in cell cylce regulation, with HIF1 and HIF2 having different roles.

Specifically, HIF2 has a permissive action. Indeed, it induces cyclin D1 and cooperates with Myc protein promoting cell proliferation. Myc is a transcription factor that repress the expression of two CDK inhibitors, p21 and p27, that are, on the contrary, directly induced in a HIF1 dependent manner [17]. The role of HIF2 has also important consequences in response to radiation therapy. Indeed, the loss of HIF2 activity promotes p53-mediated responses, increases apoptosis, and reduces clonogenic survival of irradiated and non-irradiated cells in response to radiation therapy [9].

In contrast to HIF2 effect, a non transcriptional mechanism of HIF1 on inhibition of cell cycle progression was reported in several cancer cell lines as well as in non cancer cells [17]. Specifically, HIF1 prevents kinase Cdc7 from phosphorilating the MCM complex, which is a crucial step for the transition from G1 into the S-phase. Then the duration of the G1 phase of the cell cycle is increased and the cell is induced to enter a quiescent state. It is thought that in general the arrest of cell proliferation is a mean for the cell to escape from hypoxia-induced apoptosis. Specifically, by the same means, the fact that cancer cells have the ability to enter more easily a quiescent state when environmental conditions, such as oxygen availability, are unfavorable, represents a clonal advantage over normal surrounding cells. This ability can then contribute to the appearance of highly resistant and aggressive tumor phenotypes.

Having this phenomenon in mind, Bedessem and Stephanou [7] proposed a mathematical model of the G1/S transition under hypoxia that starts from the mutually inhibitory relationship between HIF1 and cyclin-D (C_D) to include some classical ingredients of the cell cycle actors like retinoblastoma (R_b), cyclin-E (C_E), E2FA (E_2), SCF (S) as well as cell mass M. Considering that the model only takes HIF1 into account and not HIF2, and that the time scale of cell cycle is of the order of a day, an oxygen dependent equilibrium concentration H_1^{eq} is considered. Specifically, an exponentially decaying function of O_2 is used for $\hat{H}_1^{eq}(O_2)$.

Referring to Fig. 15, the core of the mathematical model writes as

$$\begin{cases} \dfrac{dC_D}{dt} = \gamma_D - \delta_{DH}\hat{H}_1^{eq}(O_2) - \delta_D C_D \\ \dfrac{dC_E}{dt} = \gamma_E M E_2 - (\delta_E + \delta_{ES} S) C_E \\ \dfrac{dS}{dt} = \gamma_S \dfrac{1-S}{K_S + 1 - S} - \delta_S \dfrac{S C_E}{K_S' + S} \\ \dfrac{dR_b}{dt} = \gamma_R - (\delta_R + \delta_{RC_D} C_D) R_b \end{cases} \tag{23}$$

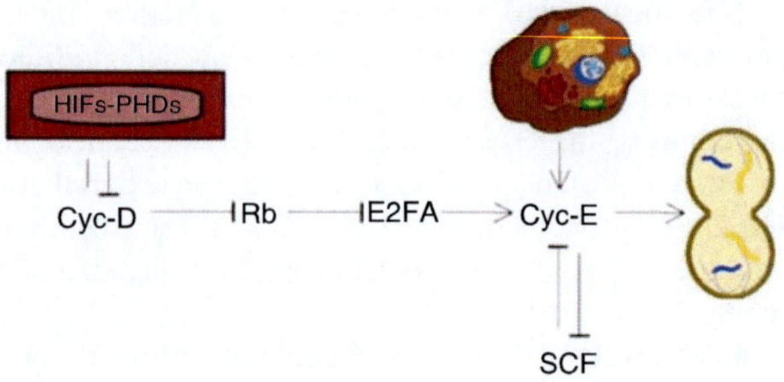

Fig. 15 Influence of HIF1 regulation on cell cycle as proposed by the model in [7]

where cell mass M grows logistically according to an independent Verhulst-type ODE and the unphosphorylated E2F follows a dynamic equilibrium described by the linear ODE

$$\frac{dE_2}{dt} = \gamma_E(E_{tot} - E_2)\,.$$

The mathematical model is able to reproduce the slowing down of the G1 phase under moderate hypoxia and the entry into quiescence of proliferating cells under severe hypoxia. They also show how the inhibition of cyclin-D by HIF-1 can induce cell quiescence.

Focusing on mesenchymal stromal cells, Zhang et al. [82] included also HIF2 in the picture and consider Myc protein as the crucial link between hypoxia sensing machinery and cell cycle in order to properly model the effect of the different HIF responses on cell cycle commitment. Furthermore, they also add a ROS-mediated protein deactivation that is assumed as a function of oxygen tension beyond a threshold oxygen level. They also explore its influence on the regulation of cell cycle progression. The mathematical model consisting of 13 equations with HIF1 and HIF2 taken again at equilibrium with a dependence from oxygen saturation that presents a maximum at 0.5% for HIF1 and at 5% for HIF2. Their results indicate that the inhibition of E2F by ROS and HIF1 under normoxia and severe hypoxia, respectively, accounts for a delay in cell cycle progression that is counteracted by HIF2, that on the contrary activates E2F.

4.4 HIF and ECM-Stiffening

Extracellular matrix (ECM) represents the non-cellular component composed by water, proteins and polysaccharides that is organized in a cell/tissue-specific manner and is crucial to confer physical stability to organs and tissues. However, its function extends beyond this role. It, for instance, enters the regulation of cell differentiation, motility and survival participating to the mechanosensing and mechanoregulation processes. Indeed, ECM-derived mechanical stimuli are converted into biochemical or electric signals that elicit cellular responses in the sense of gene expression, protein synthesis and cellular phenotype change [47].

In many solid tumors ECM is denser and stiffer thanks to an increase of collagens, fibronectin and elastin that are mainly caused by cancer associated fibroblasts (CAFs) and tumor cells themselves. These changes in composition and organization of the ECM are strictly correlated with malignant features.

It is known (see, for instance [12, 32]) that a higher ECM stiffness triggers epithelial-mesenchymal transition promoting cancer cell invasion and metastasis. In particular, Xiao et al. [77] show that HIF can have a direct involvement in the stiffening of the ECM surrounding the cell through the production of Lysyl Oxidase (LOX). In fact, LOX oxidizes lysine residues in collagens and elastin, resulting in the covalent crosslinking of ECM fibers. From the macroscopic viewpoint an increase of the links between fibers cause an increase of tensile strength and structural integrity.

Specifically, the expression of both HIF1 and LOX in epithelial ovarian cancer is significantly correlated with the tumor grade, tumor diameter and lymph node metastasis [28]. The role of HIF2 is elucidated in a study on the fibrotic disease process that characterizes the course of the thyroid-associated orbitopathy (TAO). TAO is a connective tissue disease that is associated with glycosaminoglycan deposition, adipogenesis in the orbit and fibrosis affecting the extraocular muscles [44]. Hikage et al. [25] demonstrate that HIF2 accelerates ECM deposition by inducing LOX.

Hence, abnormalities in LOX expression and activity result in connective tissue disorders and fibrotic diseases as well as in promoting cancer cell invasion and metastasis. In fact, coherently, LOX oxidase family is found to be highly expressed in invasive tumors, and is closely associated with metastasis and poor patient outcome. Conversely, inhibition of LOX family oxidase activity by pharmacological inhibitors, therapeutic antibodies or reduced LOX expression impeded tumor progression. So, the fact that inactivating LOX impaired stiff matrix, and TGF-β-mediated epithelial-mesenchymal transition, and therefore cell invasiveness in breast cancer cells, suggests LOX as a potential mediator that couples mechanotransduction to TGF-β signaling.

Conversely, HIFs also influence invasiveness of cancer cells entering the regulation of ECM degradation. In fact, HIF1 upregulates genes encoding for type IV collagen-degrading enzymes (MMP2 and MMP9), whereas membrane-bound membrane-type 1 MMP (MMP14) is upregulated in an HIF2-dependent manner [22]. Furthermore, HIF1 promotes invasiveness also leading to the upregulation of urokinase-type plasminogen activator receptor (uPAR) gene transcription that encodes for the proteolytic enzyme uPA [50]. This enzyme promotes pericellular proteolysis altering the interactions between integrins and the ECM.

Despite the importance of the relationship between HIF and ECM stiffness and its consequences, to our knowledge no mathematical model has proposed at present in the literature on this topic.

4.5 HIF and VEGF

Oxygen and nutrient supply are major drivers of angiogenesis that consists in the formation of new capillaries from existing blood vessels. It occurs in physiological (e.g., endometrial growth during menstrual cycle, wound healing) and pathological (e.g., cancer and rheumatoid arthritis) conditions. Physiological angiogenesis is a finely regulated process in which the balance between anti-angiogenic and pro-angiogenic factors is preserved. On the contrary, pathological angiogenesis is a highly chaotic process characterized by the prevalence of pro-angiogenic factors that lead to a persistent process. In tumors, angiogenesis plays a critical role in cancer progression allowing tumor growth beyond 1–2 mm in diameter. HIF1 and HIF2 are involved in different aspects of this multi-step process [84]. They transcriptionally activate genes that codify for several angiogenic factors, such as the vascular endothelial growth factor (VEGF) and its receptors VEGFR-1 (Flt-1), regulate endothelial cells proliferation and matrix degradation (mainly HIF1), promote blood vessel maturation (mainly HIF2). The impaired regulation leads to the formation of tortuous and leaky vessels that promote heterogeneous blood flow affecting tumor perfusion. Several studies address the link between hypoxia and angiogenesis, mostly focusing on VEGF signaling, however only few of them deepen HIF regulation.

Chen et al. [13] propose a model that consists in 86 reactions aimed at describing the effect of hypoxia on VEGF signaling and ATP. Specifically, they model HIF-related induction of VEGF and of its receptor that can lead downstream to the activation of ERK and Akt, two important signaling proteins involved in tumor proliferation and survival. They also consider that severe hypoxia leads to decreased levels of ATP that may impact ATP-dependent phosphorylation signaling protein such as ERK and Akt themselves. The model links the variability in tumor signalling between tumor cell-types and the intracellular concentration of some identified biomolecules and clusters of biomolecules, such as Akt and PIP3, that enters into the regulation of Akt itself.

Another study [83] focuses on the link between HIF1 and VEGF deepening the role of non-coding RNA molecules (miRNA) in modulating intracellular VEGF synthesis. This mass-action based model describes the dynamics of the interactions between the molecular components that are involved in the miRNA control of the HIF-VEGF pathway in endothelial cells. The authors do not distinguish between HIF1 and HIF2 and they consider the regulation played by FIH and the negative influence of tristetraprolin (TTP) on HIF mRNA. TTP acts primarily through post-transcriptional regulation of mRNA that can be induced by hypoxia and that regulates both VEGF and HIF. They specifically focus on two miRNAs, let7 and miR15a, considering their biogenesis pathways and they analyse different scenarios of gene over-expression/silencing in order to understand their impact on angiogenesis. In this way they identify miR15a as a crucial effector and they test *in silico* new therapeutic strategies, demonstrating that a promising way to suppress

VEGF production would be to increase AGO1 which is one of the core protein of the miRNA-induced silencing complex.

4.6 HIF and High Altitude

Shifting from cells to human populations, hypoxia also characterizes high altitude (HA) environments. The decrease in barometric pressure leads to a proportional decrease of the atmospheric oxygen partial pressure that causes a reduced oxygen content of the blood and impairs oxygenation of peripheral tissues. Hyperventilation and elevated heart rate are two of the most important immediate responses that aim at increasing oxygen supply and delivery, respectively [14]. However, physiological responses may evolve in severe pathological disease, such as high altitude pulmonary edema (HAPE) and high-altitude cerebral edema (HACE) whose risk increases with altitude and with the speed of ascent [43]. At a second time, also the number of red blood cells increases, through erythropoiesis, thanks to the production of erythropoietin (EPO), whose gene is upregulated by HIF2. In parallel, the iron intestinal absorption is increased in order to allow the body to produce more hemoglobin (Hb) to bind to the oxygen. These responses, which are due to a long-term exposure, may degenerate into pathological scenarios that globally refers to chronic mountain sickness (CMS) [73]. Indeed, an excessive increase in the concentration of red blood cells in the blood (polycythaemia) increases its viscosity that, in turn, predisposes towards pulmonary hypertension with increased risk of heart failure (cor pulmonale) [5]. The risk of developing CMS increases with the age and the duration of altitude residence, but also depends on genetic factors [73].

The comparison between ethnic groups that share the same environment, thus the same altitude, but not the same settlement timing represents a valuable tool to report a difference in highlander populations in the prevalence of CMS which is lower in long-term resident populations, such as Tibetans, than in lowlanders [60]. This might spread some light on how to control the response to hypoxic environments.

High-altitude adaptation in highlanders may be due to multiple and interconnected genes. However, a crucial role of the HIF-PHD system was demonstrated. In Tibetans, genetic variants (polymorphisms) in EPAS1 (which encodes for HIF2α) and in EGLN1 (which encodes for PHD2) were reported thanks to genome wide analysis [42]. An association was reported between these variants and the lower Hb found in Tibetans that may contribute to their HA adaptation [6, 76].

Even if the underling mechanism has not been yet elucidated, kinetic studies of a genetic variant of PHD2 have been reported to manifest a reduction in the binding affinity for the oxygen [42] that suggests a reduced HIF activation under hypoxia. Bearing in mind the pivotal role of HIFs-PHDs in tumors, one might well wonder what role these polymorphisms may have on cancer growth.

Up to now just one study [38] reports an association between some selected Tibetan variants related to EGLN1 and the risk of lung cancer, without further studying their role on lung cancer initiation or progression.

So, though it might be still premature to speculate about the effect of these genetic variants in tumors, mathematical modeling may help formulating hypotheses and making predictions.

Acknowledgments The Authors thank the Medical Student Isabella Ferrante for stimulating and fruitful discussions and for her contribution on the graphics of HIF-related reaction networks.

References

1. Aik, W.S., Chowdhury, R., Clifton, I.J., Hopkinson, R.J., Leissing, T., McDonough, M.A., Nowak, R., Schofield, C.J., Walport, L.J.: Introduction to structural studies on 2-oxoglutarate-dependent oxygenases and related enzymes. In: 2-Oxoglutarate-Dependent Oxygenases, pp. 59–94. Royal Society of Chemistry Cambridge, Cambridge (2015)
2. Appelhoff, R.J., Tian, Y.-M., Raval, R.R., Turley, H., Harris, A.L., Pugh, C.W., Ratcliffe, P.J., Gleadle, J.M.: Differential function of the prolyl hydroxylases phd1, phd2, and phd3 in the regulation of hypoxia-inducible factor. J. Biol. Chem. **279**(37), 38458–38465 (2004)
3. Arocena, M., Landeira, M., Di Paolo, A., Silva, A., Sotelo-Silveira, J., Fernández, A., Alonso, J.: Using a variant of coverslip hypoxia to visualize tumor cell alterations at increasing distances from an oxygen source. J. Cell. Physiol. **234**(10), 16671–16678 (2019)
4. Bagnall, J., Leedale, J., Taylor, S.E., Spiller, D.G., White, M.R.H., Sharkey, K.J., Bearon, R.N., Sée, V.: Tight control of hypoxia-inducible factor-α transient dynamics is essential for cell survival in hypoxia. J. Biol. Chem. **289**(9), 5549–5564 (2014)
5. Bartsch, P., Gibbs, J.S.R.: Effect of altitude on the heart and the lungs. Circulation **116**(19), 2191–2202 (2007)
6. Beall, C.M., Cavalleri, G.L., Deng, L., Elston, R.C., Gao, Y., Knight, J., Li, C., Li, J.C., Liang, Y., McCormack, M., et al.: Natural selection on epas1 (hif2α) associated with low hemoglobin concentration in tibetan highlanders. Proc. Nat. Acad. Sci. USA **107**(25), 11459–11464 (2010)
7. Bedessem, B., Stéphanou, A.: A mathematical model of hif-1α-mediated response to hypoxia on the g1/s transition. J. Math. Biosci. **248**, 31–39 (2014)
8. Berra, E., Benizri, E., Ginouvès, A., Volmat, V., Roux, D., Pouysségur, J.: Hif prolyl-hydroxylase 2 is the key oxygen sensor setting low steady-state levels of hif-1α in normoxia. EMBO J. **22**(16), 4082–4090 (2003)
9. Bertout, J.A., Majmundar, A.J., Gordan, J.D., Lam, J.C., Ditsworth, D., Keith, B., Brown, E.J., Nathanson, K.L., Simon, M.C.: Hif2α inhibition promotes p53 pathway activity, tumor cell death, and radiation responses. Proc. Nat. Acad. Sci. USA **106**(34), 14391–14396 (2009)
10. Bocharov, G., Jäger, W., Knoch, J., Neuss-Radu, M., Thiel, M.: A mathematical model of hif-1 regulated cellular energy metabolism. Vietnam J. Math. **49**(1), 119–141 (2021)
11. Brand, A., Singer, K., Koehl, G.E., Kolitzus, M., Schoenhammer, G., Thiel, A., Matos, C., Bruss, C., Klobuch, S., Peter, K., et al.: Ldha-associated lactic acid production blunts tumor immunosurveillance by t and nk cells. Cell Metab. **24**(5), 657–671 (2016)
12. Butcher, D.T., Alliston, T., Weaver, V.M.: A tense situation: forcing tumour progression. Nature Rev. Cancer **9**(2), 108–122 (2009)
13. Chen, E.P., Song, R.S., Chen, X.: Mathematical model of hypoxia and tumor signaling interplay reveals the importance of hypoxia and cell-to-cell variability in tumor growth inhibition. BMC Bioinf. **20**(1), 1–15 (2019)
14. Cogo, A.: The lung at high altitude. Multidisc. Resp. Med. **6**(1), 1–2 (2011)
15. Corbet, C., Feron, O.: Tumour acidosis: From the passenger to the driver's seat. Nature Rev. Cancer **17**(10), 577–593 (2017)

16. Coulibaly, A., Bettendorf, A., Kostina, E., Figueiredo, A.S., Velasquez Giraldo, S.Y., Bock, H.-G., Thiel, M., Lindner, H.A., Barbarossa, M.V.: Interleukin-15 signaling in hif-1α regulation in natural killer cells, insights through mathematical models. Front. Immunol. **10**(1), 2401 (2019)
17. Druker, J., Wilson, J.W., Child, F., Shakir, D., Fasanya, T., Rocha, S.: Role of hypoxia in the control of the cell cycle. Int. J. Molec. Sci. **22**(9), 4874 (2021)
18. Ehrismann, D., Flashman, E., Genn, D.N., Mathioudakis, N., Hewitson, K.S., Ratcliffe, P.J., Schofield, C.J.: Studies on the activity of the hypoxia-inducible-factor hydroxylases using an oxygen consumption assay. Biochem. J. **401**(1), 227–234 (2007)
19. Eltzschig, H.K., Carmeliet, P.: Hypoxia and inflammation. New Eng. J. Med. **364**(7), 656–665 (2011)
20. Ferrante, P., Preziosi, L., Scianna, M.: Modelling hypoxia-related inflammation scenarios in tumors. Math. Biosci. **355**, 108952 (2023)
21. Filatova, A., Seidel, S., Böğürcü, N., Gräf, S., Garvalov, B.K., Acker, T.: Acidosis acts through hsp90 in a phd/vhl-independent manner to promote hif function and stem cell maintenance in glioma acidosis promotes hif and glioma stem cells. Cancer Res. **76**(19), 5845–5856 (2016)
22. Gilkes, D.M., Semenza, G.L., Wirtz, D.: Hypoxia and the extracellular matrix: drivers of tumour metastasis. Nature Rev. Cancer **14**(6), 430–439 (2014)
23. Grivennikov, S.I., Greten, F.R., Karin, M.: Immunity, inflammation, and cancer. Cell **140**(6), 883–899 (2010)
24. Hashemzadeh, S., Shahmorad, S., Rafii-Tabar, H., Omidi, Y.: Computational modeling to determine key regulators of hypoxia effects on the lactate production in the glycolysis pathway. Sci. Rep. **10**(1), 1–8 (2020)
25. Hikage, F., Atkins, S., Kahana, A., Smith, T.J., Chun, T.-H.: Hif2α–lox pathway promotes fibrotic tissue remodeling in thyroid-associated orbitopathy. Endocrinology **160**(1), 20–35 (2019)
26. Holmquist-Mengelbier, L., Fredlund, E., Löfstedt, T., Noguera, R., Navarro, S., Nilsson, H., Pietras, A., Vallon-Christersson, J., Borg, Å, Gradin, K., Poellinger, L., Pahlman, S.: Recruitment of hif-1α and hif-2α to common target genes is differentially regulated in neuroblastoma: Hif-2α promotes an aggressive phenotype. Cancer Cell **10**(5), 413–423 (2006)
27. Honda, T., Hirakawa, Y., Mizukami, K., Yoshihara, T., Tanaka, T., Tobita, S., Nangaku, M.: A distinctive distribution of hypoxia-inducible factor-1α in cultured renal tubular cells with hypoperfusion simulated by coverslip placement. Physiol. Rep. **9**(1), e14689 (2021)
28. Ji, F., Wang, Y., Qiu, L., Li, S., Zhu, J., Liang, Z., Wan, Y., Di, W.: Hypoxia inducible factor 1α-mediated lox expression correlates with migration and invasion in epithelial ovarian cancer. Int. J. Oncol. **42**(5), 1578–1588 (2013)
29. Jiang, B.-H., Rue, E., Wang, G.L., Roe, R., Semenza, G.L.: Dimerization, dna binding, and transactivation properties of hypoxia-inducible factor 1. J. Biol. Chem. **271**(30), 17771–17778 (1996)
30. Jiang, B.-H., Semenza, G.L., Bauer, C., Marti, H.H.: Hypoxia-inducible factor 1 levels vary exponentially over a physiologically relevant range of o2 tension. Am. J. Physiol. Cell Physiol. **271**(4):C1172–C1180 (1996)
31. Kaelin, W.G., Ratcliffe, P.J.: Oxygen sensing by metazoans: the central role of the hif hydroxylase pathway. Molec. Cell **30**(4), 393–402 (2008)
32. Kass, L., Erler, J.T., Dembo, M., Weaver, V.M.: Mammary epithelial cell: influence of extracellular matrix composition and organization during development and tumorigenesis. Int. J. Biochem. Cell Biol. **39**(11), 1987–1994 (2007)
33. Kazyken, D., Lentz, S.I., Fingar, D.C.: Alkaline intracellular ph (phi) activates ampk–mtorc2 signaling to promote cell survival during growth factor limitation. J. Biol. Chem. **297**(4) (2021)
34. Kim, J.-W., Tchernyshyov, I., Semenza G.L., Dang, C.V.: Hif-1-mediated expression of pyruvate dehydrogenase kinase: a metabolic switch required for cellular adaptation to hypoxia. Cell Metab. **3**(3), 177–185 (2006)
35. Koh, M.Y., Powis, G.: Passing the baton: the hif switch. Trends Biochem. Sci. **37**(9) 364–372 (2012)

36. Koivunen, P., Hirsila, M., Kari, I.K., Myllyharju, J.: The length of peptide substrates has a marked effect on hydroxylation by the hypoxia-inducible factor prolyl 4-hydroxylases. J. Biol. Chem. **281**(39), 28712–28720 (2006)
37. Korbecki, J., Simińska, D., Gassowska-Dobrowolska, M., Listos, J., Gutowska, I., Chlubek, D., Baranowska-Bosiacka, I.: Chronic and cycling hypoxia: drivers of cancer chronic inflammation through hif-1 and nf-κb activation: A review of the molecular mechanisms. Int. J. Molec. Sci. **22**(19), 10701 (2021)
38. Lanikova, L., Reading, N.S., Hu, H., Tashi, T., Burjanivova, T., Shestakova, A., Siwakoti, B., Thakur, B.K., Pun, C.B., Sapkota, A., et al.: Evolutionary selected tibetan variants of hif pathway and risk of lung cancer. Oncotarget **8**(7), 11739 (2017)
39. Leedale, J., Herrmann, A., Bagnall, J., Fercher, A., Papkovsky, D., Sée, V., Bearon, R.N.: Modeling the dynamics of hypoxia inducible factor-1α (hif-1α) within single cells and 3d cell culture systems. Math. Biosci. **258**, 33–43 (2014)
40. Liberti, M.V., Locasale, J.W.: The warburg effect: How does it benefit cancer cells? Trends Biochem. Sci. **41**(3), 211–218 (2016)
41. Lin, Q., Yun, Z.: The hypoxia-inducible factor pathway in adipocytes: the role of hif-2 in adipose inflammation and hypertrophic cardiomyopathy. Front. Endocr. **6**, 39 (2015)
42. Lorenzo, F.R., Huff, C., Myllymäki, M., Olenchock, B., Swierczek, S., Tashi, T., Gordeuk, V., Wuren, T., Ri-Li, G., Khan, T.M. McClain, D.A., Koul, P.A., Guchhait, P., Salama, M.E., Xing, J., Semenza, G.L., Liberzon, E., Wilson, A., Simonson, T.S., Jorde, L.B., Kaelin, W.G., Koivunen, P., Prchal, J.T.: A genetic mechanism for tibetan high-altitude adaptation. Nature Genet. **46**(9), 951–956 (2014)
43. Luks, A.M., Swenson, E.R., Bärtsch, P.: Acute high-altitude sickness. Eur. Resp. Rev. **26**(143) (2017)
44. Maheshwari, R., Weis, E.: Thyroid associated orbitopathy. Indian J. Opht. **60**(2), 87 (2012)
45. McKeown, S.R.: Defining normoxia, physoxia and hypoxia in tumours-implications for treatment response. Brit. J. Radiol. **87**(1035), 20130676 (2014)
46. Metzen, E., Berchner-Pfannschmidt, U., Stengel, P., Marxsen, J., Stolze, I., Klinger, M., Huang, W.Q., Wotzlaw, C., Hellwig-Burgel, T., Jelkmann, W., Acker, H., Fandreyothers, J.: Intracellular localisation of human hif-1α hydroxylases: implications for oxygen sensing. J. Cell Sci. **116**(7), 1319–1326 (2003)
47. Miller, A.E., Hu, P., Barker, T.H.: Feeling things out: bidirectional signaling of the cell–ecm interface, implications in the mechanobiology of cell spreading, migration, proliferation, and differentiation. Adv. Healthcare Mat. **9**(8), 1901445 (2020)
48. Mu, X., Shi, W., Xu, Y., Xu, C., Zhao, T., Geng, B., Yang, J., Pan, J., Hu, S., Zhang, C., Zhang, J., Wang, C., Shen, J., Che, Y., Liu, Z., Lv, Z., Wen, H., You, Q.: Tumor-derived lactate induces m2 macrophage polarization via the activation of the erk/stat3 signaling pathway in breast cancer. Cell Cycle **17**(4), 428–438 (2018)
49. Nguyen, L.K., Cavadas, M.A.S., Scholz, C.C., Fitzpatrick, S.F., Bruning, U., Cummins, E.P., Tambuwala, M.M., Manresa, M.C., Kholodenko, B.N., Taylor, C.T., Cheong, A.: A dynamic model of the hypoxia-inducible factor 1α (hif-1α) network. J. Cell Sci. **126**(6), 1454–1463 (2013)
50. Nishi, H., Sasaki, T., Nagamitsu, Y., Terauchi, F., Nagai, T., Nagao, T., Isaka, K.: Hypoxia inducible factor-1 mediates upregulation of urokinase-type plasminogen activator receptor gene transcription during hypoxia in cervical cancer cells. Oncol. Rep. **35**(2), 992–998 (2016)
51. Papale, M., Buccarelli, M., Mollinari, C., Russo, M.A., Pallini, R., Ricci-Vitiani, L., Tafani, M.: Hypoxia, inflammation and necrosis as determinants of glioblastoma cancer stem cells progression. Int. J. Molec. Sci. **21**(8), 2660 (2020)
52. Paradise, R.K., Lauffenburger, D.A., Van Vliet, K.J.: Acidic extracellular ph promotes activation of integrin αvβ3. PLoS One **6**(1), e15746 (2011)
53. Pektas, S., Knapp, M.J.: Substrate preference of the hif-prolyl hydroxylase-2 (phd2) and substrate-induced conformational change. J. Inorg. Biochem. **126**, 55–60 (2013)
54. Piasentin, N., Milotti, E., Chignola, R.: The control of acidity in tumor cells: A biophysical model. Sci. Rep. **10**(1), 1–14 (2020)

55. Pichiule, P., Chavez, J.C., Schmidt, A.M., Vannucci, S.J.: Hypoxia-inducible factor-1 mediates neuronal expression of the receptor for advanced glycation end products following hypoxia/ischemia. J. Biol. Chem. **282**(50), 36330–36340 (2007)
56. Pressley, M., Gallaher, J.A., Brown, J.S., Tomaszewski, M.R., Borad, P., Damaghi, M., Gillies, R.J., Whelan, C.J.: Cycling hypoxia selects for constitutive hif stabilization. Sci. Rep. **11**(1), 1–14 (2021)
57. Qutub, A.A., Popel, A.S.: Three autocrine feedback loops determine hif1α expression in chronic hypoxia. J. Biochim. Biophys. Acta Molec. Cell Res. **1773**(10), 1511–1525 (2007)
58. Robey, I.F., Lien, A.D., Welsh, S.J., Baggett, B.K., Gillies, R.J.: Hypoxia-inducible factor-1α and the glycolytic phenotype in tumors. Neoplasia **7**(4), 324–330 (2005)
59. Robinson, P.J., Hack, C.E.: Mathematical model of hif-1α pathway, oxygen transport and hypoxia. Internal Report U.S. Army, AFRL-RH-WP-TR-2017-0080 (2017)
60. Ronen, R., Zhou, D., Bafna, V., Haddad, G.G.: The genetic basis of chronic mountain sickness. Physiology **29**(6), 403–412 (2014)
61. Salmond, R.J.: mtor regulation of glycolytic metabolism in t cells. Front. Cell Dev. Biol. **6**, 122 (2018)
62. Schmierer, B., Novák, B., Schofield, C.J.: Hypoxia-dependent sequestration of an oxygen sensor by a widespread structural motif can shape the hypoxic response - a predictive kinetic model. BMC Syst. Biol. **4**(1), 139 (2010)
63. Semenza, G.L.: Hydroxylation of hif-1: oxygen sensing at the molecular level. Physiology **19**(4), 176–182 (2004)
64. Semenza, G.L., Jiang, B.-H., Leung, S.W., Passantino, R., Concordet, J.-P., Maire, P., Giallongo, A.: Hypoxia response elements in the aldolase a, enolase 1, and lactate dehydrogenase a gene promoters contain essential binding sites for hypoxia-inducible factor 1. J. Biol. Chem. **271**(51), 32529–32537 (1996)
65. Stiehl, D.P., Bordoli, M.R., Abreu-Rodriguez, I., Wollenick, K., Schraml, P., Gradin, K., Poellinger, L., Kristiansen, G., Wenger, R.H.: Non-canonical hif-2α function drives autonomous breast cancer cell growth via an areg–egfr/erbb4 autocrine loop. Oncogene **31**(18), 2283–2297 (2012)
66. Stiehl, D.P., Wirthner, R., Koditz, J., Spielmann, P., Camenisch, G., Wenger, R.H.: Increased prolyl 4-hydroxylase domain proteins compensate for decreased oxygen levels: evidence for an autoregulatory oxygen-sensing system. J. Biol. Chem. **281**(33), 23482–23491 (2006)
67. Tafani, M., Schito, L., Pellegrini, L., Villanova, L., Marfe, G., Anwar, T., Rosa, R., Indelicato, M., Fini, M., Pucci, B., Russo, M.A.: Hypoxia-increased rage and p2x7r expression regulates tumor cell invasion through phosphorylation of erk1/2 and akt and nuclear translocation of nf-κb. Carcinogenesis **32**(8), 1167–1175 (2011)
68. Taneja, S., Vetter S.W., Leclerc, E.: Hypoxia and the receptor for advanced glycation end products (rage) signaling in cancer. Int. J. Mol. Sci. **22**(15), 8153 (2021)
69. Tay, S., Hughey, J.J., Lee, T.K., Lipniacki, T., Quake, S.R., Covert, M.W.: Single-cell nf-κb dynamics reveal digital activation and analogue information processing. Nature **466**(7303), 267–271 (2010)
70. Trayhurn, P.: Hypoxia and adipose tissue function and dysfunction in obesity. Physiol. Rev. **93**(1), 1–21 (2013)
71. Tuckerman, J.R., Zhao, Y., Hewitson, K.S., Tian, Y.-M., Pugh, C.W., Ratcliffe, P.J., Mole, D.R.: Determination and comparison of specific activity of the hif-prolyl hydroxylases. FEBS Lett. **576**(1–2), 145–150 (2004)
72. Vaupel, P.: Tumor microenvironmental physiology and its implications for radiation oncology. In: Seminars in Radiation Oncology, vol. 14, pp. 198–206. Elsevier, Amsterdam (2004)
73. Villafuerte, F.C., Corante, N.: Chronic mountain sickness: clinical aspects, etiology, management, and treatment. High Alt. Med Biol. **17**(2), 61–69 (2016)
74. Warburg, O.: On the origin of cancer cells. Science **123**(3191), 309–314 (1956)
75. Webb, B.A., Chimenti, M., Jacobson, M.P., Barber, D.L.: Dysregulated ph: A perfect storm for cancer progression. Nature Rev. Cancer **11**(9), 671–677 (2011)

76. Xiang, K., Peng, Y., Yang, Z., Zhang, X., Cui, C., Zhang, H., Li, M., Zhang, Y., Wu, T., Chen, H., et al.: Identification of a tibetan-specific mutation in the hypoxic gene egln1 and its contribution to high-altitude adaptation. Molec. Biol. Evol. **30**(8), 1889–1898 (2013)
77. Xiao, Q., Ge, G.: Lysyl oxidase, extracellular matrix remodeling and cancer metastasis. Cancer Microenv. **5**(3), 261–273 (2012)
78. Xu, L., Fukumura, D., Jain, R.K.: Acidic extracellular ph induces vascular endothelial growth factor (vegf) in human glioblastoma cells via erk1/2 mapk signaling pathway: Mechanism of low ph-induced vegf. J. Biol. Chem. **277**(13), 11368–11374 (2002)
79. Yu, Y., Wang, G., Simha, R., Peng, W., Turano, F., Zeng, C.: Pathway switching explains the sharp response characteristic of hypoxia response network. PLOS Comput. **3**(8), e171 (2007)
80. Zatterale, F., Longo, M., Naderi, J., Raciti, G.A., Desiderio, A., Miele, C., Beguinot, F.: Chronic adipose tissue inflammation linking obesity to insulin resistance and type 2 diabetes. Front. Physiol. **10**, 1607 (2020)
81. Zhang, N., Fu, Z., Linke, S., Chicher, J., Gorman, J.J., Visk, D., Haddad, G.G., Poellinger, L., Peet, D.J., Powell, F., Johnson, R.S.: The asparaginyl hydroxylase factor inhibiting hif-1α is an essential regulator of metabolism. Cell Metab. **11**(5), 364–378 (2010)
82. Zhang, B., Ye, H., Yang, A.: Mathematical modelling of interacting mechanisms for hypoxia mediated cell cycle commitment for mesenchymal stromal cells. BMC Syst. Biol. **12**, 35 (2018)
83. Zhao, T., Zhu, Y., Morinibu, A., Kobayashi, M., Shinomiya, K., Itasaka, S., Yoshimura, M., Guo, G., Hiraoka, M., Harada, H.: Hif-1-mediated metabolic reprogramming reduces ros levels and facilitates the metastatic colonization of cancers in lungs. Sci. Rep. **4**(1), 1–7 (2014)
84. Zimna, A., Kurpisz, M.: Hypoxia-inducible factor-1 in physiological and pathophysiological angiogenesis: applications and therapies. BioMed. Res. Int. **2015** (2015)

An Imaging-Informed Mechanical Framework to Provide a Quantitative Description of Brain Tumour Growth and the Subsequent Deformation of White Matter Tracts

Francesca Ballatore, Giulio Lucci, Andrea Borio, and Chiara Giverso

Abstract The mathematical description of brain tumours is a challenging problem, that may be fundamental to support medical observations and to build personalised therapeutic treatments for the patients. In this respect, we propose a multiphase model, based on Continuum Mechanics, where both the healthy and the diseased regions are treated as mixtures, comprising a solid and a fluid phase. Moreover, we use patient-specific imaging data to reconstruct the preferential directions for nutrient diffusion, fluid and cell motion inside the brain, since they all follow the orientation of white matter tracts. Then, given the mechanical deformation induced by the tumour onto the healthy tissue, we employ it to properly modify the preferential directions of white matter tracts. Our numerical simulations show that tumour-induced displacements and stresses may have a substantial impact on the tissue surrounding the cancer mass, even in regions distant from the tumour position. Furthermore, the model is able to highlight relevant changes in the preferential directions of nutrient diffusion and cell motion, caused by the spread of the cancer. Finally, the proposed framework may be a useful tool for the mechanical and computational modelling of other kinds of tumours growing in highly anisotropic environments and for estimating the effect of the expanding mass on the surrounding tissue.

1 Introduction

Despite the relevant advances in clinical practice supported by novel therapies and imaging techniques, the treatment of brain tumours remains not fully effective in many cases, due to cancer aggressiveness and to the intrinsically fragile nature

F. Ballatore · G. Lucci · A. Borio · C. Giverso (✉)
Politecnico di Torino, Torino, Italy
e-mail: chiara.giverso@polito.it

G. Bretti et al. (eds.), *Mathematical Models and Computer Simulations for Biomedical Applications*, SEMA SIMAI Springer Series 33,
https://doi.org/10.1007/978-3-031-35715-2_5

of brain tissue. For this reason, in the last three decades, the mathematical modelling of brain tumour growth started to attract research attention. The purpose of these models is essentially twofold: first, they can help in understanding the progression of the tumour, providing further support to medical observations; then, in silico findings coming from simulations may be employed to build personalised therapeutic strategies, which are particularly important in the case of brain cancers to maximize the efficacy of treatments while minimizing side effects for the patients. In order to obtain a realistic outcome, the interplay between models and clinical data is crucial, as well as the accurate mathematical description of the brain environment, which is known to strongly affect tumour progression [22, 47, 77, 79].

To achieve these goals, models of brain tumour growth have become increasingly refined during the years: to give an overview, in Table 1 we summarise some of the main contributions that appear in the literature. For detailed and extensive reviews on brain cancer modelling, we refer the reader to [4, 37, 47, 65].

In particular, a first distinction between models can be made according to the mathematical framework they use, which is strictly related to the scale that is considered. More specifically, microscopic and mesoscopic models provide a

Table 1 Summary of previous contributions concerning brain tumour modelling. The models are classified according to three criteria: (1) the mathematical framework employed to describe the growth of the tumour mass (first column); (2) the inclusion of tumour and tissue mechanics with quantification of deformations and stresses (third column); (3) the use of patient-specific imaging data to perform simulations (fourth column). Abbreviations: CA = Cellular Automaton; ABM = Agent-Based Model; ODE = Ordinary Differential Equations model; PDE-RD = Reaction-Diffusion equations; PDE-ARD = Advection-Reaction-Diffusion equations; PDE-KM = Kinetic model; PDE-CH = Cahn-Hilliard model; PDE-CM = Continuum-Mechanics-based model; HYB = Hybrid model; FL = Fluid; LE = Linear elasticity; NLE = Nonlinear elasticity

Type of model	References	Mechanics	Imaging
CA	[11, 12, 17, 48, 54, 55, 90]	No	No
ODE	[14, 52]	No	No
ABM	[62–64]	No	No
ABM–PDE–RD	[41, 42]	No	No
PDE–RD	[75, 83–88, 92, 94]	No	Yes (CT+MRI)
PDE–RD	[46, 53]	No	Yes (MRI+DTI)
PDE–ARD	[95]	No	No
PDE–ARD	[81]	No	Yes (DTI)
PDE–ARD–KM	[72, 82]	No	Yes (DTI)
ODE–PDE–ARD–KM	[25, 26, 33–36, 51]	No	Yes (DTI)
PDE–ARD–CM	[16, 23, 32, 49]	Yes (LE)	Yes (MRI+DTI)
PDE–CM	[9, 10, 30, 31, 58, 61]	Yes (NLE)	Yes (MRI+DTI)
PDE–CM	[66]	Yes (FL)	No
PDE–CH	[1–3, 24, 37, 59]	Yes (FL)	Yes (MRI+DTI)
HYB	[39, 56, 57, 76]	No	No
HYB	[89, 96]	No	No

description of phenomena taking place at the subcellular or cellular scale. These types of models mainly fulfill the objective of reproducing the early growth of brain tumours, accounting explicitly for interactions at the cellular level: most of them are Cellular Automata (CA), Agent-Based Models (ABM) or Ordinary Differential Equations (ODE) models. On the other hand, macroscopic models based on Partial Differential Equations (PDE) do not consider the intrinsically discrete nature of tumours, in exchange for a more flexible description performed through continuous variables. The first works [83–87, 92, 94] using reaction-diffusion equations for the migration and proliferation of gliomas paved the way for a number of subsequent studies, with an increasing level of detail. Multi-scale formulations bridging the gap between the microscopic and the macroscopic levels, grounded on Kinetic Models (KM) and their scaling [25, 26, 33–36, 51, 72, 82], also provided an interesting extension of purely diffusive, phenomenological descriptions. Another approach recently proposed to tackle the problem of brain tumour proliferation employs a Cahn-Hilliard-type (CH) equation to deal with the infiltrative nature of some brain tumours, showing a good agreement with real data [1–3, 24, 37].

As discussed above, a fundamental improvement to build reliable models for clinical applications is the inclusion of realistic imaging data, which may be used both for the estimation of parameters and for performing simulations on real geometries. In particular, the very first works [83–87, 92, 94] employed Computed Tomography (CT) and Magnetic Resonance Imaging (MRI) to extrapolate numerical values for the model parameters and to introduce a spatial distinction between white and grey matter. Then, the progress in Diffusion Tensor Imaging (DTI) allowed to account for the intrinsic anisotropy of brain tissue and became of great interest in the modelling process [53].

Nevertheless, the majority of the models mentioned so far does not take into account the mechanical impact of the growing tumour mass on the healthy brain tissue. Instead, the deformation induced by the proliferating cancer onto the surrounding areas may be harmful for the patient, since it may lead to damage in brain functionalities. In this respect, recent experimental works [21, 69, 70, 77] pointed out the preeminent role of solid stresses due to brain tumour expansion, in addition to the effects of fluid pressure. Moreover, the distribution of such stresses appears to be different even in tumours that exhibit similar imaging volumes [70, 79]. Therefore, it is important to have models that are able to capture the mechanical effects of a tumour inside the skull, to precisely evaluate the brain area affected by the cancer and the consequent possible risks for the patient. Motivated by these facts, some recent models exploited the framework of Continuum Mechanics to provide a description of the so-called "mass effect" caused by the tumour. The first and simpler biomechanical models considered the brain as a linear elastic (LE) medium [16, 23, 49], while nonlinearly elastic (NLE) constitutive equations (e.g. Neo-Hookean or Mooney-Rivlin) have been employed in successive descriptions [9, 10, 30, 31, 61].

Another relevant feature that a quantitative mechanical model allows to introduce is the modification of DTI imaging data as a consequence of tumour growth [61]. Indeed, the growing mass displaces and dislodges the surrounding white matter fibres, causing a change in the preferential directions for water diffusion and cell movement. Since for patients affected by aggressive brain cancers it is often difficult to obtain multiple imaging scans at different times, the possibility of providing a computational modification of DTI data by means of mechanics may have valuable clinical implications.

Motivated by these observations, in this chapter we propose a mechanical and computational framework which may be used to describe brain tumour growth, accounting both for a detailed mechanical representation and for the inclusion of patient-specific data. Specifically, we derive a multiphase model based on Continuum Mechanics which features the nonlinear elastic properties of brain tissue, being then able to quantify the deformation and the solid stress induced by the growing tumour. In our description, differently from previous models [61], the tumour is sharply separated from the host tissue, leading to variables of the model which are not necessarily continuous across the interface. It is therefore mandatory to specify proper interface conditions between the healthy and the cancer tissues, which also have distinct mechanical properties. Such a description is appropriate to represent solid brain tumours with a sharp separation between the cancer and the surrounding healthy area. Then, as already done in [61], we perform simulations on a realistic three-dimensional brain geometry reconstructed from MRI and DTI data and we employ the knowledge of the mechanical variables provided by our model to modify the DTI data in time, by changing the orientation of white matter fibres as a consequence of the deformation. We observe that the proposed framework may be used, properly modified, for the mechanical and computational modelling of other kinds of tumours growing in highly anisotropic environments.

In detail, the structure of the chapter is as follows. In Sect. 2 we present the mathematical model and the governing equations, describing also the process of DTI data modification following the mechanical deformation. We remark that the introduction of discontinuous variables, due to the presence of two distinct domains, requires to pay attention to the derivation of the weak form of the problem, rewritten in the Lagrangian frame. Therefore, in Sect. 3 of the present work we provide details on the derivation of the weak formulation of the problem and on its discretisation in space and time. Then, always in Sect. 3, we provide a possible estimate for all the parameters involved in the model and we define the procedure used to create the patient-specific mesh. Section 4 is devoted to the presentation of some numerical outcomes, showing the validity of the proposed framework as a proof-of-concept for the mechanical description of brain tumours and DTI modification.

2 A Multiphase Model for Brain Tumour Growth

In this section, we present a multiphase model for tumour growth and expansion, based on the theory of mixtures [6, 19, 74] and consisting of a set of mass and momentum balance equations. Even though the general framework of the proposed model could be used, in principle, to describe the development of any kind of solid tumour, we specialise it to account for brain tumour evolution in a patient-specific setting. The aim of the mathematical model presented hereafter is to evaluate the progression of this disease, in order to predict the evolution of tumour shape and to quantify the extent and the position of damaged areas. We assume that the region occupied by the tumour, denoted by $\Omega_t(t)$, is completely separated from the healthy host tissue, denoted by $\Omega_h(t)$, so that the boundary between the tumour and the surrounding environment can be described by a moving interface. Both these regions are treated as saturated domains consisting of two distinct phases, representing the cell population (labelled with subscript "s") and the interstitial fluid (labelled with subscript "ℓ"), which fill all the available space. Therefore, introducing the volumetric fraction of the cell population, ϕ_s, and the volumetric fraction of the liquid, ϕ_ℓ, the saturation constraint

$$\phi_s + \phi_\ell = 1 \tag{1}$$

holds at any time instant and at any point in the brain domain $\Omega(t) = \Omega_h(t) \cup \Omega_t(t)$. We remark that, in this description, the cellular phase represents healthy cells in $\Omega_h(t)$ and diseased cells in $\Omega_t(t)$, whereas the fluid phase resumes interstitial brain fluid, blood, and nutrients in both regions. Furthermore, we assume that the materials composing the phases are incompressible, which means that both phases of the mixture have constant true densities $\hat{\rho}_\alpha$, with $\alpha \in \{s, \ell\}$. Then, once the true density $\hat{\rho}_\alpha$ is prescribed, the partial phase density $\rho_\alpha := \hat{\rho}_\alpha \phi_\alpha$ of the material composing the α-phase is totally defined by knowing ϕ_α. Finally, since cells are mainly composed of water, we assume that the true densities of both phases are equal, i.e., $\hat{\rho}_s = \hat{\rho}_\ell$.

Throughout this chapter, we will denote by Ω^*, Ω_t^* and Ω_h^* the reference configurations of the whole brain, the tumour and the host tissue, respectively, so that $\Omega^* = \Omega_h^* \cup \Omega_t^*$ holds. It is important to underline that the tumour region Ω_t^* in the reference configuration does not evolve in time. For what concerns the differential operators, Grad and Div will be used in the following to denote the material gradient and material divergence with derivatives taken with respect to the material point in the reference configuration. Instead, the notations ∇ and $\nabla\cdot$ will denote the gradient and the divergence with respect to the spatial variable in the current configuration $\Omega(t)$. The deformation of the body from the reference configuration to the deformed one can be described using the map $\chi(\mathbf{X}, t)$, which assigns to each material point $\mathbf{X} \in \Omega^*$ its position $\mathbf{x}$ in $\Omega(t)$. Introducing the displacement field of the solid phase, defined by $\mathbf{u}_s(\mathbf{X}, t) = \chi(\mathbf{X}, t) - \mathbf{X}$, we can define the *deformation gradient tensor* of the solid phase as $\mathbb{F}_s = \mathbb{I} + \text{Grad}\, \mathbf{u}_s$, where $\mathbb{I}$ is the second order identity tensor. Moreover, the concept of evolving natural configurations is employed in the following to properly describe the mechanics of the growing body. Resorting to the modelling background proposed in [7, 8] for growing tumours, this approach consists in splitting the evolution in pure elastic deformations and deformations subsequent to anelastic distortions, such as growth and remodelling. In particular, if we assume to cut a generic particle out of the body and to relieve its state of stress while keeping the mass constant, we find the

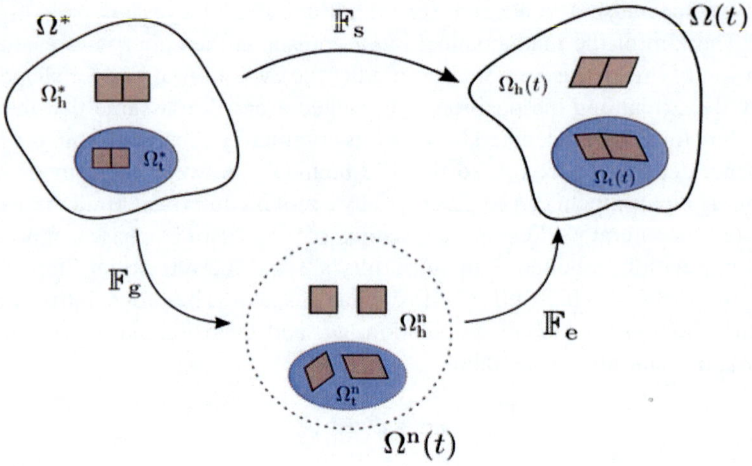

Fig. 1 Multiplicative decomposition of the deformation gradient

natural state of such particle at time t. The *natural configuration* of the body at time t is then the collection of all the particles in their natural states at time t and it is indicated by $\Omega^n(t)$. In this way it is possible to measure the deformation from the natural configuration $\Omega^n(t)$ to $\Omega(t)$ through the tensor $\mathbb{F}_e$, which is connected to the stress response of the material, while the path from the reference configuration Ω^* to the natural configuration is described by the tensor $\mathbb{F}_g$, which is directly related to growth and it is therefore named *growth tensor*. This decomposition is graphically shown in Fig. 1. To sum up, the deformation gradient $\mathbb{F}_s$ indicates how the body is deforming locally in going from the reference configuration Ω^* to $\Omega(t)$, while, in an analogous way, $\mathbb{F}_e$ tells how the body is deforming locally in going from the natural configuration $\Omega^n(t)$ to $\Omega(t)$, and $\mathbb{F}_g$ tells how the body is growing locally. The following multiplicative decomposition of the deformation gradient is therefore valid:

$$\mathbb{F}_s = \mathbb{F}_e \mathbb{F}_g \,. \tag{2}$$

Furthermore, since the deformation gradient $\mathbb{F}_s$ is invertible, it follows that $\mathbb{F}_e$ and $\mathbb{F}_g$ are invertible too. Indeed, the determinant of the deformation gradient can be expressed as

$$J_s = J_e J_g \,,$$

where $J_e = \det \mathbb{F}_e$ and $J_g = \det \mathbb{F}_g$. In particular, since we are dealing with growth processes, we have $J_g \geq 1$.

2.1 Eulerian Formulation

2.1.1 Balance Equations

Mass and Momentum Balance Laws in the Tumour Region $\Omega_{\mathrm{t}}(t)$

To derive the balance equations of our model, we firstly focus on the region occupied by the tumour. We assume that, in this region, cells proliferate since the tumour is growing. Hence, the mass and momentum balances for each phase $\alpha \in \{\mathrm{s}, \ell\}$ read

$$\frac{\partial \phi_\alpha}{\partial t} + \nabla \cdot (\phi_\alpha \mathbf{v}_\alpha) = \Gamma_\alpha \,, \tag{3}$$

$$\hat{\rho}_\alpha \phi_\alpha \left(\frac{\partial \mathbf{v}_\alpha}{\partial t} + \mathbf{v}_\alpha \cdot \nabla \mathbf{v}_\alpha \right) = \nabla \cdot \widetilde{\mathbb{T}}_\alpha + \hat{\rho}_\alpha \phi_\alpha \mathbf{b}_\alpha + \widetilde{\mathbf{m}}_\alpha \,, \tag{4}$$

where $\mathbf{v}_\alpha$ is the velocity of the α-phase, $\widetilde{\mathbb{T}}_\alpha$ is the partial Cauchy stress tensor of that phase, Γ_α is the mass growth rate and $\widetilde{\mathbf{m}}_\alpha$ represents the rate at which the α-phase exchanges momentum with the other phase. Then, the mixture is assumed to be closed with respect to mass, i.e., $\Gamma_\ell = -\Gamma_{\mathrm{s}}$, so that mass exchanges occur only among the constituents taken into account. Moreover, external body forces (such as the gravitational force) included in $\mathbf{b}_\alpha$ as well as inertial effects are negligible, since the motion of cells and interstitial fluid is very slow, when dealing with biological growth phenomena. Thus, Eq. (4) becomes

$$\nabla \cdot \widetilde{\mathbb{T}}_\alpha + \widetilde{\mathbf{m}}_\alpha = \mathbf{0} \,. \tag{5}$$

The term $\widetilde{\mathbf{m}}_\alpha$, with $\alpha \in \{\mathrm{s}, \ell\}$, can be decomposed, using thermodynamics arguments [43], into a dissipative and a non-dissipative part as $\widetilde{\mathbf{m}}_\alpha = \overline{\mathbf{m}}_{\alpha\beta} + p\nabla\phi_\alpha$, where p is the pressure of the interstitial fluid and the term $\overline{\mathbf{m}}_{\alpha\beta}$ represents the dissipative force acting on the α phase due to the other phase, denoted by subscript β. By invoking the action-reaction principle and the saturation condition (1), it holds that

$$\overline{\mathbf{m}}_{\mathrm{s}\ell} = -\overline{\mathbf{m}}_{\ell\mathrm{s}} \,. \tag{6}$$

We remark that, in defining the momentum exchange between phases, we neglected the exchange rates associated with the mass sources and sinks Γ_α, $\alpha \in \{\mathrm{s}, \ell\}$. Such an assumption is reasonable in the context of avascular tumour growth, in which the velocities of both the solid and the fluid phase are small [43].

Following standard arguments in mixture theory, the Cauchy stress associated with the α-phase of the mixture can be written as the sum of a purely hydrostatic contribution, which indicates the amount of pressure sustained by the α-phase, and an effective stress

$$\widetilde{\mathbb{T}}_\alpha = -\phi_\alpha p\mathbb{I} + \mathbb{T}_\alpha\,.$$

Moreover, we require that the effective stress of the fluid phase $\mathbb{T}_\ell$ is negligible with respect to the pressure gradient and to the interaction forces between fluid and solid phase. As a consequence, Eq. (5), specialised for the two phases, becomes

$$-\phi_s\nabla p + \nabla\cdot\mathbb{T}_s + \overline{\mathbf{m}}_{s\ell} = \mathbf{0}\,, \tag{7}$$

$$-\phi_\ell\nabla p + \overline{\mathbf{m}}_{\ell s} = \mathbf{0}\,. \tag{8}$$

The momentum balance for the mixture can then be obtained by summing (7) and (8), recalling the saturation condition (1) and the action-reaction principle (6):

$$-\nabla p + \nabla\cdot\mathbb{T}_s = \mathbf{0}\,. \tag{9}$$

Furthermore, calling μ the dynamic viscosity of the fluid component, $\mathbb{K}\,(\phi_\ell)$ the permeability tensor and taking $\overline{\mathbf{m}}_{\ell s} = -\mu\phi_\ell^2\mathbb{K}\,(\phi_\ell)^{-1}\,(\mathbf{v}_\ell - \mathbf{v}_s)$ [43], it is possible to derive from (8) the well-known Darcy's law as a momentum balance for the fluid phase

$$\mathbf{v}_\ell = \mathbf{v}_s - \frac{\mathbb{K}\,(\phi_\ell)}{\mu\phi_\ell}\nabla p\,. \tag{10}$$

To account for anisotropy in fluid motion due to the presence of white and grey matter fibres in the brain tissue, we take the permeability tensor as

$$\mathbb{K}\,(\phi_\ell) := K\,(\phi_\ell)\,\mathbb{A}\,, \tag{11}$$

where $\mathbb{A}$ denotes the Eulerian preferential directions tensor [24], whose construction will be discussed in Sect. 2.1.4.

Mass and Momentum Balance Laws in the Healthy Region $\Omega_h(t)$

In the domain occupied by the healthy tissue we assume that the proliferation of cells is compensated by natural cell death, so that the net rate of growth is equal to zero (i.e. $\Gamma_s = 0$). The closed mixture assumption implies that also the source term Γ_ℓ must be null. Hence, the mass balances in the healthy region can be written as

$$\frac{\partial\phi_\alpha}{\partial t} + \nabla\cdot(\phi_\alpha\mathbf{v}_\alpha) = 0\,, \quad \text{with}\,\alpha\in\{s,\ell\}\,.$$

As regards the momentum balance equations, they are the same as in the region occupied by the tumour, namely Eqs. (9) and (10). The differences in the mechanical properties between the healthy and the diseased tissues, affecting the stress tensor $\mathbb{T}_s$, will be provided through the constitutive equations.

2.1.2 Stress Tensor and Constitutive Equations

In order to close the system of mass and momentum balance equations and to understand how brain tumour growth influences mechanically the surrounding tissues, we have to determine an appropriate evolution law for the effective part of the Cauchy stress tensor $\mathbb{T}_s$, associated with the cellular population, both in the diseased and in the healthy region. In analogy with [7], we assume that the mechanical response is hyperelastic from the natural configuration, that is, both the healthy brain tissue and the tumour are modelled as nonlinear elastic materials.

Effective Stress Tensor in $\Omega_t(t)$

In order to fully describe the elastic response, the generalized Ogden model [71] is often considered appropriate to represent the mechanical behaviour of soft brain tissue [68]. In particular, we take into account the Mooney-Rivlin model, which represents a particular case of the generalized Ogden energy [13, 27, 68]. Let $\overline{\mathbb{C}}_e := J_e^{-\frac{2}{3}} \mathbb{C}_e$ be the isochoric part of the elastic right Cauchy-Green deformation tensor $\mathbb{C}_e := \mathbb{F}_e^T \mathbb{F}_e$. The strain energy density per unit volume of the natural configuration $\widehat{\mathcal{W}}_{sn}$ can be written as a function of the first two invariants of $\overline{\mathbb{C}}_e$ and J_e:

$$\widehat{\mathcal{W}}_{sn}\left(\overline{\mathbb{C}}_e, J_e\right) = \frac{1}{2}\mu_{1t}\left(\mathrm{I}_{\overline{\mathbb{C}}_e} - 3\right) + \frac{1}{2}\mu_{2t}\left(\mathrm{II}_{\overline{\mathbb{C}}_e} - 3\right) + \\ + \kappa_t (1 - \phi_{sn})^2 \left(\frac{J_e - 1}{1 - \phi_{sn}} - \ln \frac{J_e - \phi_{sn}}{1 - \phi_{sn}}\right), \tag{12}$$

where

$$\mathrm{I}_{\overline{\mathbb{C}}_e} := \mathrm{tr}\left(\overline{\mathbb{C}}_e\right), \qquad \mathrm{II}_{\overline{\mathbb{C}}_e} := \frac{1}{2}\left[\left(\mathrm{tr}\,\overline{\mathbb{C}}_e\right)^2 - \mathrm{tr}\left(\overline{\mathbb{C}}_e^2\right)\right].$$

The last term on the right-hand side of Eq. (12), which is different from the one employed in previous works (e.g. [61]), describes volumetric changes in the solid skeleton, occurring below the compaction point, i.e., when all pores in the structure are closed and further volume deformations are impeded due to the incompressibility of the solid phase (see [29] for further details). Furthermore, ϕ_{sn} represents the volumetric fraction of the cell phase in the natural state and it has a constant value. Finally, μ_{1t} and μ_{2t} are the material parameters of the tumour tissue whereas κ_t is the elastic parameter associated with the response of the tumour to volumetric deformations. Given $\widehat{W}_{sn}$, we can express the Cauchy stress tensor of the cellular phase as

$$\mathbb{T}_s = 2J_e^{-1}\mathbb{F}_e \frac{\partial \widehat{W}_{sn}}{\partial \mathbb{C}_e}\mathbb{F}_e^T \quad \text{in } \Omega_t(t)\,. \tag{13}$$

By working out the derivative in (13) we have

$$\begin{aligned}\frac{\partial \widehat{W}_{sn}\left(\overline{\mathbb{C}}_e, J_e\right)}{\partial \mathbb{C}_e} &= \frac{\partial \overline{\mathbb{C}}_e}{\partial \mathbb{C}_e} : \frac{\partial \widehat{W}_{sn}}{\partial \overline{\mathbb{C}}_e} + \frac{\partial J_e}{\partial \mathbb{C}_e}\frac{\partial \widehat{W}_{sn}}{\partial J_e} \\ &= J_e^{-2/3}\left(\underline{\underline{\mathbb{I}}} - \frac{1}{3}\overline{\mathbb{C}}_e^{-1} \otimes \overline{\mathbb{C}}_e\right) : \frac{\partial \widehat{W}_{sn}}{\partial \overline{\mathbb{C}}_e} + \frac{1}{2}J_e\mathbb{C}_e^{-1}\frac{\partial \widehat{W}_{sn}}{\partial J_e}\,,\end{aligned}$$

where $\underline{\underline{\mathbb{I}}}$ is the fourth-order identity tensor. For the particular choice of material constitutive relation (12), we have

$$\begin{aligned}\frac{\partial \widehat{W}_{sn}}{\partial \overline{\mathbb{C}}_e} &= \frac{\partial \widehat{W}_{sn}}{\partial \mathrm{I}_{\overline{\mathbb{C}}_e}}\mathbb{I} + \frac{\partial \widehat{W}_{sn}}{\partial \mathrm{II}_{\overline{\mathbb{C}}_e}}\left(\mathrm{I}_{\overline{\mathbb{C}}_e}\mathbb{I} - \overline{\mathbb{C}}_e\right) = \frac{1}{2}\mu_{1t}\mathbb{I} + \frac{1}{2}\mu_{2t}\left(\mathrm{I}_{\overline{\mathbb{C}}_e}\mathbb{I} - \overline{\mathbb{C}}_e\right)\,, \\ \frac{\partial \widehat{W}_{sn}}{\partial J_e} &= \kappa_t\left(1 - \phi_{sn}\right)\left(1 - \frac{1 - \phi_{sn}}{J_e - \phi_{sn}}\right)\,.\end{aligned}$$

Thus, the constitutive expression of the Cauchy stress tensor $\mathbb{T}_s$ becomes

$$\begin{aligned}\mathbb{T}_s = 2J_e^{-1}\mathbb{F}_e&\left[J_e^{-2/3}\left(\underline{\underline{\mathbb{I}}} - \frac{1}{3}\overline{\mathbb{C}}_e^{-1} \otimes \overline{\mathbb{C}}_e\right) : \left(\gamma_1\mathbb{I} + \gamma_2\overline{\mathbb{C}}_e\right)\right]\mathbb{F}_e^T + \\ &+\kappa_t\left(1 - \phi_{sn}\right)\left(1 - \frac{1 - \phi_{sn}}{J_e - \phi_{sn}}\right)\mathbb{I},\end{aligned} \tag{14}$$

where we have defined the quantities $\gamma_1 := \frac{1}{2}\mu_{1t} + \frac{1}{2}\mathrm{I}_{\overline{\mathbb{C}}_e}\mu_{2t}$ and $\gamma_2 := -\frac{1}{2}\mu_{2t}$. The constitutive expression of the Cauchy stress tensor should be accompanied by equations determining $\mathbb{F}_s$ and $\mathbb{F}_g$. The tensor $\mathbb{F}_s$ is entirely determined by the motion of the cell phase and for this reason it is not an additional unknown for the model. In fact, it satisfies

$$\dot{\mathbb{F}}_s\mathbb{F}_s^{-1} = \nabla \mathbf{v}_s.$$

So it remains to determine $\mathbb{F}_g$ by solving appropriate evolution equations. The evolution of $\mathbb{F}_g$ can be derived from Eq. (3) with $\alpha = \mathrm{s}$. It is possible to show that the following relation has to hold [43, 67]:

$$\mathrm{tr}\left(\mathbb{L}_{\mathrm{g}}\right) = \frac{\Gamma_{\mathrm{s}}}{\phi_{\mathrm{s}}}\,, \tag{15}$$

where $\mathbb{L}_{\mathrm{g}} := \dot{\mathbb{F}}_{\mathrm{g}}\mathbb{F}_{\mathrm{g}}^{-1}$ is the strain rate tensor (or velocity gradient) associated with $\mathbb{F}_{\mathrm{g}}$. In this work, we consider an isotropic growth tensor of the form

$$\mathbb{F}_{\mathrm{g}} = g\mathbb{I}\,, \tag{16}$$

with g being a scalar field to be determined. Therefore, Eq. (16) leads to $\mathrm{tr}\left(\mathbb{L}_{\mathrm{g}}\right) = 3\dot{g}g^{-1}$, which consequently yields [8]

$$\frac{\dot{g}}{g} = \frac{1}{3}\frac{\Gamma_{\mathrm{s}}}{\phi_{\mathrm{s}}}, \qquad \text{in } \Omega_{\mathrm{t}}(t)\,. \tag{17}$$

Equation (17) is an ordinary differential equation that, equipped with an initial condition, determines g uniquely. Hence, it completely determines the evolution of the growth tensor $\mathbb{F}_{\mathrm{g}}$. Of course, this is true provided that Γ_{s} is given constitutively. In particular, we assume the following constitutive equation for the latter:

$$\Gamma_{\mathrm{s}} = \nu\phi_{\mathrm{s}}\left(\phi_{\mathrm{max}} - \phi_{\mathrm{s}}\right)\left(c_n - c_0\right)_+\,, \tag{18}$$

where $(\cdot)_+$ denotes the positive part and ν is a positive coefficient. In particular, we have that the proliferation rate depends linearly on the available concentration of nutrients c_n, provided that it is greater than the hypoxia threshold c_0, below which tumour cells stop duplicating. On the other hand, as long as $c_n > c_0$, the cell phase is allowed to grow and the proliferation rate is proportional to the difference between the actual nutrients concentration and the hypoxia threshold. Moreover, the growth rate depends on the fraction of cells that is already present, since cell population grows by duplication of existing cells, and on the availability of space that can be filled by the cellular phase. Therefore proliferation reduces as ϕ_{s} approaches the maximum admissible cell volume fraction $\phi_{\mathrm{max}} \in (0, 1]$, for which contact inhibition of growth occurs. For more details about the definition of Γ_{s}, see [24, 61] and references therein.

Effective Stress Tensor in $\Omega_{\mathrm{h}}(t)$

In the host healthy tissue, as stated before, the net source term Γ_{s} is null, since the death of healthy cells is compensated by proliferation. This implies that, in principle, the multiplicative decomposition (2) is not needed in $\Omega_{\mathrm{h}}(t)$. However, for simplicity, in order to have all quantities defined on both the tumour and the healthy tissue, it is possible to apply a fictitious multiplicative decomposition of the deformation gradient $\mathbb{F}_{\mathrm{s}}$, by taking $\mathbb{F}_{\mathrm{g}} = \mathbb{I}$ in $\Omega_{\mathrm{h}}(t)$. Then, even though the constitutive mechanical model for the healthy tissue might be taken as totally different from the one describing the tumour, we assume that the solid phase is described by a Mooney-Rivlin strain energy density function $\widehat{\mathcal{W}}_{\rm sn}$ as in (12) also in the healthy domain. Nevertheless, even assuming the same functional form, the material parameters could be different, i.e., we could use $\mu_{1\rm h}$, $\mu_{2\rm h}$ and $\kappa_{\rm h}$ different from $\mu_{1\rm t}$, $\mu_{2\rm t}$ and $\kappa_{\rm t}$, respectively.

2.1.3 Nutrients

The rate of tumour growth $\Gamma_{\rm s}$ is influenced by many different factors, but of course the amount of nutrients plays a fundamental role, because it strongly affects the cells capability to duplicate. Consequently, it is necessary to introduce in the model an equation describing nutrients evolution in the domain. We assume that they are transported by the fluid phase and they can diffuse into it. On the other side, they are taken by the growing tumour and uniformly supplied by blood vessels. We introduce the hypothesis that the nutrients absorbed by the healthy tissue are immediately replaced by the vasculature, whereas the nutrients uptake by the tumour tissue is not negligible. Following these assumptions, we can write the mass balance equation governing the concentration of available nutrients c_n in $\Omega(t)$, normalising it with respect to the physiological concentration taken at the border of the brain:

$$\frac{\partial}{\partial t}(\phi_\ell c_n) + \nabla \cdot (\phi_\ell c_n \mathbf{v}_\ell) = \nabla \cdot (\phi_\ell \mathbb{D} \nabla c_n) + \Gamma_\ell c_n + G_n\,, \tag{19}$$

where $\mathbb{D}$ is the Eulerian diffusion tensor (discussed later in Sect. 2.1.4), the term $\Gamma_\ell c_n$ is related to the variation of the nutrients amount due to absorption/production of the liquid in which the chemical is dissolved, and G_n is the chemical source term occurring without net variation of the liquid amount. In particular, we will consider the form

$$G_n = \begin{cases} -\zeta \phi_{\rm s} \phi_\ell c_n + S_n (1 - c_n)\, \phi_\ell & \text{in } \Omega_{\rm t}(t) \\ 0 & \text{in } \Omega_{\rm h}(t) \end{cases}. \tag{20}$$

The expression of G_n in the tumour domain describes the fact that nutrients are consumed by the tumour with a constant rate ζ. Furthermore, nutrients are supplied at a rate S_n as far as their concentration is below the physiological value, whereas above the physiological value they are absorbed. The consumption and the delivery of nutrients is also weighted with a factor ϕ_ℓ to describe the fact that if there is a higher availability of fluid phase, then a greater uptake or supply of nutrients can be provided. On the other hand, in the healthy region we assume that production and absorption of nutrients are reciprocally balanced. Using standard calculus techniques and recalling the mass balance equation of the fluid phase, with $\Gamma_\ell = 0$

in $\Omega_{\text{h}}(t)$, we can rewrite Eq. (19) in the tumour and in the healthy domain as

$$\frac{\partial c_n}{\partial t} + \mathbf{v}_\ell \cdot \nabla c_n = \frac{1}{\phi_\ell} \nabla \cdot (\phi_\ell \mathbb{D} \nabla c_n) + [-\zeta \phi_{\text{s}} c_n + S_n (1 - c_n)] \qquad \text{in } \Omega_{\text{t}}(t)\,, \tag{21a}$$

$$\frac{\partial c_n}{\partial t} + \mathbf{v}_\ell \cdot \nabla c_n = \frac{1}{\phi_\ell} \nabla \cdot (\phi_\ell \mathbb{D} \nabla c_n) \qquad \text{in } \Omega_{\text{h}}(t)\,. \tag{21b}$$

2.1.4 Diffusion Tensor $\mathbb{D}$ and Preferential Directions Tensor $\mathbb{A}$

We still need to provide a definition for the diffusion tensor $\mathbb{D}$, appearing in Eq. (19), and for the tensor of preferential directions $\mathbb{A}$, which affects the permeability tensor $\mathbb{K}$ in Eq. (11). Since the displacement induced by the tumour modifies the direction of brain fibres in the surrounding environment and therefore alters the directions along which diffusion and fluid motion happen, we take advantage of the mechanical description included in our model to progressively modify these tensors as time evolves. The diffusion tensor at the initial time instant $\mathbb{D}_0$ can be constructed from DTI imaging data, through a computational processing summarised in Sect. 3.4. Indeed, DTI scan quantifies the diffusion of water inside the brain and for this reason it seems appropriate to employ such data to describe nutrients diffusion and to determine the orientation of nerve tracts and other structures inside the brain. Thus, we can write

$$\mathbb{D}_0 = \lambda_1^0 \mathbf{e}_1^0 \otimes \mathbf{e}_1^0 + \lambda_2^0 \mathbf{e}_2^0 \otimes \mathbf{e}_2^0 + \lambda_3^0 \mathbf{e}_3^0 \otimes \mathbf{e}_3^0\,,$$

where we call $\lambda_1^0 > \lambda_2^0 > \lambda_3^0$ the decreasing order eigenvalues and $\mathbf{e}_1^0, \mathbf{e}_2^0, \mathbf{e}_3^0$ the corresponding orthogonal eigenvectors of the initial tensor $\mathbb{D}_0$, taken from DTI images. The construction of $\mathbb{A}_0$, i.e. the initial value of tensor $\mathbb{A}$, is also performed using these data: the aim is to evaluate the preferential directions identified by the presence of white matter tracts. In particular, it is assumed that $\mathbb{A}_0$ has the same eigenvectors as the diffusion tensor, while the eigenvalues are properly rescaled to enhance anisotropy along the preferential directions, as described in [1, 2, 53]. The tensor $\mathbb{A}_0$ is then defined as

$$\mathbb{A}_0 = \frac{1}{\text{A}_{av}} \widehat{\mathbb{A}}_0, \quad \widehat{\mathbb{A}}_0 = a_1^0(r) \lambda_1^0 \mathbf{e}_1^0 \otimes \mathbf{e}_1^0 + a_2^0(r) \lambda_2^0 \mathbf{e}_2^0 \otimes \mathbf{e}_2^0 + \lambda_3^0 \mathbf{e}_3^0 \otimes \mathbf{e}_3^0,$$

$$\text{A}_{av} = \frac{1}{3} \operatorname{tr}(\widehat{\mathbb{A}}_0),$$

where r is the tuning parameter of anisotropy and $a_i^0(r)$ are functions of r given by

$$a_1^0(r) = r c_l^0 + r c_p^0 + c_s^0\,, \quad a_2^0(r) = c_l^0 + r c_p^0 + c_s^0\,, \tag{22}$$

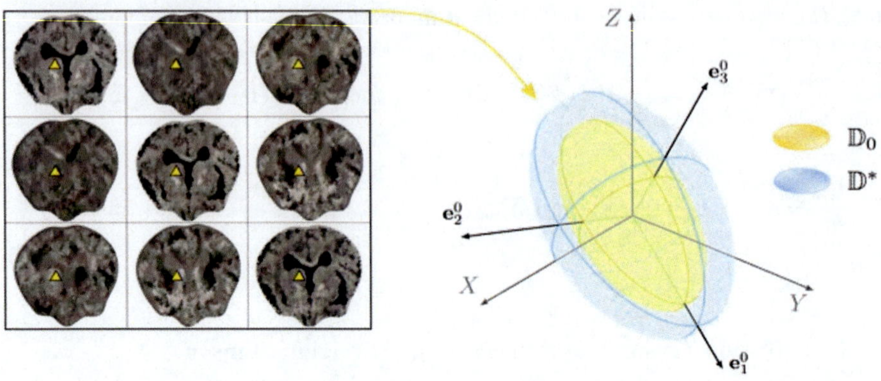

Fig. 2 Computational reconstruction and modification of the components of the diffusion tensor taken from DTI data. On the left, a sample of the six components of the initial tensor $\mathbb{D}_0$ built using medical imaging data is shown. In particular, a diffusion value is assigned to each brain mesh cell, with higher values appearing in black. For a representative cell marked by the yellow triangle, on the right we sketch the diffusion ellipsoid of the initial and modified tensors in the reference configuration. Specifically, the smaller (yellow) ellipsoid represents the preferential directions and values of diffusion at the initial time instant, i.e. the eigenvectors and eigenvalues of $\mathbb{D}_0$, respectively. The initial eigenvectors are then modified according to the deformation of the tissue to obtain the current tensor $\mathbb{D}$, as described in Sect. 2.1.4. In the figure, we sketch (in blue) the pullback $\mathbb{D}^* := J_s \mathbb{F}_s^{-1} \mathbb{D} \mathbb{F}_s^{-T}$ of the modified diffusion tensor. We observe that both $\mathbb{D}_0$ and $\mathbb{D}^*$ are defined in the reference configuration and have the same eigenvectors, but different eigenvalues

being c_l^0, c_p^0, c_s^0 the linear, planar and spherical anisotropy coefficients, respectively, defined as [53]:

$$c_l^0 = \frac{\lambda_1^0 - \lambda_2^0}{\lambda_1^0 + \lambda_2^0 + \lambda_3^0}, \qquad c_p^0 = \frac{2(\lambda_2^0 - \lambda_3^0)}{\lambda_1^0 + \lambda_2^0 + \lambda_3^0}, \qquad c_s^0 = \frac{3\lambda_3^0}{\lambda_1^0 + \lambda_2^0 + \lambda_3^0}.$$

Hence, the case $r = 1$ corresponds to a situation where the anisotropy is not increased, and therefore the tensor $\mathbb{A}_0$ is simply given by a normalisation of the diffusion tensor $\mathbb{D}_0$. Instead, if $r > 1$, then anisotropy is enhanced according to the values of the anisotropic coefficients, as given by Eq. (22).

The modification of tensors $\mathbb{D}_0$ and $\mathbb{A}_0$ is then done considering only the reorientation of the preferential directions and not their extension or compression. For this reason, the deformation gradient $\mathbb{F}_s$ is used to deform the eigenvectors and the deformed eigenvectors are normalized to account only for changes in the direction (see Fig. 2). Hence, for the modified diffusion tensor we can use the expression

$$\mathbb{D} = \lambda_1^0 \frac{\mathbb{F}_s \mathbf{e}_1^0 \otimes \mathbb{F}_s \mathbf{e}_1^0}{|\mathbb{F}_s \mathbf{e}_1^0|^2} + \lambda_2^0 \frac{\mathbb{F}_s \mathbf{e}_2^0 \otimes \mathbb{F}_s \mathbf{e}_2^0}{|\mathbb{F}_s \mathbf{e}_2^0|^2} + \lambda_3^0 \frac{\mathbb{F}_s \mathbf{e}_3^0 \otimes \mathbb{F}_s \mathbf{e}_3^0}{|\mathbb{F}_s \mathbf{e}_3^0|^2},$$

where we observe that

$$\left|\mathbb{F}_s \mathbf{e}_i^0\right|^2 = \mathbb{F}_s \mathbf{e}_i^0 \cdot \mathbb{F}_s \mathbf{e}_i^0 = \mathbf{e}_i^0 \cdot \mathbb{C}_s \mathbf{e}_i^0 , \quad i = 1, 2, 3 , \quad \mathbb{C}_s = \mathbb{F}_s^T \mathbb{F}_s .$$

The preferential directions tensor $\mathbb{A}$ can be derived in a similar way to $\mathbb{D}$, i.e. deforming the eigenvectors through the deformation gradient and normalizing them:

$$\mathbb{A} = \frac{1}{A_{av}} \widehat{\mathbb{A}} , \qquad A_{av} = \frac{1}{3} \operatorname{tr}(\widehat{\mathbb{A}}) ,$$

$$\widehat{\mathbb{A}} = a_1^0(r)\lambda_1^0 \frac{\mathbb{F}_s \mathbf{e}_1^0 \otimes \mathbb{F}_s \mathbf{e}_1^0}{\mathbf{e}_1^0 \cdot \mathbb{C}_s \mathbf{e}_1^0} + a_2^0(r)\lambda_2^0 \frac{\mathbb{F}_s \mathbf{e}_2^0 \otimes \mathbb{F}_s \mathbf{e}_2^0}{\mathbf{e}_2^0 \cdot \mathbb{C}_s \mathbf{e}_2^0} + \lambda_3^0 \frac{\mathbb{F}_s \mathbf{e}_3^0 \otimes \mathbb{F}_s \mathbf{e}_3^0}{\mathbf{e}_3^0 \cdot \mathbb{C}_s \mathbf{e}_3^0} .$$

2.1.5 Interface Conditions at the Boundary Between the Tumour and the Healthy Tissue

Since the material interface $\partial\Omega_t(t)$ between the tumour and the healthy tissue moves with the tumour cells with velocity $\mathbf{v}_s|_{\partial\Omega_t(t)}$, we have to satisfy the following interface conditions on the two sides of the boundary, in order to guarantee the continuity of the normal displacement, the normal stress, chemical concentration and fluxes at the interface:

$$[\![\mathbf{v}_s \cdot \mathbf{n}]\!]|_{\partial\Omega_t(t)} = 0 , \tag{23a}$$

$$[\![\phi_\ell \left(\mathbf{v}_\ell - \mathbf{v}_s\right) \cdot \mathbf{n} d\Sigma]\!]|_{\partial\Omega_t(t)} = 0 , \tag{23b}$$

$$[\![p]\!]|_{\partial\Omega_t(t)} = 0 , \tag{23c}$$

$$[\![c_n]\!]|_{\partial\Omega_t(t)} = 0 , \tag{23d}$$

$$[\![\mathbb{T}\mathbf{n} d\Sigma]\!]|_{\partial\Omega_t(t)} = \mathbf{0} , \tag{23e}$$

$$[\![(\phi_\ell c_n \left(\mathbf{v}_\ell - \mathbf{v}_s\right) - \phi_\ell \mathbb{D} \nabla c_n) \cdot \mathbf{n} d\Sigma]\!]|_{\partial\Omega_t(t)} = 0 , \tag{23f}$$

where $[\![\cdot]\!]|_{\partial\Omega_t(t)}$ denotes the jump across the interface, $\mathbf{n}$ is the unit normal vector to $\partial\Omega_t(t)$ pointing outwards and $d\Sigma$ is the area element at the interface. In particular we underline that, by combining the continuity across the interface of the total stress $\mathbb{T} = -p\mathbb{I} + \mathbb{T}_s$, prescribed by Eq. (23e), and the continuity of the pressure p, prescribed by Eq. (23c), it follows that also the effective stress $\mathbb{T}_s$ is continuous. Furthermore, due to the presence in biological tissues of cell-cell and cell-extracellular matrix adhesion molecules, it is physically reasonable to assume not only the continuity of the velocity $\mathbf{v}_s$ along the normal direction, but also that there are not breakages and rotations between the tumour and the healthy tissue. This hypothesis requires that $\mathbf{v}_s$ is continuous also along the unit tangential component $\boldsymbol{\tau}$:

$$[\![\mathbf{v}_s \cdot \boldsymbol{\tau}]\!]|_{\partial\Omega_t(t)} = 0 .$$

This assumption leads us to say that the displacement field $\mathbf{u}_s$ is continuous along $\partial\Omega_t(t)$ and that the areas $d\Sigma$ deform in the same way at the interface, but it does not imply that also $\mathbb{F}_s$ and J_s are continuous. We observe, as well, that the cell volumetric fraction ϕ_s is in general discontinuous at the tumour interface since it is related to the inverse of the Jacobian J_s. In the end, removing $d\Sigma$ in (23b), (23e) and (23f) for the assumption made above, the interface conditions that we impose are

$$[\![\mathbf{u}_s]\!]|_{\partial\Omega_t(t)} = \mathbf{0} , \tag{24a}$$

$$[\![\phi_\ell (\mathbf{v}_\ell - \mathbf{v}_s) \cdot \mathbf{n}]\!]|_{\partial\Omega_t(t)} = 0 , \tag{24b}$$

$$[\![p]\!]|_{\partial\Omega_t(t)} = 0 , \tag{24c}$$

$$[\![c_n]\!]|_{\partial\Omega_t(t)} = 0 , \tag{24d}$$

$$[\![\mathbb{T}_s \mathbf{n}]\!]|_{\partial\Omega_t(t)} = \mathbf{0} , \tag{24e}$$

$$[\![(\phi_\ell c_n (\mathbf{v}_\ell - \mathbf{v}_s) - \phi_\ell \mathbb{D} \nabla c_n) \cdot \mathbf{n}]\!]|_{\partial\Omega_t(t)} = 0 . \tag{24f}$$

2.2 *Lagrangian Formulation of the Model*

To approach the numerical implementation of the model, our aim is to rewrite the equations derived in Sect. 2.1 using a Lagrangian description of motion. In this way, all the quantities of interest are considered in terms of material coordinates. Henceforth, we will then use a superscript * to denote any material element. Furthermore, we will use the same symbols to denote the variables in the spatial and material description, omitting the explicit spatial dependence. We recall the following equalities, which will be useful in successive computations:

$$d\boldsymbol{\Sigma} = J_s \mathbb{F}_s^{-\mathrm{T}} d\boldsymbol{\Sigma}^*, \qquad dV = J_s dV^* , \tag{25}$$

where $d\boldsymbol{\Sigma} = \mathbf{n} d\Sigma$ and dV represent the infinitesimal element of area and volume in spatial coordinates, respectively, $d\boldsymbol{\Sigma}^* = \mathbf{N} d\Sigma^*$ and dV^* denote the infinitesimal element of area and volume in material coordinates, while $\mathbf{N}$ denotes the unit normal vector to $\partial\Omega_t^*$ pointing outwards [44].

Firstly, we integrate Eq. (3) for $\alpha = \mathrm{s}$ over the tumour domain $\Omega_t(t)$ to obtain

$$\int_{\Omega_t(t)} \left[\frac{\partial \phi_s}{\partial t} + \nabla \cdot (\phi_s \mathbf{v}_s) \right] dV = \int_{\Omega_t(t)} \Gamma_s \, dV .$$

Using Reynolds' transport theorem, see for example [44], and moving the volume elements to the reference configuration by means of (25), we obtain

$$\frac{d}{dt}\int_{\Omega_t^*} \phi_s J_s \, dV^* = \int_{\Omega_t^*} \Gamma_s J_s \, dV^* \,,$$

which locally becomes

$$\overline{J_s\phi_s}^{\,\cdot} = J_s \Gamma_s \,. \tag{26}$$

For what concerns Eq. (3) for $\alpha = \ell$, integrating over the tumour domain gives

$$\int_{\Omega_{t(t)}} \left[\frac{\partial \phi_\ell}{\partial t} + \nabla \cdot (\phi_\ell \mathbf{v}_\ell)\right] dV = -\int_{\Omega_{t(t)}} \Gamma_s \, dV \,.$$

Since the interface does not move with the fluid, we have to make use of the generalized Reynolds' transport theorem [44] which, together with the divergence theorem and Eq. (25), yields

$$\frac{d}{dt}\int_{\Omega_t^*} \phi_\ell J_s \, dV^* - \int_{\Omega_t^*} \mathrm{Div}\left[J_s \phi_\ell \mathbb{F}_s^{-1} (\mathbf{v}_s - \mathbf{v}_\ell)\right] dV^* = -\int_{\Omega_t^*} \Gamma_s J_s \, dV^* \,,$$

which localized gives

$$\overline{J_s\phi_\ell}^{\,\cdot} + \mathrm{Div}\left[J_s \phi_\ell \mathbb{F}_s^{-1} (\mathbf{v}_\ell - \mathbf{v}_s)\right] = -\Gamma_s J_s \,. \tag{27}$$

Then, if we recall that ϕ_{sn} is the volumetric fraction in the natural state and it is a constant quantity, using (15) we can rewrite (26) as

$$J_s \phi_s = J_g \phi_{sn} \quad \Rightarrow \quad J_s \phi_s = g^3 \phi_{sn} \,.$$

As regards the momentum balance of the solid phase, if we integrate (9) over the tumour domain and we remember that $\mathbb{T} = -p\mathbb{I} + \mathbb{T}_s$ is the Cauchy stress tensor of the mixture, we obtain

$$\int_{\Omega_{t(t)}} \nabla \cdot \mathbb{T} \, dV = \mathbf{0} \,.$$

Introducing the first Piola-Kirchhoff stress tensor $\mathbb{P} := J_s \mathbb{T}\mathbb{F}_s^{-T}$, the latter becomes

$$\int_{\Omega_t^*} \mathrm{Div}\, \mathbb{P} \, dV^* = \mathbf{0} \quad \Longrightarrow \quad \mathrm{Div}\, \mathbb{P} = \mathbf{0} \,.$$

Recalling that $\mathbb{P} = J_s \mathbb{T}\mathbb{F}_s^{-T} = -J_s p \mathbb{F}_s^{-T} + \mathbb{P}_s$, we have

$$\mathrm{Div}\left[-J_s p \mathbb{F}_s^{-T} + \mathbb{P}_s\right] = \mathbf{0}\,.$$

In order to rewrite (10) using the Lagrangian formulation, we integrate over a surface

$$\int_S \phi_\ell \left(\mathbf{v}_\ell - \mathbf{v}_s\right) \cdot d\mathbf{\Sigma} = -\int_S \frac{\mathbb{K}(\phi_\ell)}{\mu} \nabla p \cdot d\mathbf{\Sigma}\,.$$

Moving the integrals to the reference configuration, we get

$$\int_{S^*} \left[\frac{\mathbb{K}}{\mu} \mathbb{F}_s^{-T}\, \mathrm{Grad}\, p + \phi_\ell \left(\mathbf{v}_\ell - \mathbf{v}_s\right)\right] \cdot J_s \mathbb{F}_s^{-T} d\mathbf{\Sigma}^* = 0\,.$$

Let us assume that all the involved quantities are regular, we have then the local form

$$\mathbf{v}_\ell - \mathbf{v_s} = -\mathbb{F}_s \frac{\mathbb{K}^*}{J_s \mu \phi_\ell}\, \mathrm{Grad}\, p\,, \tag{28}$$

defining the tensor pullback $\mathbb{K}^* := J_s \mathbb{F}_s^{-1} \mathbb{K} \mathbb{F}_s^{-T}$.

In the light of Eq. (28), it is then convenient to further reformulate the mass balances by summing up Eqs. (26) and (27). Using the saturation condition and the closed mixture assumption, the mass balance for the mixture therefore reads

$$\dot{J}_s = \mathrm{Div}\left[\frac{\mathbb{K}^*}{\mu}\, \mathrm{Grad}\, p\right].$$

Referring to the nutrients balance Eq. (21a), integrating it over the tumour domain and recalling the closed mixture assumption, we obtain

$$\int_{\Omega_t(t)} \left[\frac{\partial\left(\phi_\ell c_n\right)}{\partial t} + \nabla \cdot \left(\phi_\ell c_n \mathbf{v}_\ell\right)\right] dV =$$
$$\int_{\Omega_t(t)} \nabla \cdot \left(\phi_\ell \mathbb{D} \nabla c_n\right) dV - \int_{\Omega_t(t)} \left(\Gamma_s c_n - G_n\right) dV\,,$$

that localized by means of the generalized Reynolds' transport theorem and the Gauss theorem leads to

$$\overline{J_s \phi_\ell c_n}^{\,\cdot} - \mathrm{Div}\left[J_s \phi_\ell c_n \mathbb{F}_s^{-1} \left(\mathbf{v}_s - \mathbf{v}_\ell\right)\right] - \mathrm{Div}\left[J_s \phi_\ell \mathbb{F}_s^{-1} \mathbb{D} \mathbb{F}_s^{-T}\, \mathrm{Grad}\, c_n\right] =$$
$$-\Gamma_s c_n J_s + G_n J_s\,.$$

If we define $\mathbb{D}^* := J_s \mathbb{F}_s^{-1} \mathbb{D} \mathbb{F}_s^{-T}$ and we recall the mass balance of the fluid phase (27), substituting Darcy's law in the reference configuration (28) we can rewrite it as

$$J_s \phi_\ell \dot{c}_n - \frac{\mathbb{K}^*}{\mu} \operatorname{Grad} p \cdot \operatorname{Grad} c_n - \operatorname{Div}\left[\phi_\ell \mathbb{D}^* \operatorname{Grad} c_n\right] = J_s G_n \,. \tag{29}$$

In conclusion, the set of equations in Lagrangian form in the tumour reference domain Ω_t^* is

$$\dot{J}_s = \operatorname{Div}\left[\frac{\mathbb{K}^*}{\mu} \operatorname{Grad} p\right], \tag{30a}$$

$$\phi_s = J_s^{-1} g^3 \phi_{sn} \,, \tag{30b}$$

$$\phi_s + \phi_\ell = 1 \,, \tag{30c}$$

$$\operatorname{Div}\left[-J_s p \mathbb{F}_s^{-T} + \mathbb{P}_s\right] = \mathbf{0} \,, \tag{30d}$$

$$\frac{\dot{g}}{g} = \frac{1}{3} \frac{\Gamma_s}{\phi_s} \,, \tag{30e}$$

$$J_s \phi_\ell \dot{c}_n - \frac{\mathbb{K}^*}{\mu} \operatorname{Grad} p \cdot \operatorname{Grad} c_n - \operatorname{Div}\left[\phi_\ell \mathbb{D}^* \operatorname{Grad} c_n\right] = J_s G_n \,. \tag{30f}$$

A similar reasoning and analogous computations can be used to derive the Lagrangian equations in the healthy tissue reference domain, so that we end up with the following set of equations in Ω_h^*

$$\dot{J}_s = \operatorname{Div}\left[\frac{\mathbb{K}^*}{\mu} \operatorname{Grad} p\right], \tag{31a}$$

$$\phi_s = J_s^{-1} g^3 \phi_{sn} \,, \tag{31b}$$

$$\phi_s + \phi_\ell = 1 \,, \tag{31c}$$

$$\operatorname{Div}\left[-J_s p \mathbb{F}_s^{-T} + \mathbb{P}_s\right] = \mathbf{0} \,, \tag{31d}$$

$$\dot{g} = 0 \,, \tag{31e}$$

$$J_s \phi_\ell \dot{c}_n - \frac{\mathbb{K}^*}{\mu} \operatorname{Grad} p \cdot \operatorname{Grad} c_n - \operatorname{Div}\left[\phi_\ell \mathbb{D}^* \operatorname{Grad} c_n\right] = 0 \,. \tag{31f}$$

The systems (30) and (31) allow to determine all the unknown fields, namely, the displacement field $\mathbf{u}_s(\mathbf{X}, t)$ and the scalar fields $p(\mathbf{X}, t)$, $\phi_s(\mathbf{X}, t)$, $\phi_\ell(\mathbf{X}, t)$, $g(\mathbf{X}, t)$ and $c_n(\mathbf{X}, t)$, $\forall \mathbf{X} \in \Omega^* = \Omega_t^* \cup \Omega_h^*$ and $\forall t \in (0, T)$, if we provide proper interface, initial and boundary conditions.

Interface Conditions The interface conditions derived in Sect. 2.1.5, i.e. Eqs. (24a)–(24f), need to be reformulated in Lagrangian coordinates, by making use of the relations (25). Then, the set of interface conditions we obtain are the following:

$$[\![\mathbf{u}_\mathrm{s}]\!]|_{\partial\Omega_\mathrm{t}^*} = \mathbf{0}\,, \tag{32a}$$

$$[\![\frac{\mathbb{K}^*}{\mu}\,\mathrm{Grad}\,p \cdot \mathbf{N}]\!]|_{\partial\Omega_\mathrm{t}^*} = 0\,, \tag{32b}$$

$$[\![\mathbb{P}_\mathrm{s}\mathbf{N}]\!]|_{\partial\Omega_\mathrm{t}^*} = \mathbf{0}\,, \tag{32c}$$

$$[\![p]\!]|_{\partial\Omega_\mathrm{t}^*} = 0\,, \tag{32d}$$

$$[\![c_n]\!]|_{\partial\Omega_\mathrm{t}^*} = 0\,, \tag{32e}$$

$$[\![\phi_\ell \mathbb{D}^*\,\mathrm{Grad}\,c_n \cdot \mathbf{N}]\!]\big|_{\partial\Omega_\mathrm{t}^*} = 0\,. \tag{32f}$$

Boundary Conditions Before imposing the boundary conditions, it is important to remark that $\partial\Omega_\mathrm{h}^* = \partial\Omega_\mathrm{t}^* \cup \partial\Omega_\mathrm{out}^*$ is the boundary of the healthy domain that is composed by the interface with the tumour $\partial\Omega_\mathrm{t}^*$ and by the external boundary corresponding to the cranial skull $\partial\Omega_\mathrm{out}^*$. In our simulations for tumour growth in the brain, we consider the following boundary conditions on $\partial\Omega_\mathrm{out}^*$:

$$\mathbf{u}_\mathrm{s} = \mathbf{0} \qquad \text{on } \partial\Omega_\mathrm{out}^*\,, \quad \forall t \in (0, T)\,, \tag{33a}$$

$$p = 0 \qquad \text{on } \partial\Omega_\mathrm{out}^*\,, \quad \forall t \in (0, T)\,, \tag{33b}$$

$$c_n = 1 \qquad \text{on } \partial\Omega_\mathrm{out}^*\,, \quad \forall t \in (0, T)\,. \tag{33c}$$

We impose a null Dirichlet boundary condition for the displacement $\mathbf{u}_\mathrm{s}$ and for the pressure p. For the nutrients concentration we suppose that the brain boundary is sufficiently far from the tumour. Therefore, we can assume that, on the boundary, the oxygen concentration is maintained constant at the physiological value of 1 by the vasculature. When the tumour grows close to the boundary, the boundary conditions proposed in [61] should be applied.

Initial Conditions At the beginning of the tumour growth process we assume that the displacement and the pressure are equal to zero. Furthermore, we take the scalar field g, related to the growth component of the deformation gradient, as equal to 1 everywhere in the domain at $t = 0$. We also assume that the volumetric fraction of the cell phase is initially equal to the constant volumetric fraction in the natural state ϕ_sn. Finally, in order to obtain the initial nutrients concentration $c_n^0(\mathbf{X})$, we solve the steady version of the nutrients governing equation, neglecting advection:

$$-\,\mathrm{Div}\,[\phi_\ell \mathbb{D}_0\,\mathrm{Grad}\,c_n] = J_\mathrm{s} G_n\,.$$

In conclusion, we have the following set of initial conditions:

$$\begin{aligned}
\mathbf{u}_s(\mathbf{X}, 0) &= \mathbf{0} && \forall \mathbf{X} \in \Omega^*, \\
p(\mathbf{X}, 0) &= 0 && \forall \mathbf{X} \in \Omega^*, \\
g(\mathbf{X}, 0) &= 1 && \forall \mathbf{X} \in \Omega^*, \\
\phi_s(\mathbf{X}, 0) &= \phi_{sn} && \forall \mathbf{X} \in \Omega^*, \\
c_n(\mathbf{X}, 0) &= c_n^0(\mathbf{X}) && \forall \mathbf{X} \in \Omega^*.
\end{aligned}$$

3 Numerical Implementation

In this section, we discuss how the Lagrangian model for brain tumour growth, equipped with proper boundary and initial conditions, is solved through numerical simulations. First of all, we derive the weak formulation of the Lagrangian model. Then, we discretise in time and space the weak formulation and we assess the values of the parameters that appear in the system. Finally, we describe how we manage to generate the patient-specific mesh that has been used for the computation.

3.1 Weak Formulation of the Lagrangian Model

We will derive now a weak formulation of our Lagrangian model. We first write the weak form in each domain Ω_t^* and Ω_h^* separately and then we extend the weak form to the whole domain $\Omega^* = \Omega_t^* \cup \Omega_h^*$. At this point, we define the test functions space that meets the Dirichlet conditions we impose on the external boundary for p (33b) and c_n (33c), recalling that p and c_n are continuous functions over Ω^*:

$$H^1_{0,\partial\Omega^*_{\text{out}}}(\Omega^*) := \left\{ q \in H^1(\Omega^*) \colon q = 0 \text{ on } \partial\Omega^*_{\text{out}} \right\}.$$

Furthermore, we establish the vector test functions space that meets the Dirichlet conditions we impose on the external boundary for the continuous vector function $\mathbf{u}_s$ (33a):

$$\boldsymbol{H}^1_{0,\partial\Omega^*_{\text{out}}}(\Omega^*) := \left\{ \boldsymbol{q} \in \boldsymbol{H}^1(\Omega^*) \colon \boldsymbol{q} = \mathbf{0} \text{ on } \partial\Omega^*_{\text{out}} \right\}.$$

Then, starting from Eq. (30a), we multiply each side by a test function $q_t \in H^1_{0,\partial\Omega^*_{\text{out}}}(\Omega^*)$ and we integrate the whole equation over the Lagrangian tumour domain:

$$\int_{\Omega_t^*} \dot{J}_s q_t \, dV^* = \int_{\Omega_t^*} \mathrm{Div}\left[\frac{\mathbb{K}^*}{\mu} \,\mathrm{Grad}\, p\right] q_t \, dV^* .$$

Integrating by parts the second order derivatives, we obtain

$$\int_{\Omega_t^*} \dot{J}_s q_t \, dV^* = -\int_{\Omega_t^*} \mathrm{Grad}\, q_t \cdot \frac{\mathbb{K}^*}{\mu} \mathrm{Grad}\, p \, dV^* + \int_{\partial\Omega_t^*} q_t \frac{\mathbb{K}^*}{\mu} \mathrm{Grad}\, p \cdot \mathbf{N} d\Sigma^* .$$

In the healthy domain we take as test function $q_h \in H^1_{0,\partial\Omega^*_{out}}(\Omega^*)$ and we find

$$\int_{\Omega_h^*} \dot{J}_s q_h \, dV^* = -\int_{\Omega_h^*} \mathrm{Grad}\, q_h \cdot \frac{\mathbb{K}^*}{\mu} \mathrm{Grad}\, p \, dV^* + \int_{\partial\Omega_t^*} q_h \frac{\mathbb{K}^*}{\mu} \mathrm{Grad}\, p \cdot \mathbf{N} d\Sigma^* ,$$

since the test function q_h is required to vanish on the boundary $\partial\Omega^*_{out}$ because it belongs to $H^1_{0,\partial\Omega^*_{out}}(\Omega^*)$. Summing up the equations in the healthy and tumour domain taking $q \in H^1_{0,\partial\Omega^*_{out}}(\Omega^*)$ we have

$$\int_{\Omega^*} \dot{J}_s q \, dV^* = -\int_{\Omega^*} \mathrm{Grad}\, q \cdot \frac{\mathbb{K}^*}{\mu} \mathrm{Grad}\, p \, dV^* - \int_{\partial\Omega_t^*} \left[\!\left[q \frac{\mathbb{K}^*}{\mu} \mathrm{Grad}\, p \right]\!\right] \cdot \mathbf{N} d\Sigma^* .$$

Since the test function q belongs to $H^1_{0,\partial\Omega^*_{out}}(\Omega^*)$ and so it is continuous inside the domain, thanks to interface condition (32b) we finally have

$$\int_{\Omega^*} \dot{J}_s q \, dV^* = -\int_{\Omega^*} \mathrm{Grad}\, q \cdot \frac{\mathbb{K}^*}{\mu} \mathrm{Grad}\, p \, dV^* , \tag{34}$$

for all test functions $q \in H^1_{0,\partial\Omega^*_{out}}(\Omega^*)$.

For what concerns the momentum balance, we multiply (30d) by a vector test function $\boldsymbol{q}_t \in \boldsymbol{H}^1_{0,\partial\Omega^*_{out}}(\Omega^*)$ and then we integrate over the tumour reference domain, obtaining

$$\int_{\Omega_t^*} \mathrm{Div}\left[-J_s p \mathbb{F}_s^{-T} + \mathbb{P}_s\right] \cdot \boldsymbol{q}_t \, dV^* = 0 .$$

Using tensor integration by parts, we get

$$-\int_{\Omega_t^*} \left(-J_s p \mathbb{F}_s^{-T} + \mathbb{P}_s\right) : \mathrm{Grad}\, \boldsymbol{q}_t \, dV^* + \int_{\partial\Omega_t^*} \left(-J_s p \mathbb{F}_s^{-T} + \mathbb{P}_s\right) \mathbf{N} \cdot \boldsymbol{q}_t \, d\Sigma^* = 0 .$$

If we do the same in the healthy domain and sum the two equations, with $\boldsymbol{q} \in \boldsymbol{H}^1_{0,\partial\Omega^*_{\text{out}}}(\Omega^*)$, the weak formulation on the whole domain is

$$-\int_{\Omega^*}\left(-J_{\text{s}}p\mathbb{F}_{\text{s}}^{-\text{T}}+\mathbb{P}_{\text{s}}\right):\text{Grad}\,\boldsymbol{q}\,dV^*-\int_{\partial\Omega^*_{\text{t}}}\left[\!\left[\left(-J_{\text{s}}p\mathbb{F}_{\text{s}}^{-\text{T}}+\mathbb{P}_{\text{s}}\right)\mathbf{N}\cdot\boldsymbol{q}\right]\!\right]d\Sigma^*=0\,.$$

Recalling that the displacement is taken continuous in all directions (32a), the areas deform in the same way at the interface. For this reason, the first relation in (25) implies $[\![J_{\text{s}}\mathbb{F}_{\text{s}}^{-\text{T}}\mathbf{N}]\!]|_{\partial\Omega^*_{\text{t}}}=\mathbf{0}$. Looking at this condition and at the interface conditions (32c) and (32d), recalling that $\boldsymbol{q} \in \boldsymbol{H}^1_{0,\partial\Omega^*_{\text{out}}}(\Omega^*)$, the jump vanishes and we are left with

$$-\int_{\Omega^*}\left(-J_{\text{s}}p\mathbb{F}_{\text{s}}^{-\text{T}}+\mathbb{P}_{\text{s}}\right):\text{Grad}\,\boldsymbol{q}\,dV^*=0\,. \tag{35}$$

We need then a weak formulation for the equation of the nutrients. In order to derive it, we multiply (30f) by a test function $q_{\text{t}} \in H^1_{0,\partial\Omega^*_{\text{out}}}(\Omega^*)$ and we integrate by parts, obtaining

$$\int_{\Omega^*_{\text{t}}}\left(J_{\text{s}}\phi_\ell\dot{c}_n-\frac{\mathbb{K}^*}{\mu}\text{Grad}\,p\cdot\text{Grad}\,c_n\right)q_{\text{t}}\,dV^*+\int_{\Omega^*_{\text{t}}}\phi_\ell\,\text{Grad}\,q_{\text{t}}\cdot\mathbb{D}^*\,\text{Grad}\,c_n\,dV^*+$$
$$-\int_{\partial\Omega^*_{\text{t}}}q_{\text{t}}\phi_\ell\mathbb{D}^*\,\text{Grad}\,c_n\cdot\mathbf{N}d\Sigma^*=\int_{\Omega^*_{\text{t}}}J_{\text{s}}G_nq_{\text{t}}\,dV^*\,.$$

We follow the same approach in the healthy domain and then we sum the two equations. Taking $q \in H^1_{0,\partial\Omega^*_{\text{out}}}(\Omega^*)$, the test function vanishes on the external boundary and, recalling the interface condition (32f), we finally have

$$\int_{\Omega^*}\left(J_{\text{s}}\phi_\ell\dot{c}_n-\frac{\mathbb{K}^*}{\mu}\text{Grad}\,p\cdot\text{Grad}\,c_n\right)q\,dV^*+\int_{\Omega^*}\phi_\ell\,\text{Grad}\,q\cdot\mathbb{D}^*\,\text{Grad}\,c_n\,dV^*=$$
$$=\int_{\Omega^*}J_{\text{s}}G_nq\,dV^*\,. \tag{36}$$

We remark that, given the pressure p and the displacement $\mathbf{u}_{\text{s}}$ obtained by solving (34) and (35), Eq. (36) represents a linear variational problem to be solved for the unknown c_n.

3.2 Discrete Formulation of the Continuous Variational Problems

We need now to introduce a time and spatial discrete formulation of the continuous variational problems (34), (35) and (36). We make use of linear tetrahedron $\mathbb{P}_1$ elements, so we introduce the following finite element spaces:

$$
\begin{aligned}
\boldsymbol{V}_h &:= \left\{ \boldsymbol{q}_h \in \left[C^0\left(\overline{\Omega^*}\right)\right]^3 : \boldsymbol{q}_h\big|_K \in [\mathbb{P}_1(K)]^3 \ \forall K \in \mathcal{T}_h,\ \boldsymbol{q}_h = \boldsymbol{0} \text{ on } \partial\Omega^*_{\text{out}} \right\} \\
&\subset \boldsymbol{H}^1_{0,\partial\Omega^*_{\text{out}}}\left(\Omega^*\right), \\
W_{h0} &:= \left\{ q_h \in C^0\left(\overline{\Omega^*}\right) : q_h|_K \in \mathbb{P}_1(K)\ \forall K \in \mathcal{T}_h,\ q_h = 0 \text{ on } \partial\Omega^*_{\text{out}} \right\} \\
&\subset H^1_{0,\partial\Omega^*_{\text{out}}}\left(\Omega^*\right), \\
W_{h1} &:= \left\{ q_h \in C^0\left(\overline{\Omega^*}\right) : q_h|_K \in \mathbb{P}_1(K)\ \forall K \in \mathcal{T}_h,\ q_h = 1 \text{ on } \partial\Omega^*_{\text{out}} \right\} \\
&\subset H^1\left(\Omega^*\right),
\end{aligned}
$$

where $\mathcal{T}_h$ is a decomposition of the domain Ω^* into tetrahedra K conforming to the tumour boundary.

For what concerns the time discretization, given N time instants on the interval $(0, T)$, $\Delta t := T/N$ is the time step and we use a superscript k to denote the value of a quantity at time $t_k = k\Delta t$. In order to simplify the notation, we will drop the superscript $k+1$ to denote the value of a quantity of interest at the next time step.

Then, we can define the fully discrete variational problem, summing (34) and (35) to rewrite them into a single nonlinear variational problem.

Thus, we can formulate the problem as follows

for $k = 1, \ldots, N$, given $\left(\mathbf{u}_h^k, p_h^k, c_h^k\right) \in \boldsymbol{V}_h \times W_{h0} \times W_{h1}$ find $(\mathbf{u}_h, p_h, c_h) \in \boldsymbol{V}_h \times W_{h0} \times W_{h1}$ such that $\forall(\boldsymbol{v}_h, w_h, q_h) \in \boldsymbol{V}_h \times W_{h0} \times W_{h0}$ it holds

$$
\begin{aligned}
&\left(J_{\text{s}}\left(\mathbf{u}_h\right), w_h\right) + \Delta t \left(\operatorname{Grad} w_h, \frac{\mathbb{K}^*}{\mu} \operatorname{Grad} p_h\right) - \left(\mathbb{P}\left(\mathbf{u}_h, p_h\right), \operatorname{Grad} \boldsymbol{v}_h\right) \\
&= \left(J_{\text{s}}^k\left(\mathbf{u}_h^k\right), w_h\right), \\
&\left(J_{\text{s}}\left(\mathbf{u}_h\right) c_h, q_h\right) - \Delta t \left(\frac{\mathbb{K}^*}{\mu \phi_\ell} \operatorname{Grad} p_h \cdot \operatorname{Grad} c_h, q_h\right) \\
&+ \Delta t \left(\operatorname{Grad} q_h, \mathbb{D}^* \operatorname{Grad} c_h\right) = \\
&= \left(J_{\text{s}}\left(\mathbf{u}_h\right) c_h^k, q_h\right) + \Delta t \left(J_{\text{s}}\left(\mathbf{u}_h\right) \frac{G_n\left(c_h\right)}{\phi_\ell}, q_h\right),
\end{aligned}
$$

where for simplicity we have denoted by $(\cdot,\cdot)$ the standard scalar product on the spaces $L^2(\Omega^*)$, $L^2(\Omega^*;\mathbb{R}^3)$ and $L^2(\Omega^*;\mathbb{R}^{3\times 3})$ when appropriate.

The last step is to introduce a proper discretisation of the other equations involved, namely the ordinary differential equation for g (17), the saturation condition (30c) and the relation (30b). Let g_h, $\phi_{s,h}$ and $\phi_{\ell,h}$ be piecewise-constant functions approximating g, ϕ_s and ϕ_ℓ, respectively. Regarding (17), it can be discretised in time using the explicit Euler method, only in the nodes which belong to the tumour domain Ω_t^*:

$$g_h = g_h^k\left(1 + \Delta t\frac{\Gamma_s(c_h^k, \phi_{s,h}^k)}{3\phi_{s,h}^k}\right).$$

Equation (30b) is discretized as

$$\phi_{s,h} = J_s^{-1}(\mathbf{u}_h)g_h^3\phi_{sn}.$$

Once we have computed $\phi_{s,h}$, we can derive $\phi_{\ell,h}$ using the saturation condition

$$\phi_{\ell,h} = 1 - \phi_{s,h}.$$

Furthermore, a time-step control was made, which allows us to ensure the convergence of the numerical simulation.

3.3 Parameters Estimation

Before performing numerical simulations, we have to assess the values of the parameters that appear in the system. It is important to remark that the choice of the parameters is fundamental to have a realistic outcome for the model. On the other hand, when working in the field of mathematical biomedicine, it is often difficult to have precise estimations of the parameters involved. In this section, we review the literature in order to assign a value or a range of admissible values to the parameters introduced in our model.

Firstly, we deal with the mechanical parameters μ_{1h} and μ_{2h} that appear in the Mooney-Rivlin energy density for the healthy tissue. In the article of Balbi et al. [13], the authors propose as mean values for the material parameters $\mu_{1h} = 3.06\cdot 10^{-4}$ MPa and $\mu_{2h} = 5.94\cdot 10^{-4}$ MPa, which we choose as references. For what concerns instead the Mooney-Rivlin parameters in the diseased tissue, we will consider them as ten times greater than the healthy ones, that means $\mu_{1t} = 3.06\cdot 10^{-3}$ MPa and $\mu_{2t} = 5.94\cdot 10^{-3}$ MPa.

The volumetric moduli κ_t and κ_h penalise volumetric changes in the solid skeleton. Unfortunately, it is difficult to estimate their numerical value, since in the majority of the works available in the literature, both on the experimental and on the modelling sides, the brain is not considered as a mixture. We take as a reference the work by Prevost et al. [73], who estimate a range of $2 \cdot 10^2 - 2 \cdot 10^4$ Pa for the volumetric modulus. Therefore, taking into account that the brain is very soft, we choose $\kappa_t = 1.389 \cdot 10^{-3}$ MPa and $\kappa_h = 1.389 \cdot 10^{-4}$ MPa.

Then, we have to provide values of the parameters involved in the growth rate Γ_s proposed in Eq. (18). The growth parameter ν is estimated using typical proliferation times for glioma cells in vitro, which range between 24 and 48 hours: this corresponds to values of 0.5–1 day^{-1}. As underlined in [24, 61], proliferation is strongly affected by the availability of nutrients, allowing also smaller values to be appropriate for ν. Having said that, in the following we will fix a value of $\nu = 0.5$ day^{-1}. For what concerns the maximum cell volume fraction $\phi_{\max}$, since we are not modelling the formation of calcifications and necrotic regions, we assume that a minimum amount of extracellular liquid is always present in the tissue to keep cells alive. Consequently, $\phi_{\max} < 1$. In the numerical simulations we set $\phi_{\max} = 0.95$.

The values for the hypoxia threshold c_0 which are found in the literature [1, 39, 40, 89] are quite different and cover a range from 0.15 to 0.5. We will consider $c_0 = 0.30$ in simulations, as done by Agosti et al. in [1]. Concerning the nutrients consumption rate ζ that appears in Eq. (20), we follow the approach by Colombo et al. [24] and so we know it can be estimated as $\zeta = 8640$ day^{-1}. For the estimation of the nutrients supply rate S_n, we rely on the value of 10^4 day^{-1} proposed in [20], as done also in [1, 24]. We need then to give an estimate of the cell volumetric fraction in the natural state ϕ_{sn}, which is a constant given from the outset. Specifically, we consider a value of $\phi_{sn} = 0.40$, in accordance with the fact that the extra-cellular space, which is complementary to the solid volume fraction, amounts approximately at 61% [18].

It is also necessary to introduce and estimate the function $K(\phi_\ell)$ which appears in the permeability tensor expression (11). For the saturation condition (1), it can be equivalently expressed as a function of ϕ_s. As a functional form for such a permeability function, we will consider the exponential Holmes-Mow expression [50], which is frequently used in the modelling of soft tissues [28]

$$K(\phi_\ell) = \tilde{K}(\phi_s) = k_0 \left[\frac{\phi_{sn}(1-\phi_s)}{\phi_s(1-\phi_{sn})}\right]^{\alpha_0} e^{\frac{m}{2}\left(\frac{\phi_{sn}^2-\phi_s^2}{\phi_s^2}\right)},$$

in which α_0 and m are model parameters, whereas k_0 is a reference permeability value taken in the natural state. In particular, for what concerns the value of k_0, in the literature it is often estimated the ratio $k_c := \frac{k_0}{\mu}$, where μ is the dynamic viscosity of the fluid phase. Following this definition and given the spatial and temporal scales of our model, such a ratio has units mm^2/(MPa · day). Values found in the literature cover quite a wide range: Mascheroni et al. [67] consider a value of $4.2 \cdot 10^4$ mm^2/(MPa · day) for the fluid phase in a mixture model for brain tumour spheroids. On the other hand, the values $4.31 \cdot 10^5 - 6.47 \cdot 10^5$mm^2/(MPa · day) were proposed by Basser [15] for the permeability of white

Table 2 List of the values of the model parameters employed in the numerical simulations

Parameter	Description	Value
μ_{1h}	Mooney-Rivlin material parameter (healthy)	$3.06 \cdot 10^{-4}$ MPa
μ_{2h}	Mooney-Rivlin material parameter (healthy)	$5.94 \cdot 10^{-4}$ MPa
μ_{1t}	Mooney-Rivlin material parameter (tumour)	$3.06 \cdot 10^{-3}$ MPa
μ_{2t}	Mooney-Rivlin material parameter (tumour)	$5.94 \cdot 10^{-3}$ MPa
κ_h	Volumetric modulus (healthy)	$1.389 \cdot 10^{-4}$ MPa
κ_t	Volumetric modulus (tumour)	$1.389 \cdot 10^{-3}$ MPa
ν	Cell proliferation constant	0.5 day^{-1}
c_0	Hypoxia threshold	0.30
ζ	Nutrients consumption rate	8640 day^{-1}
S_n	Nutrients supply rate	10^4 day^{-1}
ϕ_{sn}	Cell volume fraction in the natural state	0.40
ϕ_{max}	Maximum cell volume fraction	0.95
k_c	Hydraulic conductivity	$2.17 \cdot 10^5\ \text{mm}^2 \cdot \text{MPa}^{-1} \cdot \text{day}^{-1}$
α_0	Holmes-Mow permeability parameter	0.0848
m	Holmes-Mow permeability parameter	4.638

and grey matter, respectively. We consider an intermediate value, that corresponds to $k_c = 2.17 \cdot 10^5\ \text{mm}^2/(\text{MPa} \cdot\ \text{day})$, as done also in [61]. Furthermore, the values $\alpha_0 = 0.0848$ and $m = 4.638$ are considered, as it is usually done for the Holmes-Mow permeability in soft tissues [28].

In Table 2 we report the complete list of all the used parameters, together with the values employed in the simulations.

3.4 Mesh Preparation

The last step needed to perform the numerical simulations is the introduction of the computational brain mesh, containing the values of $\mathbb{D}_0$ and, consequently, the values of $\mathbb{A}_0$. The mesh was constructed using the Magnetic Resonance Imaging (MRI) data of a single patient, acquired in the context of normal clinical practice by the Istituto Neurologico Carlo Besta in Milan (Italy). The main advantages of MRI lie in its efficiency in detecting brain tumours and its capability to highlight the different tissue types composing the brain. Nevertheless, MRI does not provide any information concerning the micro-structural architecture of the brain and the preferential direction of water diffusion. Therefore, the values of $\mathbb{D}_0$ should be derived from Diffusion Tensor Imaging (DTI) data. In this technique the water diffusivity within each voxel of the brain is described through a symmetric and positive definite tensor, whose diagonal elements are proportional to the apparent diffusivity along the three measurement directions, while the other elements are the correlation terms between molecular displacements in directions perpendicular to the measurement direction at a given time. The capacity of DTI to determine the anisotropic diffusion of water molecules provides a way to identify and visualise the orientation of white matter tracts and, consequently, the direction of cell migration.

First, we performed the segmentation of MRI grey-scale images, i.e. the partitioning into several image segments, by assigning a label to every pixel in order to reconstruct the boundary of the brain and eventually distinguish different brain tissues. This can be done using specific software packages, such as *Slicer3D* [78]. Once the segmentation has been performed, the computational mesh was constructed using *Tetgen* [45], which is able to generate tetrahedral meshes of any 3D polyhedral domain. This software enables us to construct a tumour-conformal mesh, in order to clearly separate the healthy domain from the tumour domain. In the present work, this latter is built as a sphere of radius 7 mm located in the healthy tissue of the patient, in order to use DTI data that are not corrupted by the presence of the tumour. Nevertheless, a similar procedure could be possibly implemented in those cases in which the tumour boundary is reconstructed from MRI, identifying the cancerous region during the segmentation process. Then, the mesh has been refined along the tumour boundary, where the variables of our problem show the greatest variations. Finally, we employ data from DTI in order to build the diffusion tensor $\mathbb{D}_0$. In order to include these data in our model, the six images coming from DTI and corresponding to the six independent components of the diffusion tensor need to be aligned with the ones from MRI (i.e., they should be shifted, rotated and eventually rescaled in order to perfectly overlay with the MR images). This process can be done using the software *FSL* (*FMRIB Software Library*) [38]. Once all the images are aligned, it is possible to integrate the six components of the tensor $\mathbb{D}_0$ into the computational mesh built upon MRI data, using the scripts implemented in *VMTK* software library [93]. A Z-normal slice of all components of $\mathbb{D}_0$ is reported in Fig. 2 on the left. For further details about the construction of the computational mesh, we refer the reader to [1, 24, 61].

4 Numerical Simulations in the Brain

We apply the model presented in Sect. 2.2 to describe the growth of a tumour with an initial radius of 7 mm for a period of 80 days. The simulations have been performed using the discretisation method described in Sect. 3.2, implemented in the software *FEniCS* [5, 60, 91], which provides a high-level Python and C++ interface for solving PDEs through the Finite Element Method and allows to quickly translate weak formulations into finite element code. In Fig. 3 we show the results in terms of magnitude of the displacement vector $\|\mathbf{u}_s\|$, pressure p and nutrient concentration c_n, while in Fig. 4 we plot the scalar field g, the cell volume fraction ϕ_s and the volumetric solid Cauchy stress $\sigma := -\frac{1}{3}\text{tr}(\mathbb{T}_s)$, along three sagittal, axial, and coronal planes centered within the tumour.

All these results highlight the anisotropic evolution of the tumour inside the healthy tissue. in particular, for what concerns the displacement, it presents an

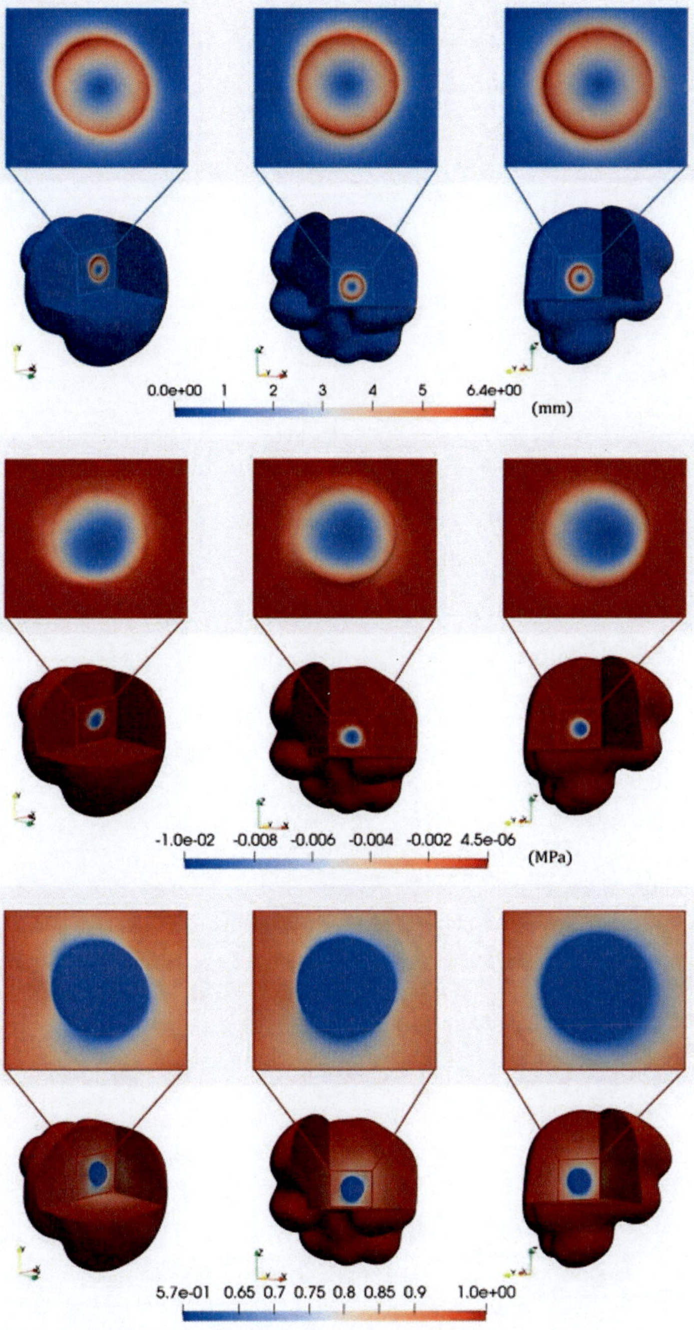

Fig. 3 Comparison between the displacement magnitude $\|\mathbf{u}_s\|$, the pressure p and nutrients c_n after $t = 80$ days of tumour growth in the brain, clipped along three different planes (XY, XZ and YZ respectively)

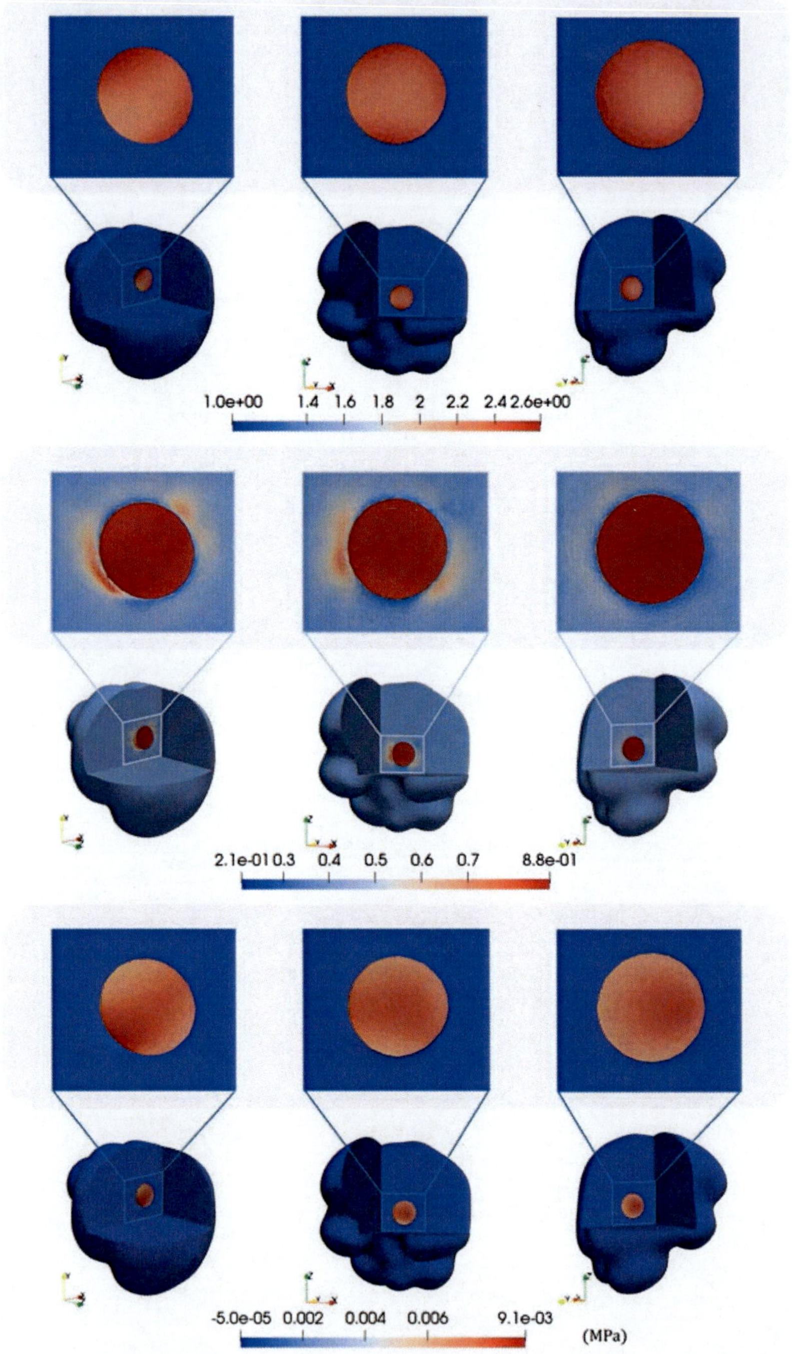

Fig. 4 Comparison between the scalar field g, the cell volume fraction ϕ_s and the stress measure σ after $t = 80$ days of tumour growth in the brain, clipped along three different planes (XY, XZ and YZ respectively)

Table 3 Maximum and minimum displacement values along each direction at time $t = 80$ days

	X	Y	Z
Min	−5.58 mm	−5.30 mm	−5.85 mm
Max	5.53 mm	6.16 mm	5.93 mm

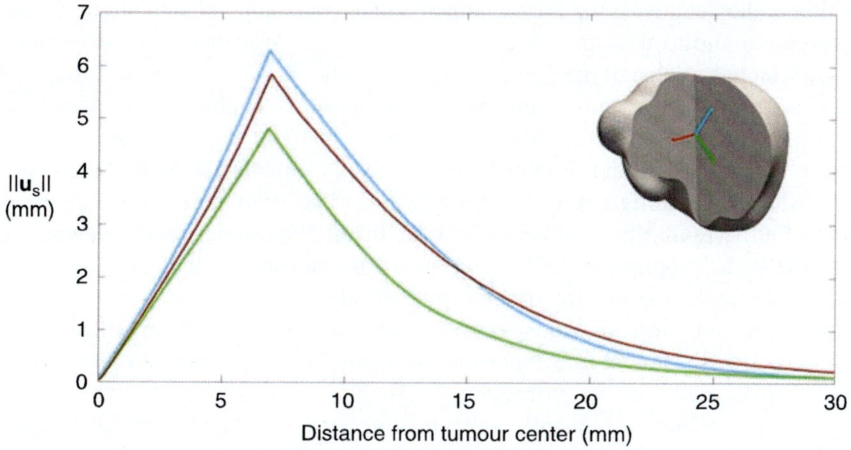

Fig. 5 Comparison of the displacement along three different rays originating from the tumour center

evident anisotropic behaviour and the maximum of its magnitude reaches 6.4 mm. Hence, at the end of the simulation, the final radius is almost doubled compared to the initial one of 7 mm. In Table 3 we report the maximum and the minimum displacement values along all directions of the axes. Furthermore, the final volume is significantly increased, as it reaches the value of about 11.5 cm^3, while the initial one was eight times smaller. Using these data, it is possible to compute an average velocity of radial expansion (VRE), which is based on the variation of the radius during growth. Considering the tumour as a sphere, even if it is not exactly the case due to anisotropy, we obtain a VRE value of about 31.94 mm/year, which is biologically feasible and shows good agreement with experimental measurements reported in [80]. Furthermore, the magnitude of displacement along different representative rays originating from the tumour center is shown in Fig. 5. First of all, it is possible to notice the anisotropic behaviour of the growth, which is enhanced along certain directions. In addition, the displacement has a peak at the tumour boundary and then it undergoes a rapid decrease. However, the graph shows a slight displacement even at 30 mm from the tumour center, meaning that tumour growth affects the surrounding brain tissue up to that distance. Moreover, for what concerns the pressure in Fig. 3, negative values are shown in the tumour zone since the fluid is consumed during the uncontrolled cellular proliferation. Regarding the concentration of nutrients, it decreases inside the tumour and close to its boundary, while it is maintained at the physiological value of 1 far from the cancer region. We observe that, for the chosen set of parameters, the concentration of nutrients is never above the physiological threshold. The volumetric fraction of the cell phase ϕ_s (Fig. 4) increases in an anisotropic way, but it does not reach the value of ϕ_{max}, for which contact inhibition of growth occurs. We remark that, since growth in the tumour region is assumed isotropic, the anisotropic distribution of the cell volume fraction in the healthy tissue is determined by the motion of cells due to tissue solid compression and to fluid pressure. This value is thus influenced by many factors, among which the value of the permeability, the preferential directions of fibres in the tissue, the growth term and the mechanical properties of healthy and tumour tissue. For the parameters set in the numerical simulations and the patient-specific DTI data, in regions where the tumour boundary displacements are higher, the healthy cells dislocate circumferentially and fluids tend to accumulate close to the tumour border. Furthermore, the positive value of σ inside the tumour region denotes that the tissue is in compression in this area, while the negative values of σ around the cancer mass underline that the tissue is in traction.

In conclusion, we show some results related to the DTI data modification due to tumour growth. In fact, the expansion of the mass and the induced displacement alter the fibre tracts in the surroundings. Thanks to our model, we can quantify these changes, which significantly affect the diffusion and the preferential directions tensor. In order to compare the differences between $\mathbb{D}_0$ and $\mathbb{D}$ at the last time step ($t = 80$ days), we plot in Fig. 6 the Frobenius norm

$$\frac{\|\mathbb{D} - \mathbb{D}_0\|_F}{\|\mathbb{D}_0\|_F}.$$

It can be noted that variations in the diffusion data are non-uniform in the tumour area, with highest variations occurring close to the boundary of the growing cancer,

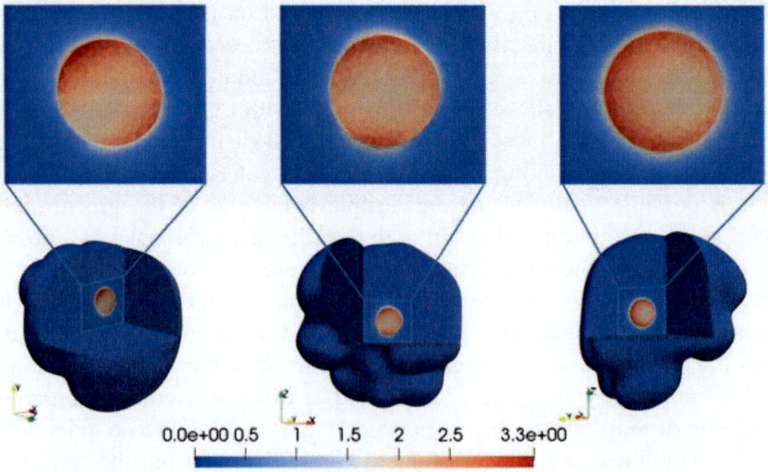

Fig. 6 Frobenius norm about DTI data modification due to tumour growth

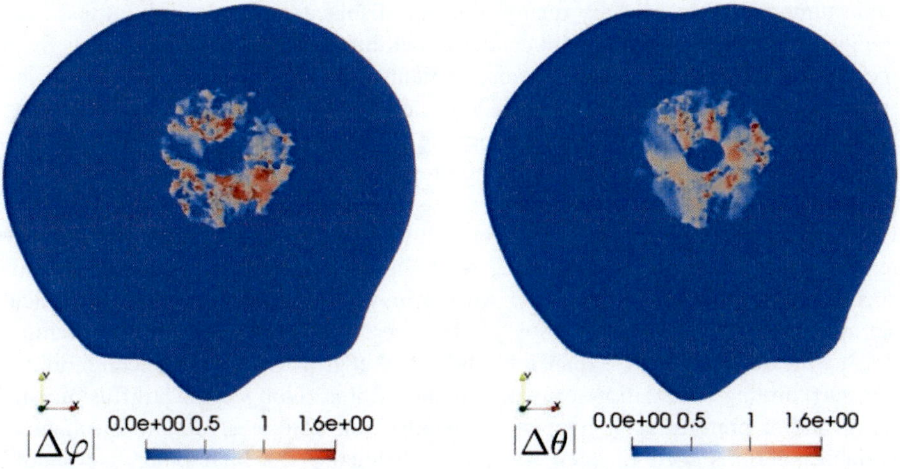

Fig. 7 Variation during tumour growth of the eigenvector associated with the greatest eigenvalue of the diffusion tensor $\mathbb{D}$, quantified in terms of absolute variations of the polar angle $|\Delta\varphi|$ and the azimuthal angle $|\Delta\theta|$

where the tissue experiences the most relevant displacements. Moreover, we also investigate the variation in the eigenvector direction $\mathbf{e}_1^0$ associated with the greatest eigenvalue λ_1^0 of the diffusion tensor $\mathbb{D}_0$. With the aim of reconstructing how much the principal direction is deformed, we express $\mathbf{e}_1^0$ and the eigenvector associated with the greatest eigenvalue of $\mathbb{D}$ using the spherical coordinates (r, φ, θ), where $\varphi \in [0, \pi]$ is the polar angle and $\theta \in [-\pi, \pi]$ is the azimuthal angle. Then we compute the absolute variations of the angles φ and θ between the two eigenvectors, identifying among themselves the polar and azimuthal angles differing by multiples of π. Furthermore, we have properly rescaled the angles variation between 0 and $\frac{\pi}{2}$. As we can notice in Fig. 7, there exist regions in which the two angles defining the spherical coordinates of the two eigenvectors have significant variations. Hence, this leads to relevant changes in the arrangement of fibres and in the preferential direction of nutrients diffusion.

5 Conclusions and Future Developments

The investigation of brain tumour growth through mathematical models has attracted a great interest in recent years. However, the quantification of the mechanical impact due to a neoplastic mass inside the brain is still a challenging problem, which has not been addressed extensively. To shed light on this issue, we have proposed a mathematical and computational framework, grounded on Continuum Mechanics and mixture theory, which is able to incorporate medical imaging data. In particular, our model is able to provide a detailed description of brain tumour mass effect, by quantifying the displacements, the solid stress, and the fluid pressure associated with cancer growth in a realistic brain geometry. Such mechanical quantities are usually very difficult to be evaluated in vivo. However, their knowledge is fundamental to understand how the growing tumour affects the surrounding tissues, as well as to capture the brain area influenced by the cancer.

An additional feature embedded in our framework is the possibility of incorporating DTI data to account for real diffusion patterns and for the anisotropy of white matter fibres inside the brain. This allows to build diffusion and permeability tensors based on neuroimaging data, which is a relevant step towards a patient-specific personalisation of the model. Moreover, we have shown that our mechanical set-up can be advantageously employed to describe how such tensors are modified due to the tumour mass expansion. Indeed, the growth of the cancer modifies the surrounding orientation of white matter fibres, along which diffusion and fluid motion preferentially happen. A detailed knowledge of all the mechanical variables permits then to keep track of such changes, which otherwise should be evaluated directly through several imaging scans at different instants of time. Finally, we have provided a numerical algorithm able to represent discontinuous deformation gradients, through the use of a mesh conforming to the sharp host-tumour material interface. Simulations have been performed using the Python-based platform FEniCS, which allows to implement a weak formulation of the equations in a flexible way.

Our results on a prototype brain mesh, reconstructed from real medical data, suggest the feasibility of the model to reproduce anisotropic growth patterns. Moreover, a quantification of tumour-induced displacements and solid stresses clearly highlights the substantial mechanical impact that a proliferating mass may exert onto the surrounding healthy tissue. In particular, our simulations show a displacement magnitude due to tumour growth of about 6.4 mm after 80 days, which may have harmful effects on a patient. Moreover, a displacement of a few millimetres can be found also in regions even 20 mm distant from the tumour boundary. As a consequence, even distal areas of the brain might be affected by the growing mass and withstand unnatural deformations. Finally, we have highlighted how the diffusion tensor and the preferential directions of diffusion are altered by the spread of the cancer.

Although the framework presented in the chapter is focused on brain tumours, its mathematical foundations remain very general. Therefore, it could also be adapted to other tissues exhibiting anisotropic structures, which may undergo changes subsequent to tumour proliferation.

However, there are still some issues that should be addressed in future research developments. First of all, the prediction of the model should be validated by simulating the growth of a tumour located in the diagnosed region and by comparing the in silico evolution with the clinical one. This task would require the acquisition of different images during the clinical evolution of the pathology, possibly starting from a tumour in the early stage, so that the DTI data will not be corrupted by the growing mass yet. Indeed, the problem of reconstructing DTI coefficients, and consequently a realistic anisotropy inside and close to the tumour region, is not trivial, since medical imaging data are often isotropic in the diseased region due to fibres disruption. Then, it would be interesting to model therapies, in order to evaluate their effects in the treatment of brain tumours. In addition, simulations of resection may take advantage of the proposed mechanical description, to account for deformations happening when the tumour mass is surgically removed. Finally, another possible improvement might be represented by the inclusion of an anisotropic growth tensor.

Acknowledgments This work was supported by the MUR, through the project PRIN2017 n. 2017KL4EF3 and through the grant Dipartimenti di Eccellenza 2018–2022 project n. E11G18000350001. GL acknowledges the support of the National Group of Mathematical Physics (GNFM–INdAM) through the INdAM–GNFM project *Progetto Giovani 2023*, n. CUPE53C22001930001. AB acknowledges the support of the INdAM Research group GNCS through the grant *Progetti di Ricerca 2022*.

Computational resources were provided by HPC@POLITO (http://hpc.polito.it). The neuroimaging data used in this study are gently provided by Dr. Francesco Acerbi and Dr. Alberto Bizzi (Istituto Neurologico Carlo Besta, Milan, Italy).

The authors are indebted to Prof. Pasquale Ciarletta for helpful discussions and support on the work.

References

1. Agosti, A., Cattaneo, C., Giverso, C., Ambrosi, D., Ciarletta, P.: A computational framework for the personalized clinical treatment of glioblastoma multiforme. ZAMM **98**, 2307–2327 (2018)
2. Agosti, A., Giverso, C., Faggiano, E., Stamm, A., Ciarletta, P.: A personalized mathematical tool for neuro-oncology: a clinical case study. Int. J. Non-Linear Mech. **107**, 170–181 (2018)
3. Agosti, A., Ciarletta, P., Garcke, H., Hinze, M.: Learning patient-specific parameters for a diffuse interface glioblastoma model from neuroimaging data. Math. Methods Appl. Sci. **43**(15), 8945–8979 (2020)
4. Alfonso, J.C.L., Talkenberger, K., Seifert, M., Klink, B., Hawkins-Daarud, A., Swanson, K.R., Hatzikirou, H., Deutsch, A.: The biology and mathematical modelling of glioma invasion: a review. J. R. Soc. Interface **14**, 1–20 (2017)
5. Alnaes, M.S., Blechta, J., Hake, J., Johansson, A., Kehlet, B., Logg, A., Richardson, C., Ring, J., Rognes, M.E., Wells, G.N.: The FEniCS project version 1.5. Arch. Numer. Softw. **3**, 9–23 (2015)
6. Ambrosi, D., Preziosi, L.: On the closure of mass balance models for tumor growth. Math. Models Methods Appl. Sci. **12**(5), 737–754 (2002)
7. Ambrosi, D., Mollica, F.: On the mechanics of a growing tumor. Int. J. Eng. Sci. **40**, 1297–1316 (2002)
8. Ambrosi, D., Mollica, F.: The role of stress in the growth of a multicell spheroid. J. Math. Biol. **48**, 477–499 (2004)
9. Angeli, S., Stylianopoulos, T.: Biphasic modeling of brain tumor biomechanics and response to radiation treatment. J. Biomech. **49**, 1524–1531 (2016)
10. Angeli, S., Emblem, K.E., Due-Tonnessen, P., Stylianopoulos, T.: Towards patient-specific modeling of brain tumor growth and formation of secondary nodes guided by DTI-MRI. NeuroImage Clin. **20**, 664–673 (2018)
11. Aubert, M., Badoual, M., Fereol, S., Christov, C., Grammaticos, B.: A cellular automaton model for the migration of glioma cells. Phys. Biol. **3**(2), 93–100 (2006)

12. Aubert, M., Badoual, M., Christov, C., Grammaticos, B.: A model for glioma cell migration on collagen and astrocytes. J. R. Soc. Interface **5**, 75–83 (2008)
13. Balbi, V., Trotta, A., Destrade, M., Annaidh, A.: Poynting effect of brain matter in torsion. Soft Matter **15**, 5147–5153 (2019)
14. Banerjee, S., Khajanchi, S., Chaudhuri, S.: A mathematical model to elucidate brain tumor abrogation by immunotherapy with T11 target structure. PLoS One **10**(5), e0123611 (2015)
15. Basser, P.J.: Interstitial pressure, volume, and flow during infusion into brain tissue. Microvasc. Res. **44**, 143–165 (1992)
16. Bondiau, P.-Y., Clatz, O., Sermesant, M., Marcy, P.-Y., Delingette, H., Frenay, M., Ayache, N.: Biocomputing: numerical simulation of glioblastoma growth using diffusion tensor imaging. Phys. Med. Biol. **53**, 879–893 (2008)
17. Bottger, K., Hatzikirou, H., Chauviere, A., Deutsch, A.: Investigation of the migration/proliferation dichotomy and its impact on avascular glioma invasion. Math. Model. Nat. Phenom. **7**, 105–135 (2012)
18. Bruehlmeier, M., Roelcke, U., Blauenstein, P., Missimer, J., Schubiger, P.A., Locher, J.T., Pellikka, R., Ametamey, S.M.: Measurement of the extracellular space in brain tumors using 76br-bromide and PET. J. Nucl. Med. **44**, 1210–1218 (2003)
19. Byrne, H., Preziosi, L.: Modelling solid tumour growth using the theory of mixtures. Math. Med. Biol. **20**, 341–366 (2003)
20. Chatelain, C., Balois, T., Ciarletta, P., Ben Amar, M.: Emergence of microstructural patterns in skin cancer: a phase separation analysis in a binary mixture. New J. Phys. **13**, 115013 (2011)
21. Chauhan, V.P., Boucher, Y., Ferrone, C.R., Roberge, S., Martin, J.D., Stylianopoulos, T., Bardeesy, N., DePinho, R.A., Padera, T.P., Munn, L.L., Jain, R.K.: Compression of pancreatic tumor blood vessels by hyaluronan is caused by solid stress and not interstitial fluid pressure. Cancer Cell **1**(26), 14–15 (2014)
22. Cheng, G., Tse, J., Jain, R.K., Munn, L.L.: Micro-environmental mechanical stress controls tumor spheroid size and morphology by suppressing proliferation and inducing apoptosis in cancer cells. PLoS One **4**(2), 1–11 (2009)
23. Clatz, O., Sermesant, M., Bondiau, P., Delingette, H., Warfield, S.K., Malandain, G., Ayache, N.: Realistic simulation of the 3-D growth of brain tumors in MR images coupling diffusion with biomechanical deformation. IEEE Trans. Med. Imaging **24**, 1334–1346 (2005)
24. Colombo, M.C., Giverso, C., Faggiano, E., Boffano, C., Acerbi, F., Ciarletta, P.: Towards the personalized treatment of Glioblastoma: Integrating patient-specific clinical data in a continuous mechanical model. PLoS One **10**, e0132887 (2015)
25. Conte, M., Gerardo-Giorda, L., Groppi, M.: Glioma invasion and its interplay with nervous tissue and therapy: a multiscale model. J. Theor. Biol. **486**, 110088 (2020)
26. Conte, M., Surulescu, C.: Mathematical modeling of glioma invasion: acid- and vasculature mediated go-or-grow dichotomy and the influence of tissue anisotropy. Appl. Math. Comput. **407**, 126305 (2021)
27. Destrade, M., Gilchrist, M.D., Murphy, J.G., Rashid, B., Saccomandi, G.: Extreme softness of brain matter in simple shear. Int. J. Non-Linear Mech. **75**, 54–58 (2015)
28. Di Stefano, S., Carfagna, M., Knodel, M.M., Hashlamoun, K., Federico, S., Grillo, A.: Anelastic reorganisation of fibre-reinforced biological tissues. Comput. Visualiz. Sci. **20**, 95–109 (2019)
29. Ehlers, W., Eipper, G.: Finite elastic deformations in liquid-saturated and empty porous solids. Transp. Porous Media **34**, 179–191 (1999)
30. Ehlers, W., Wagner, A.: Multi-component modelling of human brain tissue: a contribution to the constitutive and computational description of deformation, flow and diffusion processes with application to the invasive drug-delivery problem. Comput. Methods Biomech. Biomed. Eng. **18**, 861–879 (2015)
31. Ehlers, W., Rehm, M.M., Schroder, P., Stohr, D., Wagner, A.: Multiphasic modelling and computation of metastatic lung-cancer cell proliferation and atrophy in brain tissue based on experimental data. Biomech. Model. Mechanobiol. **21**, 277–315 (2022)

32. Elazab, A., Anter, A.M., Bai, H., Hu, Q., Hussain, Z., Ni, D., Wang, T., Lei, B.: An optimized generic cerebral tumor growth modeling framework by coupling biomechanical and diffusive models with treatment effects. Appl. Soft Comput. **80**, 617–627 (2019)
33. Engwer, C., Wenske, M.: Estimating the extent of glioblastoma invasion. J. Math. Biol. **82**, 10 (2021)
34. Engwer, C., Hillen, T., Knappitsch, M., Surulescu, C.: Glioma follow white matter tracts: A multiscale DTI-based model. J. Math. Biol. **71**, 551–582 (2015)
35. Engwer, C., Hunt, A., Surulescu, C.: Effective equations for anisotropic glioma spread with proliferation: a multiscale approach and comparisons with previous settings. Math. Med. Biol. **33**, 435–459 (2016)
36. Engwer, C., Knappitsch, M., Surulescu, C.: A multiscale model for glioma spread including cell-tissue interactions and proliferation. Math. Biosci. Eng. **13**, 443–460 (2016)
37. Falco, J., Agosti, A., Vetrano, I.G., Bizzi, A., Restelli, F., Broggi, M., Schiariti, M., DiMeco, F., Ferroli, P., Ciarletta, P., Acerbi, F.: In silico mathematical modelling for glioblastoma: a critical review and a patient-specific case. J. Clin. Med. **10**, 2169 (2021)
38. FMRIB Software Library. https://fsl.fmrib.ox.ac.uk/fsl/fslwiki
39. Frieboes, H.B., Lowengrub, J.S., Wise, S., Zheng, X., Macklin, P., Bearer, E.L., Cristini, V.: Computer simulations of glioma growth and morphology. NeuroImage **37**, S59–S70 (2007)
40. Gerlee, P., Anderson, A.R.A.: An evolutionary hybrid cellular automaton model of solid tumour growth. J. Theor. Biol. **246**, 583–603 (2007)
41. Gerlee, P., Nelander, S.: The impact of phenotypic switching on glioblastoma growth and invasion. PLoS Comput. Biol. **8**, e1002556 (2012)
42. Gerlee, P., Nelander, S.: Travelling wave analysis of a mathematical model of glioblastoma growth. Math. Biosci. **276**, 75–81 (2016)
43. Giverso, C., Scianna, M., Grillo, A.: Growing avascular tumours as elasto-plastic bodies by the theory of evolving natural configurations. Mech. Res. Commun. **68**, 31–39 (2015)
44. Gurtin, M.E., Fried, E., Anand, L.: The Mechanics and Thermodynamics of Continua. Cambridge University Press (2010)
45. Hang, S.: TetGen, a Delaunay-based quality tetrahedral mesh generator. Assoc. Comput. Mach. **41**(2), 0098–3500 (2015)
46. Hathout, L., Patel, V., Wen, P.: A 3-dimensional DTI MRI-based model of GBM growth and response to radiation therapy. Int. J. Oncol. **49**, 1081–1087 (2016)
47. Hatzikirou, H., Deutsch, A., Schaller, C., Simon, M., Swanson, K.: Mathematical modelling of glioblastoma tumour development: a review. Math. Models Methods Appl. Sci. **15**(11), 1779–1794 (2005)
48. Hatzikirou, H., Basanta, D., Simon, M., Schaller, K., Deutsch, A.: 'Go or grow': the key to the emergence of invasion in tumour progression?. Math. Med. Biol. **29**, 49–65 (2012)
49. Hogea, C., Davatzikos, C., Biros, G.: An image-driven parameter estimation problem for a reaction–diffusion glioma growth model with mass effects. J. Math. Biol. **56**, 793–825 (2008)
50. Holmes, M.H., Mow, V.C.: The nonlinear characteristics of soft gels and hydrated connective tissues in ultrafiltration. J. Biomech. **23**(11), 1145–1156 (1990)
51. Hunt, A., Surulescu, C.: A multiscale modeling approach to glioma invasion with therapy. Vietnam J. Math. **45**, 221–240 (2016)
52. Iarosz, K.C., Borges, F.S., Batista, A.M., Baptista, M.S., Siqueira, R.A.N., Viana, R.L., Lopes, S.R.: Mathematical model of brain tumour with glia–neuron interactions and chemotherapy treatment. J. Theor. Biol. **368**, 113–121 (2015)
53. Jbabdi, S., Mandonnet, E., Duffau, H., Capelle, L., Swanson, K.R., Pélégrini-Issac, M., Guillevin, R., Benali, H.: Simulation of anisotropic growth of low-grade gliomas using diffusion tensor imaging. Magn. Resonance Med. **54**, 616–624 (2005)
54. Kansal, A.R., Torquato, S., Harsh IV, G.R., Chiocca, E.A., Deisboeck, T.S.: Simulated brain tumor growth dynamics using a three-dimensional cellular automaton. J. Theor. Biol. **203**, 367–382 (2000)
55. Khain, E., Katakowski, M., Hopkins, S., Szalad, A., Zheng, X., Jiang, F., Chopp, M.: Collective behavior of brain tumor cells: the role of hypoxia. Phys. Rev. E **83**, 031920 (2011)

56. Kim, Y., Powathil, G., Kang, H., Trucu, D., Kim, H., Lawler, S., Chaplain, M.: Strategies of eradicating glioma cells: a multi-scale mathematical model with miR451-AMPK-mTOR control. PLoS One **10**, e0114370 (2015)
57. Kim, Y., Kang, H., Powathil, G., Kim, H., Trucu, D., Lee, W., Lawler, S., Chaplain, M.: Role of extracellular matrix and microenvironment in regulation of tumor growth and LAR-mediated invasion in glioblastoma. PLoS One **13**, e0204865 (2018)
58. Kyriacou, S.K., Davatzikos, C., Zinreich, S.J., Bryan, R.N.: Nonlinear elastic registration of brain images with tumor pathology using a biomechanical model. IEEE Trans. Med. Imaging **18**(7), 580–592 (1999)
59. Lipkova, J., Menze, B., Wiestler, B., Koumoutsakos, P., Lowengrub, J.S.: Modelling glioma progression, mass effect and intracranial pressure in patient anatomy. J. R. Soc. Interface **19**, 20210922 (2022)
60. Logg, A., Mardal, K.-A., Wells, G.N., et al.: Automated Solution of Differential Equations by the Finite Element Method. Springer (2012)
61. Lucci, G., Agosti, A., Ciarletta, P., Giverso, C.: Coupling solid and fluid stresses with brain tumour growth and white matter tract deformations in a neuroimaging-informed model. Biomech. Model. Mechanobiol. **21**, 1483–1509 (2022)
62. Mansury, Y., Deisboeck, T.S.: The impact of "search precision" in an agent-based tumor model. J. Theor. Biol. **224**, 325–337 (2003)
63. Mansury, Y., Deisboeck, T.S.: Modelling tumours as complex biosystems: an agent-based approach. In: Complex Systems Science: The Emergence of a New Approach in BioMedicine. Kluwer (2004)
64. Mansury, Y., Kimura, M., Lobo, J., Deisboeck, T.S.: Emerging patterns in tumor systems: Simulating the dynamics of multicellular clusters with an agent-based spatial agglomeration model. J. Theor. Biol. **219**, 343–370 (2002)
65. Martirosyan, N.L., Rutter, E.M., Ramey, W.L., Kostelich, E.J., Kuang, Y., Preul, M.C.: Mathematically modeling the biological properties of gliomas: a review. Math. Biosci. Eng. **12**, 879–905 (2015)
66. Mascheroni, P., Stigliano, S., Carfagna, M., Boso, D.P., Preziosi, L., Decuzzi, P., Schrefler, B.A.: Predicting the growth of glioblastoma multiforme spheroids using a multiphase porous media model. Biomech. Model. Mechanobiol. **15**, 1215–1228 (2016)
67. Mascheroni, P., Carfagna, M., Grillo, A., Boso, D.P., Schrefler, B.A.: An avascular tumor growth model based on porous media mechanics and evolving natural states. Math. Mech. Solids **23**, 686–712 (2018)
68. Mihai, L.A., Budday, S., Holzapfel, G.A., Kuhl, E., Goriely, A.: A family of hyperelastic models for human brain tissue. J. Mech. Phys. Solids **106**, 60–79 (2017)
69. Nia, H.T., Liu, H., Seano, G., Datta, M., Jones, D., Rahbari, N., Incio, J., Chauhan, V.P., Jung, K., Martin, J.D., Askoxylakis, V., Padera, T.P., Fukumura, D., Boucher, Y., Hornicek, F.J., Grodzinsky, A.J., Baish, J.W., Munn, L.L., Jain, R.K.: Solid stress and elastic energy as measures of tumour mechanopathology. Nat. Biomed. Eng. **1**, 0004 (2017)
70. Nia, H.T., Datta, M., Seano, G., Zhang, S., Ho, W.W., Roberge, S., Huang, P., Munn, L.L., Jain, R.K.: In vivo compression and imaging in mouse brain to measure the effects of solid stress. Nat. Protoc. **15**(8), 2321–2340 (2020)
71. Ogden, R.W.: Large deformation isotropic elasticity - on the correlation of theory and experiment for incompressible rubberlike solids. Proc. R. Soc. Lond. **326**, 565–584 (1972)
72. Painter, K.J., Hillen, T.: Mathematical modelling of glioma growth: the use of diffusion tensor imaging (DTI) data to predict the anisotropic pathways of cancer invasion. J. Theor. Biol. **323**, 25–39 (2013)
73. Prevost, T.P., Balakrishnan, A., Suresh, S., Socrate, S.: Biomechanics of brain tissue. Acta Biomater. **7**, 83–95 (2011)
74. Rajagopal, K.R., Tao, L.: Mechanics of Mixtures. World Scientific Publishing Co Pte Ltd (1995)
75. Rockne, R., Alvord Jr., E.C., Rockhill, J.K., Swanson, K.R.: A mathematical model for brain tumor response to radiation therapy. J. Math. Biol. **58**, 561–578 (2009)

76. Sander, L.M., Deisboeck, T.S.: Growth patterns of microscopic brain tumours. Phys. Rev. E **66**, 051901 (2002)
77. Seano, G., Nia, H.T., Emblem, K.E., Datta, M., Ren, J., Krishnan, S., Kloepper, J., Pinho, M.C., Ho, W.W., Ghosh, M., Askoxylakis, V., Ferraro, G.B., Riedemann, L., Gerstner, E.R., Batchelor, T.T., Wen, P.Y., Lin, N.U., Grodzinsky, A.J., Fukumura, D., Huang, P., Baish, J.W., Padera, T.P., Munn, L.L., Jain, R.K.: Solid stress in brain tumours causes neuronal loss and neurological dysfunction and can be reversed by lithium. Nat. Biomed. Eng. **3**(3), 230–245 (2019)
78. Slicer3D, image computing platform. https://www.slicer.org/
79. Steed, T.C., Treiber, J.M., Brandel, M.G., Patel, K.S., Dale, A.M., Carter, B.S., Chen, C.C.: Quantification of glioblastoma mass effect by lateral ventricle displacement. Sci. Rep. **8**, 2827 (2018)
80. Stensjøen, A.L., Solheim, O., Kvistad, K.A., Haberg, A.K., Salvesen, Ø., Berntsen, E.M.: Growth dynamics of untreated glioblastomas in vivo. Neuro-Oncology **17**(10), 1402–1411 (2015)
81. Suveges, S., Hossain-Ibrahim, K., Steele, J.D., Eftimie, R., Trucu, D.: Mathematical modelling of glioblastomas invasion within the brain: A 3D multi-scale moving-boundary approach. Mathematics **9**, 2214 (2021)
82. Swan, A., Hillen, T., Bowman, J.C., Murtha, A.D.: A patient-specific anisotropic diffusion model for brain tumour spread. Bull. Math. Biol. **80**, 1259–1291 (2018)
83. Swanson, K.R., Alvord Jr., E.C., Murray, J.D.: A quantitative model for differential motility of gliomas in grey and white matter. Cell Prolif. **33**, 317–329 (2000)
84. Swanson, K.R., Alvord Jr., E.C., Murray, J.D.: Virtual brain tumours (gliomas) enhance the reality of medical imaging and highlight inadequacies of current therapy. Br. J. Cancer **86**, 14–18 (2002)
85. Swanson, K.R., Alvord Jr., E.C., Murray, J.D.: Quantifying efficiency of chemotherapy of brain tumors with homogeneous and heterogeneous drug delivery. Acta Biotheor. **50**, 223–237 (2002)
86. Swanson, K.R., Bridge, C., Murray, J.D., Alvord Jr., E.C.: Virtual and real brain tumors: using mathematical modeling to quantify glioma growth and invasion. J. Neurol. Sci. **216**, 1–10 (2003)
87. Swanson, K.R., Alvord Jr., E.C., Murray, J.D.: Virtual resection of gliomas: effect of extent of resection on recurrence. Math. Comput. Model. **37**, 1177–1190 (2003)
88. Swanson, K.R., Rockne, R.C., Claridge, J., Chaplain, M.A., Alvord Jr., E.C., Anderson, A.R.: Quantifying the role of angiogenesis in malignant progression of gliomas: in silico modeling integrates imaging and histology. Cancer Res. **71**, 7366–7375 (2011)
89. Tanaka, M.L., Debinski, W., Puri, I.K.: Hybrid mathematical model of glioma progression. Cell Prolif. **42**, 637–646 (2009)
90. Tektonidis, M., Hatzikirou, H., Chauvière, A., Simon, M., Schaller, K., Deutsch, A.: Identification of intrinsic in vitro cellular mechanisms for glioma invasion. J. Theor. Biol. **287**, 131–147 (2011)
91. The FEniCS Project. https://fenicsproject.org/ (2021)
92. Tracqui, P., Cruywagen, C.G., Woodward, D.E., Bartoo, G.T., Murray, J.D., Alvord Jr., E.C.: A mathematical model of glioma growth: the effect of chemotherapy on spatiotemporal growth. Cell Prolif. **28**, 17–31 (1995)
93. VMTK, The Vascular Modeling Toolkit. http://www.vmtk.org/
94. Woodward, D.E., Cook, J., Tracqui, P., Cruywagen, C.G., Murray, J.D., Alvord Jr., E.C.: A mathematical model of glioma growth: the effect of extent of surgical resection. Cell Prolif. **29**, 269–288 (1996)
95. Yangjin, K., Hyejin, J., Othmer, H.: The role of the tumor microenvironment in glioblastoma: a mathematical model. IEEE Trans. Biomed. Eng. **64**, 519–527 (2017)
96. Zhang, L., Strouthos, C.G., Wang, Z., Deisboeck, T.S.: Simulating brain tumor heterogeneity with a multiscale agent-based model: Linking molecular signatures, phenotypes and expansion rate. Math. Comput. Model. **49**, 307–319 (2009)

A Multi-Scale Immune System Simulator for the Onset of Type 2 Diabetes

Maria Concetta Palumbo and Filippo Castiglione

Abstract Type 2 diabetes mellitus (T2D), the most common form of diabetes, is a chronic metabolic disorder characterized by hyperglycemia, insulin resistance, and insulin deficiency. Although genetic predisposition determines in part the susceptibility to T2D, an unhealthy diet and a sedentary lifestyle are commonly recognised as essential drivers for the onset of the disease. Indeed, considerable evidence suggests that regular exercise and appropriate nutrition bring undeniable health benefits by reducing the risk of developing T2D or delaying its onset.

The literature dealing with mathematical modelling for diabetes is abundant and in the view of a growing more personalized medicine the benefits of having tools to represent different virtual patient populations are clear.

In this study, we describe a multi-scale computational model of the human metabolic and inflammatory status that is determined by the individual dietary and activity habits. It includes a description of the immune activation and inflammation, a model for the food intake, stomach emptying and gut absorption of all the three macronutrients (proteins, carbohydrates, fats), a component to account for the effects of physical activity on the hormones' regulation and the inflammatory state of the individual, and finally, a characterization of energy intake-expenditure balance. All these pieces are merged into a single integrated simulation tool to provide a helpful aid that can be used proactively to prevent the onset of the disease. Moreover, this model turns out to help design virtual cohorts of patients to conduct in silico studies.

M. C. Palumbo (✉)
Institute for Applied Computing (IAC) "Mauro Picone", National Research Council of Italy, Rome, Italy
e-mail: c.palumbo@iac.cnr.it

F. Castiglione
Institute for Applied Computing (IAC) "Mauro Picone", National Research Council of Italy, Rome, Italy

Biotech Research Center, Technology Innovation Institute, Masdar City, Abu Dhabi, UAE

G. Bretti et al. (eds.), *Mathematical Models and Computer Simulations for Biomedical Applications*, SEMA SIMAI Springer Series 33,
https://doi.org/10.1007/978-3-031-35715-2_6

1 Introduction

Type 2 diabetes mellitus is a chronic, non-communicable disease characterized by high levels of sugar in the blood, preventing glucose homeostasis. This pathology consists in progressively losing the capability to respond to insulin, a condition termed insulin resistance, with a consequent increase in the blood sugar level, causing hyperglycemia [1]. The number of people with diabetes mellitus has quadrupled in the past three decades, and diabetes mellitus is one of the major causes of death. About 90% of adults with diabetes mellitus have T2D. Most patients with T2D have at least one complication, and cardiovascular diseases are the leading cause of morbidity and mortality in these patients. T2D results from the interaction between genetic, environmental and behavioural risk factors. Proper nutrition, regular physical activity, maintaining a normal weight, not smoking and drinking alcohol in moderation, are factors helping to manage T2D and are common non-pharmacological treatment options [2–5].

Mathematical models have helped describing glucose homeostasis. However, computational models able to represent metabolic homeostasis for the prevention of type 2 diabetes in normal life conditions are quite limited, since they do not embrace the simultaneous detailing of these three factors: exercise, food intake, immunological phenomena [6–16].

We have recently developed an integrated, multi-scale and patient-specific model (hereinafter named M-T2D), whose goal is to simulate the main metabolic and inflammatory processes undergoing the transition to unhealthy phenotypes of type 2 diabetes. M-T2D embraces, to a certain degree of sophistication, different levels of description from the molecular/cellular to organs and the whole body [17, 18] and consists, among other things, in the extension of an existing model for fuel homeostasis proposed by Kim in [19]. The improvements to the original model involve a better characterization of the exercise, the description of the ingestion of mixed meals, the computation of the energy balance leading to gaining/losing weight, the detailing of the immunological scenario of the subject, the characterization based on the subjects-specific features to achieve greater generalization and user-customization.

The description of physical exercise is based on the use of a relative exercise intensity, affecting the metabolic model by means of the description of epinephrine role in the definition of the insulin concentration [20]. In fact, physical exercise stimulates the release of epinephrine (aka adrenaline) into the circulation from the adrenal medulla. Type, duration and intensity of exercise influence the changes in epinephrine concentartion, as well as factors such as age, body weight and fitness status. In our model, a set of differential equations describing the dynamics of epinephrine due to exercise is implemented, in which the dependence on exercise intensity is explicitly given in terms of the relative exercise intensity (the percentage of maximal oxygen uptake or consumption, utilization, termed PVO_{2max}). Since adrenaline released during exercise impacts plasma insulin, we have introduced a contribution accounting for the release/clearance of adrenaline into the equation describing the insulin dynamics.

Exercise also stimulates the secretion of Interleukin-6 (IL-6) [21] by the skeletal muscle, whereas, in resting conditions, IL-6 is mainly produced by the adipose tissue [22, 23], giving evidence of its dual nature as an adipokine (i.e., adipose tissue-derived cytokine) and as a myokine (i.e., muscle-derived cytokine). During exercise, IL-6 plays a central role in glucose homeostasis, since it stimulates the production and secretion of Glucagon-like Peptide-1 (GLP-1) from intestinal L-cells and pancreas, causing an enhanced insulin response. We integrated a previously validated system model of the effects of exercise on plasma IL-6 dynamics in healthy humans, combining the contributions of both adipose tissue and skeletal muscle [24].

Meals consumed every day include different nutrients, such as carbohydrates, proteins and fats, whose description is, till now, lacking. M-T2D provides a description of the food intake, stomach emptying and gut absorption of all the three macronutrients [18].

We integrate into M-T2D a description of a user-specific energy balance between physical exercise and food ingestion, which determines the excess caloric intake leading to weight gain/loss.

Obesity is one of the main risk factors leading to the onset of several health problems, including but not limited to insulin resistance and T2D [25]. The hallmark of obesity is a state of systemic low-grade chronic inflammation leaded by metabolic tissues, such as adipose tissue, liver and muscle [26–28] and accompanied by elevated levels of circulating markers of inflammation, such as pro-inflammatory cytokines and chemokines. To model the effect of a high calorie diet on the interplay between adipose tissue and inflammation we use adipogenesis due to excessive calorie intake to induce the secretion of inflammatory cytokines such as tumor necrosis factor α (TNFα, or TNFa), interleukin-6 (IL-6), and interleukin-1β (IL-1β, or IL-1b) [29]. These cytokines create the conditions for the exacerbation of the inflammation due to the activation of a pro-inflammatory biased immune response.

To describe the inflammatory status of the subject, we employed a model of the immune system response previously used to describe immune-related pathologies [30, 31]. The model accounts for the major hallmarks of the immune response by representing the main actors of the cellular and humoral immune response in a discrete sequence of events. To model the inflammation, a series of events culminating into a full blown immune activity fueling the systemic inflammation, is triggered by the proliferating adipose compartment providing, at the same time, the already-mentioned initial level of inflammatory cytokines [29] but also the necessary antigenic peptide stimulation.

M-T2D is customizable, as it includes patient specific features such as anthropometric measures (age, sex, height, body weight, fitness status), the basal concentrations of some metabolites or hormones (i.e. glucose, epinephrine), calories ingested during the meals, and some characteristics of the moderate-intensity physical activities performed (the target value T_v of exercise intensity and the duration of the session T_{ex}). Moreover, the immune inflammation modelling component contains patient-specific parameters such as basal concentration of systemic inflammatory cytokines and blood leukocyte counts.

M-T2D covers a time frame from hours to years. As a consequence, it can be viewed as a tool to forecast the effects of adhering to different lifestyles for both self-monitoring and clinical decision support systems. Likewise, it can be used to simulate physiological reasonable metabolic and inflammatory responses to stimuli to represent different cohorts of virtual patients (i.e. in-silico clinical trials).

2 Mathematical Models

In the following sections, we describe the components of M-T2D unifying several ordinary differential equations and an agent-based model into a single simulation tool (see Fig. 1).

In particular, Sect. 2.1 describes the metabolic model of fuel homeostasis; in Sect. 2.2 we elucidate the hormonal glucagon/insulin model, which describe the changes in the plasma concentrations of insulin and glucagon caused by meals ingestion or by the epinephrine production during the physical activity; the model describing how exercise modulates the epinephrine secretion/elimination is illustrated in Sect. 2.3; Sect. 2.4 elucidates the model of food intake, stomach emptying and macronutrient absorption, meals intake and exercise cause an intake/expenditure of calories; Sect. 2.5 detail the model of energy balance and body weight gain/loss as a consequence of excess/deficit of calories; the excess of caloric intake leads to a gain in weight caused by the increase in size of adipocytes which store fat and, at the same time, trigger the release of inflammatory signals: these details are given in Sect. 2.6); finally, the release of Interleukin-6 (IL-6), one of the most crucial cytokine having a central role in glucose homeostasis, is modeled from both the adipose tissue and the skeletal muscle while exercising, and this is described in Sect. 2.7.

The input of the model is summarised in Table 1. It includes the patterns of the physical activities performed by the individual, expressed as the time interval of the exercise (start–end, in minutes) and its intensity. Another input consists is the nutritional pattern, namely the amounts of carbohydrates, proteins and fats (in grams) ingested during each meal period (expressed in minutes).

All components of M-T2D use a time scale of 1 min per step, while in the immune system and inflammation model a time step corresponds to 8 h of real life because the immune processes develop in a much longer time frame. By analysing the results at a short time scale, one can analyse how M-T2D behaves in response to challenge tests, i.e. minutes-to-days. On the other hand, M-T2D can give an idea of the effects of a long-term behaviour subject to the user provided input of nutritional and exercise patterns which last weeks or even months.

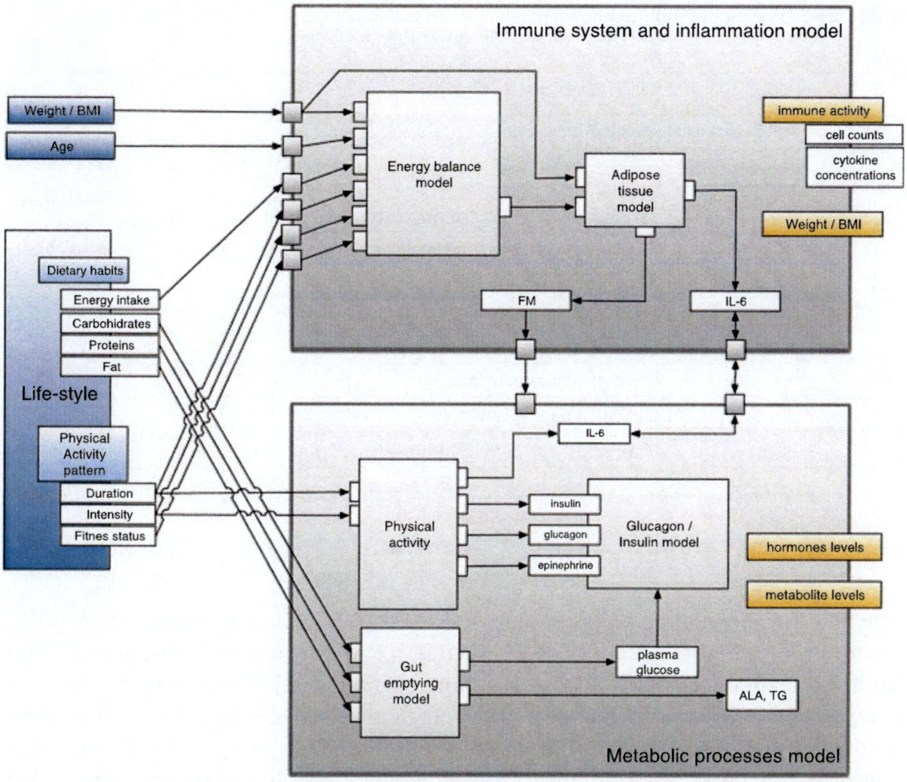

Fig. 1 Model architecture. Input variables are shown on the left (blue boxes). The output is represented by the yellow boxes at the right. The overall model includes (1) the glucagon-insulin model (Sect. 2.2), and, (2) the physical activity model (Sect. 2.3), (3) the gut emptying model (Sect. 2.4), (4) the energy balance model (see Sect. 2.5), (5) the adipose tissue model (Sect. 2.6). The energy balance module computes the calorie excess. This is used in input by the adipose tissue model which computes fat mass (FM) and the secreted IL-6. These two variables are used by the metabolic processes models. Also, the physical activity model uses the specification of the physical exercise (duration/intensity) to give in output the IL-6 generated during the activity. The gut emptying model transforms the input in glucose, alanine (ALA) and triglycerides (TG). The glucagon/insulin model computes the levels of plasma glucagon and insulin on the basis of the levels of glucose and epinephrine

2.1 *The Model of Metabolism*

Seven tissues/compartments compose the metabolic model, inspired by the work by Kim [19]: brain, heart, liver, gastrointestinal tract, skeletal muscle, adipose tissue and other tissues. Each tissue compartment is then described by dynamic mass balances and twenty-five major cellular reactions between twenty-two metabolites, reported in Table 2.

Table 1 List of model input parameters

Parameter	Meaning
S	Sex (male/female)
A	Age in years
BW	Body weight in kg
H	Height in m
T_v	Target value of exercise intensity in percentage of VO_{2max}
$fitness\ status$	Cardio-respiratory fitness classification from "poor" to "superior" [32]
$t_{ex}^{start}, t_{ex}^{end}$	Start/end of the exercise session in min
t_{meal}^{start}	Meal start in min
D_{cho}	Dose of carbohydrates in grams
D_{prot}	Dose of proteins in grams
D_{fat}	Dose of fats in grams
$C_{a,g}^{*}$	fasting glucose level in mmol/l

2.2 *The Hormonal Glucagon/Insulin Model*

The multi-scale model of glucose homeostasis M-T2D, inspired by Kim [19], incorporates the hormonal model by Saunders et al., in which both glucagon and insulin are produced and glucose regulation is achieved by altering the balance between the two hormones [33]. Any change in the glucagon-to-insulin ratio modulates the flux rates of the metabolites in the different tissues. For what concerns the description of the physical activity, the basic assumption is that the pancreatic insulin secretion is affected by the exercise-induced change in the hormone epinephrine, whose contribution to the insulin dynamics is included as follows

$$\frac{dC_I}{dt} = C_I(t) \cdot \big[\psi \cdot [h - k_3 \cdot (C_G(t) - C_{G,0}) - k_4 \cdot (C_I(t) - C_{I,0}) \\ -k_5 \cdot (C_E(t) - C_{E,0})] - D\big]$$

with the following equation describing the glucagon

$$\frac{dC_G}{dt} = C_G(t) \cdot \big[\phi \cdot [h - k_1 \cdot (C_G(t) - C_{G,0}) - k_2 \cdot (C_I(t) - C_{I,0})] - D\big]$$

in which $C_I(t)$, $C_G(t)$, $C_E(t)$ (in pmol/l) are the insulin, glucagon and the epinephrine blood concentrations, $C_{I,0}$, $C_{G,0}$, $C_{E,0}$ are their basal values. Food consumption and exercise represent two different operation regimes, thus we have estimated two distinct sets of unknown parameters $(k_1, k_2, k_3, k_4, k_5, D)$ for the glucagon/insulin model [17, 18].

Table 2 Organs/tissues, metabolites and reactions in the computational model of fuel homeostasis

List of organs	List of metabolites	List of reactions
1. Brain	1. GLC: Glucose	1. Glycolysis I
2. Heart	2. PYR: Pyruvate	2. Glycolysis II
3. Skeletal muscle	3. LAC: Lactate	3. Glycolysis III
4. Liver	4. ALA: Alanine	4. Gluconeogenesis I
5. GI tract	5. GLR: Glycerol	5. Gluconeogenesis II
6. Adipose tissue	6. FFA: Free fatty acids	6. Gluconeogenesis III
	7. TG: Triglycerides	7. Glycogenesis
	8. O2: Oxygen	8. Glycogenolysis
	9. CO2: Carbon dioxide	9. Pyruvate reduction
	10. G6P: Glucose-6-phosphate	10. Lactate oxidation
	11. GLY: Glycogen	11. Glycerol phosphorylation
	12. GAP: Glyceraldehyde-3-phosphate	12. GAP reduction
	13. GRP: Glycerol-3-phosphate	13. Glycerol 3-P oxidation
	14. AcoA: Acetyl coenzyme A	14. Alanine formation
	15. CoA: Coenzyme A	15. Alanine utilization
	16. NAD+: Aldehyde dehydrogenase	16. Pyruvate oxidation
	17. NADH: Nicotinamide adenine dinucleotide	17. Fatty acids oxidation
	18. ATP: Adenosine triphosphate	18. Fatty acids synthesis
	19. ADP: Adenosine diphosphate	19. Lipolysis
	20. Pi: Phosphate	20. Triglycerides synthesis
	21. PCR: Phosphocreatine	21. TCA cycle
	22. CR: Creatine	22. Oxidative phosphorylation
		23. Phosphocreatine breakdown
		24. Phosphocreatine synthesis
		25. ATP hydrolysis

Since meals intake and physical activity represent different operation regimes, the meals and the exercise periods cannot overlap. We fixed the duration of the meal period, also including the gastric emptying and the gut absorption, to 180 min. We chose to model the condition between meals and exercise using the same parameters as for the meal intake, but setting the food quantities to zero.

2.3 The Model of the Physical Exercise

The description of the exercise consists in modelling the dynamics of oxygen uptake (VO_2) on the basis of the study by Lenart and Roy [34, 35], as already done in previous works [17, 24]. Changes in oxygen consumption during the exercise session (performed in the range of moderate-intensity workload, above the lactate threshold) and the subsequent recovery phase are described through the following linear first-order differential equation:

$$\frac{dPVO_{2max}(t)}{dt} = -0.8 \cdot PVO_{2max}(t) + 0.8 \cdot u(t) \tag{1}$$

with

$$u(t) = \begin{cases} 0 & 0 \le t < t_{ex}^{start} \text{ and } t > t_{ex}^{end} \\ T_v & t_{ex}^{start} \le t \le t_{ex}^{end}. \end{cases}$$

The suprabasal oxygen consumption $PVO_{2max}(t)$ is expressed as a percentage of the maximum value ($\%VO_{2max}$), T_v is the target value of exercise intensity and $u(t)$ describes the input as a step function assuming value T_v for the entire duration of the exercise.

The model of epinephrine secretion and elimination (based on the PVO_{2max} as proposed by Kildegaard and colleagues in [20]) describes the changes in epinephrine concentration due to exercise as follows

$$\frac{dC_E(t)}{dt} = \frac{1}{V_d} \cdot (f_1 + f_2 + f_3) \cdot BW - k \cdot C_E(t) \tag{2}$$

in which $C_E(t)$ (in nmol/l) is the epinephrine concentration, V_d (in l) is the volume of distribution, f_1 is a constant representing a basal epinephrine secretion (in nmol/kg/min), k is the epinephrine elimination constant (in min^{-1}). The term f_2 (in nmol/kg/min) accounts for the epinephrine contribution depending on the arterial glucose level (expressed by $Ca, g(t)$ in mmol/l) and f_3 (nmol/kg/min) is the contribution from the physical activity

$$f_2 = f_2(C_{a,g}(t)) = c_1 / \left(1 + e^{c_2 \cdot (Ca,g(t) - c_3)}\right) \tag{3}$$

$$f_3 = f_3(PVO_{2max}(t)) = d_1 / \left(1 + e^{d_2 \cdot (d_3 - PVO_{2max}(t))}\right). \tag{4}$$

We have computed the epinephrine elimination constant k imposing the steady-state condition in Eq. (2). Consequently, Eqs. (3) and (4) are calculated in the absence of physical activity (i.e. $T_v = 0$, $PVO_{2max}(t) = 0$), with the glucose and epinephrine concentrations corresponding to their fasting value $C^*_{a,g}$ and $C_{E,0}$, thus obtaining

$$f_2^* = f_2(C^*_{a,g}) = c_1 / \left(1 + e^{c_2 \cdot (Ca,g^* - c_3)}\right)$$

$$f_3^* = f_3(0) = d_1 / \left(1 + e^{d_2 \cdot d_3}\right)$$

$$k = \frac{BW}{V_d \cdot C_{E,0}} \cdot (f_1 + f_2^* + f_3^*).$$

2.4 The Model of Food Intake, Stomach Emptying and Macronutrient Absorption

Meal consumption comprises the following processes: nutrient intake, stomach emptying, absorption of macronutrients carbohydrates (represented as glucose in the model), proteins (represented as amino acids), and fats (as triglycerides). Considering that proteins represent a minor contribution to energy metabolism compared to carbohydrates and lipids, amino acids are represented solely by alanine, as also done in [19]. We have converted the quantities (grams) of carbohydrates (CHO), proteins, and fats ingested into their millimole equivalent dose of glucose (D_{glu}), alanine (D_{ala}), and triglycerides (D_{tg}).

To formulate the equations introducing stomach emptying and gut nutrient absorption into the model, we followed Dalla Man et al. [36] and Elashoff et al. [37]: the amount of nutrient n (where n stands for glucose from CHO, alanine from proteins, and triglycerides from fats) emptied in the duodenum increases following a power exponential function given by

$$q_{duo}(t, n) = D_n \cdot (1 - e^{(-(K_{e,n} \cdot t)^{\beta_n})})$$

with $K_{e,n}$ being a kinetic constant, and β_n a shape factor depending on the meal type in which the nutrient is present. The gastric emptying rate $G_{emp,n}(t)$ for nutrient n is

$$G_{emp,n}(t) = D_n \cdot \beta_n \cdot K_{e,n}^{\beta_n} \cdot t^{(\beta_n - 1)} \cdot e^{-(K_{e,n} \cdot t)^{\beta_n}}.$$

The equations describing first-order kinetics of glucose absorption and rate of appearance are given below:

$$\begin{cases} dy_{n(t)}/dt = -K_{abs,n} \cdot y_n(t) + G_{emp,n}(t) \\ Ra_n(t) = F_{abs,n} \cdot K_{abs,n} \cdot y_n(t) \end{cases}$$

in which $y_n(t)$ is the amount of nutrient n in the gut (in mmol), $K_{abs,n}$ (in min^{-1}) represents the rate constant of intestinal absorption, Ra_n (expressed in mmol/min) is the rate of appearance of the nutrient in the gut and $F_{abs,n}$ describes the fraction of intestinal absorption which actually appears in plasma. Finally, we add a quantity equal to $Ra_n / V_{eff,n,GI}$ to the concentration of the nutrient in the model of the GI organ. $V_{eff,n,GI}$ (expressed in l) describes the effective distribution volume of the nutrient in the gut compartment, according to the original model in [19]. Further details on the model dealing with the food intake are available in [18].

2.5 Modeling Total Daily Energy Balance and Body Weight

Total daily energy balance ($TDEB$) can be roughly expressed as the difference between "energy in" ($TDEI$, i.e. Total Daily Energy Intake) and "energy out" ($TDEE$, i.e. Total Daily Energy Expenditure)

$$TDEB = TDEI - TDEE. \tag{5}$$

The $TDEI$ (in kcal/day) is the energy content of the daily meals. The energy values are 4.0 kcal/g (17 kJ/g) for protein, 9.0 kcal/g (37 kJ/g) for fat and 4.0 kcal/g (17 kJ/g) for carbohydrates, on the basis of the Atwater general factor system [38]. $TDEE$ (in kcal/day) is the sum of the Resting Energy Expenditure (REE), the Activity Energy Expenditure (AEE [39] and the Thermic Effect of Food (TEF).

$$TDEE = REE + AEE + TEF \tag{6}$$

Mifflin and coworkers in [40] have estimated REE, considering weight, height, age, and sex as follows

$$\begin{cases} REE_{female}(kcal) = 10 \cdot BW + 6.25 \cdot H - 5 \cdot A - 161 \\ REE_{male}(kcal) = 10 \cdot BW + 6.25 \cdot H - 5 \cdot A + 5. \end{cases}$$

We evaluate AEE by computing the milliliters of oxygen consumed during each physical activity session. Given T_v as the target value of exercise intensity, the oxygen consumption in ml $\cdot$ min^{-1} is given by $T_v \cdot VO_{2max}$. To determine the kilocalories consumed per minutes, we multiplied this value by the following conversion factor $5 \cdot 10^{-3} \cdot kcal$ per milliliter of oxygen consumed [41, 42]. Thus, taking into account the exercise duration,

$$AEE(kcal) = \frac{T_v}{100} \cdot VO_{2max} \cdot BW \cdot 5 \cdot 10^{-3} \cdot (t_{ex}^{end} - t_{ex}^{start}).$$

We have computed VO_{2max} considering the tables of Heyward in [32] that indicate the values of VO_{2max} for males and females on the basis of age and fitness status. TEF is the amount of energy consumed due to the digesting and processing food [42–44]. In healthy subjects with a mixed diet, TEF represents about 10% of $TDEI$ [44]. Consequently, Eqs. (5) and (6) become

$$TDEE = REE + AEE + 0.1 \cdot TDEI$$

$$TDEB = 0.9 * TDEI - REE - AEE$$

According to the regression equations provided by Westerterp in [45], sex, age, body weight, and height determine the Fat Free Mass (FFM) as follows,

$$FFM = \alpha + \beta \mathrm{A} + \gamma \mathrm{H} + \delta \mathrm{BW}$$

where the parameters $\alpha, \beta, \gamma, \delta$ depend on the gender and are respectively $\alpha = -12.47$ kg, $\beta = -0.074$ kg/years, $\gamma = 27.392$ kg/m, and $\delta = 0.218$ for female and $\alpha = -18.36$ kg, $\beta = -0.105$ kg/years, $\gamma = 34.009$ kg/m, and $\delta = 0.292$ for male. The value for the Fat Mass (FM) can be obtained from the following simple compartmental relation between fat and body weight [46]

$$FM = BW - FFM.$$

Changes in FM and FFM depend on the $TDEB$ as follows

$$\begin{cases} \frac{\Delta FM}{\Delta t} = \frac{fef(t) \cdot TDEB}{fm_{ed}} \\ \frac{\Delta FFM}{\Delta t} = \frac{(1 - fef(t)) \cdot TDEB}{ffm_{ed}} \end{cases}$$

where the fat energy factor (fef) determines which part of the energy balance is transformed to FM, fat mass energy density (fm_{ed}) fat-free mass energy density (ffm_{ed}) [45].

2.6 *Modeling the Effect of a Calorie Excess on the Adipocytes*

The population of adipocytes represents a form of a reservoir for fat, and the source of inflammatory signals. The relationship between adipocyte diameter and Body Mass Index ($\mathrm{BMI} = \mathrm{BW}/\mathrm{H}^2$) is determined as in Eq. (7) and is used to set the initial distribution of the volume of adipocytes as estimated in [47]:

$$\phi(\mathrm{BMI}) = \phi_c - \lambda e^{-\delta \mathrm{BMI}} \tag{7}$$

where, respectively for female and male, $\phi_c = 123$ and $120\,\mu$m, $\lambda = 198.445$ and $74.905\,\mu$m and $\delta = 0.061$ and 0.049 m^2/kg [47, 48]. Omental fat constitutes the most of visceral adipose tissue (VAT, i.e. the main spot of the inflammation), hence we assume the relation in Eq. (7) is valid for all adipocytes in VAT.

During periods of excessive calorie intake, visceral adipocyte swelling accommodates for the storing of the energy surplus. The growth continues until a "critical" volume (or diameter ϕ_c) is reached. Beyond this value, adipogenesis (generation of new adipocytes) from precursor cells begins [49–51]. The model of adipocyte swelling and adipogenesis is described in [29]. Adipogenesis induces secretion of inflammatory cytokines such as TNFα, IL-6, and IL-1β, that create the triggering condition for the further exacerbation of the inflammation due to the activation of a pro-inflammatory biased immune response.

2.7 The Model of IL-6 Release

It is well-known that IL-6 production due to exercise depends on the duration and the intensity of the physical activity [52]. The model of IL-6 release is described in [24] by the following ordinary differential equations and depends on the oxygen consumption introduced in Eq. (1) of Sect. 2.3:

$$\begin{cases} \frac{dIL6_m(t)}{dt} = SR_{ex} \cdot PVO_{2max}(t) - k_m \cdot IL6_m(t) & IL6_m(0) = 0 \\ \frac{dIL6_p(t)}{dt} = k_m \cdot IL6_m(t) - k_e \cdot IL6_p(t) + \frac{Ra_{IL6}}{V} & IL6_p(0) = IL6_{basal}. \end{cases}$$

$IL6_m(t)$ and $IL6_p(t)$ represent, respectively, IL-6 concentration in the muscle compartment and in the plasma. The first term on the right hand side in the first differential equation, accounts for muscle IL-6 increase from stationary conditions in response to exercise, described as linearly dependent on PVO_{2max} through the secretion rate SR_{ex}. In the second equation, the first term of the right-hand side describes the increase in the plasma IL-6 from its basal value ($IL6_{basal}$) due to skeletal muscles. The second term represents IL-6 removal from circulation after exercise operated by the hepatosplanchnic viscera. The third term accounts for the IL-6 production rate during non-perturbed conditions, which is represented by the adipose tissue contribution (Ra_{IL6}) normalized to the distribution volume V. (further information regarding the choice of the parameters k_e, k_m, SR_{ex} can be found in [24]). The value of the inflammatory marker IL-6 (C_{IL-6}) provided by the model of inflammation described below, updates the adipose tissue contribution Ra_{IL6}

$$Ra_{IL6}(t+1) = Ra_{IL6}(t) + C_{IL6}$$

2.8 The Model of Inflammation

The model of the immune system response is a derivation of [53]. Being it an agent-based model (ABM), it represents all entities individually [54, 55]. It is a polyclonal model that makes use of bit strings to represent the "binding site" of cells and molecules (e.g., lymphocytes receptors, B-cell receptors, T-cell receptors, Major Histocompatibility Complexes (MHC), antigen peptides and epitopes, immuno-complexes, etc.). The model represents several primary and secondary immune compartments playing a critical role in the immune response: generic tissue (e.g., epithelial and adipose tissue), lymphoid tissue, thymus gland and bone marrow.

These components are bundled together in a complex yet parsimonious description of immunity focused at the mesoscopic (i.e. cellular) level. The ABM includes the major classes of cells of the lymphoid lineage (T helper lymphocytes or Th, cytotoxic T lymphocytes or CTLs, B lymphocytes, antibody-producer plasma cells, PLB and natural killer cells, NK) and some of the myeloid lineage i.e. macrophages (MA, divided in M1 and M2), and dendritic cells. Helper T-cells are further divided in the following phenotypes: Th0, Th1, Th2, Th17 and T regulatory cells (Tregs). B and plasma B-cells are also divided into two phenotypes B-1, B-2 according to their antibody isotype IgM or IgG respectively. B-2 cells are further subdivided into those producing IgG1 or IgG2. All these entities interact following a set of rules describing the different phases of the recognition and response of the immune system against a pathogen. In particular, the model accounts for phagocytosis, antigen presentation, cytokine release, cytotoxicity, antibody secretion, and cell activation from inactive or anergic states to active states. Additionally, this model reproduces the gene-regulation mechanisms leading to macrophage differentiation during the different stages of inflammation [30]. The model simulates the innate immunity and an elaborate form of adaptive immunity (including both humoral and cytotoxic immune responses).

The interactions among the cells determine their functional behaviour. We have coded the interactions as probabilistic rules defining the transition of each cell entity from one state to another. Each interaction requires cell entities to be in a specific state choosing in a set of possible states (e.g.,, naïve, active, resting, duplicating) that are dependent on the cell type. Once this condition is fulfilled, the interaction probability is directly related to the effective level of binding between ligands and receptors. Unlike the many immunological models, the present one does not only simulate the cellular level of the inter-cellular interactions but also the intra-cellular processes of antigen uptake and presentation. We have modelled both the cytosolic and endocytic pathways. In the model, endogenous antigen is fragmented and combined with MHC class I molecules for presentation on the cell surface to CTLs' receptors, whereas the exogenous antigen is degraded into smaller parts (i.e., peptides), which are then bound to MHC class II molecules for presentation to the T helpers' receptors. At variance with classical cellular automata models, there is no correlation among entities residing on different sites at a fixed time step, and the deterministic character of automata dynamics is replaced by a stochastic behaviour. However, at the end of each time step entities diffuse from site to site introducing spatial correlations.

The number of cells populating the small volume of the body represented by the model is calculated according to generic leukocyte formulas.

A stochastic execution of the automaton rules enact the immune reaction to antigens. Upon recognition of immunogenic peptides (whatever the source), antigen-presenting cells stimulate effector lymphoid cells to initiate clone expansion and production of antibodies. For the purpose of simulating the inflammation underlying T2D, the hypertrophic adipocytes presenting the so-called MHC-I-restricted immunopeptidomes [56] provide the stimulus. The resulting cytotoxic response, provides the fuel for the progression of obesity-induced visceral adipose tissue chronic inflammation that is critically linked to T2D.

The architecture of the combined model, with the relationships among the various components and through specified key variables, is shown in the Fig. 1.

3 Results

3.1 *Setting the Parameters for the Glucagon/Insulin Model*

The computational model proposed simulates the responses of hormones and metabolites during different stimuli due to everyday activities, such as food intake or exercise. To allow this diversity, we have modified the hormonal glucagon/insulin model, also estimating the parameters when required. For what concerns the simulation of the physical activity, we did not modify the parameters k_1, k_2, k_3, k_4, D provided by Kim in [19], introducing only a new parameter accounting for the exercise regime, namely k_5. The parameter k_5 is estimated as in [19] by fitting plasma insulin and glucagon concentrations mean data, collected by Hirsch and colleagues [57] and obtained from thirteen healthy young men before, during and after 60 min of exercise at $T_v = 60$. Dealing with the regime of meal intake and absorption, for the estimation procedure we used experimental data of glucose, glucagon, and insulin plasma concentrations from a 50g oral glucose tolerance test (OGTT) study conducted by Knop and colleagues in [58], in which blood samples were drawn from 10 healthy subjects before, during, and 4h after the test. Table 3 summarizes the estimated parameters of the glucagon/insulin model for both regimes, obtained by applying the procedure in Sect. 2.2.

Table 3 Estimated parameters of the glucagon/insulin model

Parameter	Value for meal	Value for exercise	Units
k_1	0.01569	0.2707	$pM^{-1} \cdot min^{-1}$
k_2	0.00014	0.0535	$pM^{-1} \cdot min^{-1}$
k_3	3.08666	0.1507	$pM^{-1} \cdot min^{-1}$
k_4	0.07095	0.0309	$pM^{-1} \cdot min^{-1}$
k_5	0	$3.6 \cdot 10^{-5}$	pM^{-1}
D	0.00650	0.1	min^{-1}

3.2 Simulating Different Lifestyle Scenarios

The ability of M-T2D to reproduce the dynamics of different metabolites and hormones with a good level of physiological coherence has already been proved in [17, 18] in the case of single or multiple experimental testing conditions. However, the simulator can also reproduce periods spanning months or even years, and different lifestyle scenarios leading to different outcomes. In this Section, we show how M-T2D can reproduce different lifestyle scenarios.

Unhealthy habits such as an unbalanced diet and physical inactivity can lead to an increase in the risk of developing T2D, described as a rise in the level of the fasting glucose; notwithstanding, a change in the habits can lead to the restoration of the level of the fasting glucose. The two simulations here reported, deal with a male subject (40 years old, 80 kg, 1.7 m). According to Sect. 2.5, the $TDEI$ for this subject is about 2250 kcal/day, which we divide into three meals, each consisting of 90 g of CHO, 30 g of proteins and 30 g of fats.

In the first simulation, the virtual subject undergoes a 12-weeks of overfeeding, in which it eats 3375 kcal/day ($2250 \cdot 1.5$ kcal/day). Then, he starts a diet for 16 weeks, consisting of a 10% reduction of his daily energy requirement of 2250 kcal/day (2025 kcal/day). Some results of the simulations are shown in Figs. 2 and 3.

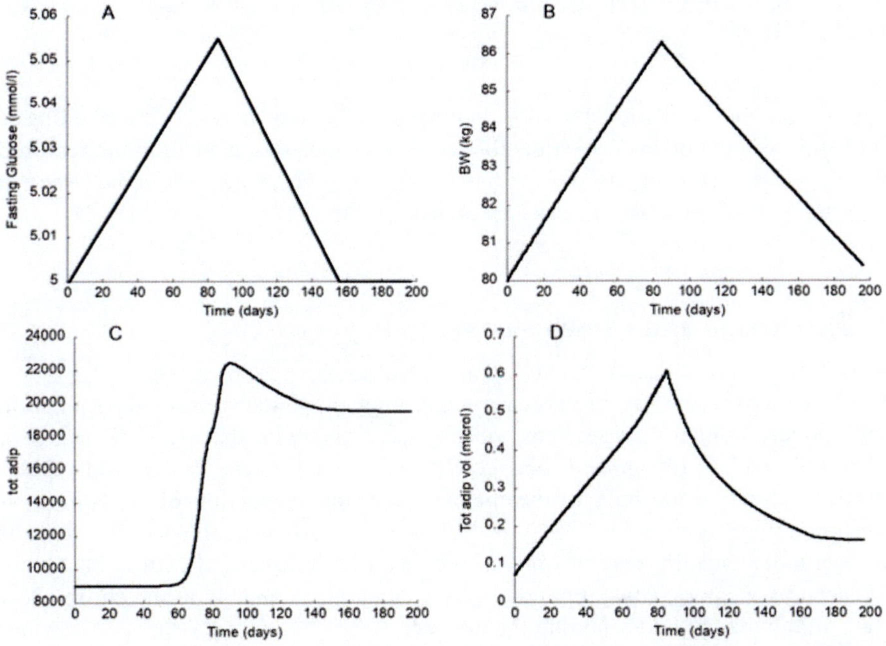

Fig. 2 First simulation (only diet intervention). Trends of: (**a**) fasting glucose; (**b**) body weight; (**c**) Total number of adipocytes (expressed as the number of adipocytes per 10microl); (**d**) Volume of adipocytes

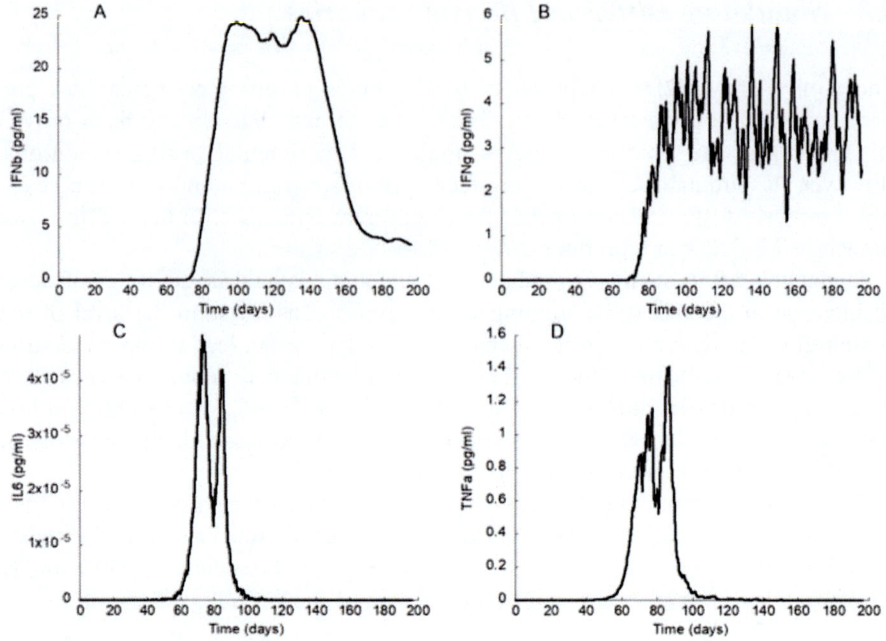

Fig. 3 Inflammatory markers for the first simulation (only diet intervention). (**a**) IFNβ; (**b**) IFNγ; (**c**) IL6; (**d**) TNFα

In the second simulation, the subject undergoes the same period of overfeeding as in the first simulation andsubsequently he assumes meals equivalent to the required $TDEI$ while performing three sessions of 60 min of physical exercise per week at $T_v = 60\%$. The results of this case are shown in Figs. 4 and 5.

4 Discussion and Conclusions

There is a growing body of evidence suggesting the association of inflammation with modern human diseases (e.g., obesity, cardiovascular disease, T2D, cancer). Also, it is widely recognized that weight management through diet and regular physical activity contribute to the management and prevention of a *plethora* of noncommunicable diseases. For type 2 diabetes in particular, the accumulation of fat over an extended period of time, as the result of an excess of caloric intake not balanced by a caloric consumption to physical activity, can lead to the chronic low grade inflammation underpinning the disease.

We have developed an integrated, multi-scale and patient-specific model with the aim of simulating the metabolic and inflammatory processes involved in the onset of T2D from reactions at the cellular level up to whole-body activities across all of the major body organs.

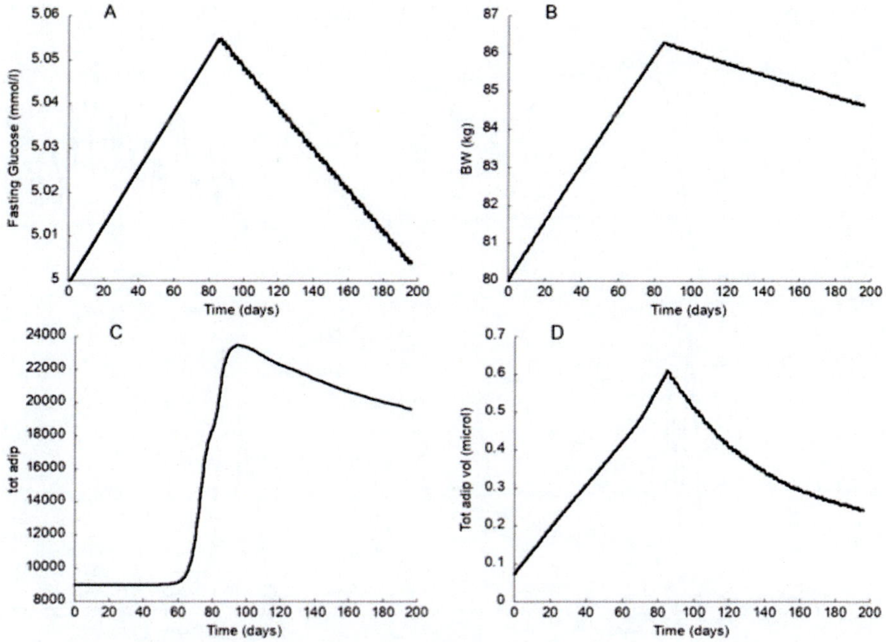

Fig. 4 Second simulation (normal eating and physical exercise). Trends of: (**a**) fasting glucose; (**b**) body weight; (**c**) Total number of adipocytes (expressed as the number of adipocytes per 10microl); (**d**) Volume of adipocytes

In Figs. 2, 3, 4, and 5 we show that a prolonged period of overfeeding (the first 12 weeks in both simulations) results in an increase in the body weight (Panel b in Figs. 2 and 4), determined by the rise in the number and volume of adipocytes (Panels c and d in Figs. 2 and 4). As a consequence of this harmful lifestyle, a cytotoxic response is triggered, leading to the rise of the inflammatory markers, as evident from Fig. 3 and 5 and the level of the plasma fasting glucose grows (Panel a in Figs. 2 and 4).

To revert this progression, diet and physical activity are essential, giving rise to the decreasing trends and restoring the levels of fasting glucose. The intervention based solely on diet results in larger, more rapid reductions of weight. Indeed, losing weight depends more on a diet than on exercise. However, regularly exercising provides more benefits than reducing nutrition alone. In fact, in this kind of intervention, the fat mass decreases while the fat-free mass augments. In contrast, during a diet, also the lean mass decreases. An effective way to maintain fat-free mass and the resting energy expenditure during a hypo energetic diet intervention would therefore consist following a more structured diet plan, incorporating an exercise program. Although the literature remains controversial regarding the benefits of exercise on body weight loss, physical activity is the primary factor impacting body weight maintenance after the interventions [59, 60].

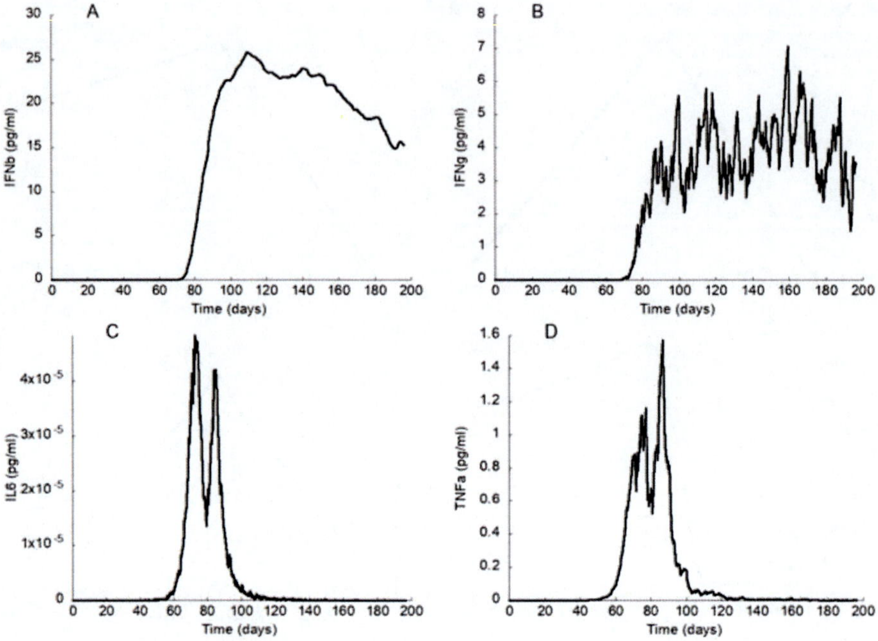

Fig. 5 Inflammatory markers for the second simulation (normal eating and physical exercise). (**a**) IFNβ; (**b**) IFNγ; (**c**) IL6; (**d**) TNFα

As illustrated in Table 1, M-T2D depends on user-specific parameters and customized exercise and food content patterns, thus enabling it to forecast personalized outcomes.

This computational model can represent viable support in the generation of *in-silico clinical trial* simulations and can provide a valuable tool for the decision-making process requiring quasi-realistic scenarios. Indeed, the ability to follow the dynamic changes of metabolites, hormones, physical features, and inflammatory markers due to the combined action of food intake and exercise can translate into a subject-specific self-diagnostic eHealth application and medical devices in the view of more predictive, preventive, personalized and participative medicine [61].

References

1. Galicia-Garcia, U., Benito-Vicente, A., Jebari, S., Larrea-Sebal, A., Siddiqi, H., Uribe, K.B., et al.: Pathophysiology of type 2 diabetes mellitus. Int. J. Mol. Sci. **21**(17), 6275 (2020)
2. Colberg, S.R., Sigal, R.J., Fernhall, B., Regensteiner, J.G., Blissmer, B.J., Rubin, R.R., et al.: Exercise and type 2 diabetes: the American College of Sports Medicine and the American Diabetes Association: joint position statement. Diabetes Care **33**(12), e147–e167 (2010)

3. Asif, M.: The prevention and control the type-2 diabetes by changing lifestyle and dietary pattern. J. Educ. Health Promot. **3** (2014)
4. Chen, L., Magliano, D.J., Zimmet, P.Z.: The worldwide epidemiology of type 2 diabetes mellitus–present and future perspectives. Nature Rev. Endocrinol. **8**(4), 228–236 (2012)
5. Zheng, Y., Ley, S.H., Hu, F.B.: Global aetiology and epidemiology of type 2 diabetes mellitus and its complications. Nature Rev. Endocrinol. **14**(2), 88–98 (2018)
6. Adamu, I.I., Garba, E.J.D., Haruna, Y.: Mathematical model for the dynamics of glucose regulatory system under the combined use of dieting and physical exercise. Ozean J. Appl. Sci. **5**(3), 229–241 (2012)
7. Svitra, D., Basov, I., Vilkytė, R.: Modelling of glycaemia dynamics: impact of physical exercises. Nonlinear Anal. Model. Control **15**(2), 213–232 (2010)
8. Breton, M.D.: Physical activity-the major unaccounted impediment to closed loop control. J. Diabetes Sci. Technol. **2**(1), 169–174 (2008). https://doi.org/10.1177/193229680800200127
9. Dalla Man, C., Breton, M.D., Cobelli, C.: Physical activity into the meal glucose-insulin model of type 1 diabetes: in silico studies. J. Diabetes Sci. Technol. **3**(1), 56–67 (2009). https://doi.org/10.1177/193229680900300107
10. Balakrishnan, N.P., Samavedham, L., Rangaiah, G.P.: Personalized hybrid models for exercise, meal, and insulin interventions in Type 1 diabetic children and adolescents. Ind. Eng. Chem. Res. **52**(36), 13020–13033 (2013)
11. Resalat, N., El Youssef, J., Tyler, N., Castle, J., Jacobs, P.G.: A statistical virtual patient population for the glucoregulatory system in type 1 diabetes with integrated exercise model. PLoS One **14**(7), e0217301 (2019)
12. Islam, M.J., Hoque, A.S.M.L.: Virtual diabetic patient with physical activity dynamics. Comput. Methods Program. Biomed. **213**, 106485 (2022)
13. Deichmann, J., Bachmann, S., Pfister, M., Szinnai, G., Kaltenbach, H.M.: A Comprehensive Model of Glucose-Insulin Regulation Including Acute and Prolonged Effects of Physical Activity in Type 1 Diabetes. bioRxiv (2021)
14. Sarkar, J., Dwivedi, G., Chen, Q., Sheu, I.E., Paich, M., Chelini, C.M., et al.: A long-term mechanistic computational model of physiological factors driving the onset of type 2 diabetes in an individual. PLoS One **13**(2), e0192472 (2018)
15. Kurata, H.: Virtual metabolic human dynamic model for pathological analysis and therapy design for diabetes. Iscience **24**(2), 102101 (2021)
16. López-Palau, N.E., Olais-Govea, J.M.: Mathematical model of blood glucose dynamics by emulating the pathophysiology of glucose metabolism in type 2 diabetes mellitus. Sci. Rep. **10**(1), 1–11 (2020)
17. Palumbo, M.C., Morettini, M., Tieri, P., Diele, F., Sacchetti, M., Castiglione, F.: Personalizing physical exercise in a computational model of fuel homeostasis. PLoS Comput. Biol. **14**(4), e1006073–e1006073 (2018)
18. Palumbo, M.C., Morettini, M., Tieri, P., de Graaf, A., Krishnan, S., Castiglione, F.: A computational model of the effects of macronutrients absorption and physical exercise on hormonal regulation and metabolic homeostasis Comput. Biol. Med. **163**, 107158 (2023)
19. Kim, J., Saidel, G.M., Cabrera, M.E.: Multi-scale computational model of fuel homeostasis during exercise: effect of hormonal control. Ann. Biomed. Eng. **35**(1), 69–90 (2007). https://doi.org/10.1007/s10439-006-9201-x
20. Kildegaard, J., Christensen, T.F., Johansen, M.D., Randløv, J., Hejlesen, O.K.: Modeling the effect of blood glucose and physical exercise on plasma adrenaline in people with type 1 diabetes. Diabetes Technol. Ther. **9**(6), 501–507 (2007)
21. Pedersen, B., Steensberg, A., Fischer, C., Keller, C., Keller, P., Plomgaard, P., et al.: Searching for the exercise factor: is IL-6 a candidate? J. Muscle Res. Cell Motil. **24**(2), 113–119 (2003)
22. Mohamed-Ali, V., Goodrick, S., Rawesh, A., Katz, D.R., Miles, J.M., Yudkin, J.S., et al.: Subcutaneous adipose tissue releases interleukin-6, but not tumor necrosis factor-alpha, in vivo. J. Clin. Endocrinol. Metab. **82**(12), 4196–4200 (1997)
23. Lyngsø, D., Simonsen, L., Bülow, J.: Interleukin-6 production in human subcutaneous abdominal adipose tissue: the effect of exercise. J. Physiol. **543**(1), 373–378 (2002)

24. Morettini, M., Palumbo, M.C., Sacchetti, M., Castiglione, F., Mazzá, C.: A system model of the effects of exercise on plasma Interleukin-6 dynamics in healthy individuals: Role of skeletal muscle and adipose tissue. PLoS One **12**(7), e0181224 (2017). https://doi.org/10.1371/journal.pone.0181224
25. Hotamisligil, G.S.: Inflammation and metabolic disorders. Nature **444**(7121), 860–867 (2006)
26. Gregor, M.F., Hotamisligil, G.S.: Inflammatory mechanisms in obesity. Ann. Rev. Immunol. **29**, 415–445 (2011). https://doi.org/10.1146/annurev-immunol-031210-101322
27. Faloia, E., Michetti, G., De Robertis, M., Luconi, M.P., Furlani, G., Boscaro, M.: Inflammation as a link between obesity and metabolic syndrome. J. Nutr. Metab. (2012). https://doi.org/10.1155/2012/476380
28. Rodríguez-Hernández, H., Simental-Mendía, L.E., Rodríguez-Ramírez, G., Reyes-Romero, M.A.: Obesity and inflammation: epidemiology, risk factors, and markers of inflammation. Int. J. Endocrinol. (2013). https://doi.org/10.1155/2013/678159
29. Prana, V., Tieri, P., Palumbo, M.C., Mancini, E., Castiglione, F.: Modeling the effect of high calorie diet on the interplay between adipose tissue, inflammation, and diabetes. Comput. Math. Methods Med. (2019). https://doi.org/10.1155/2019/7525834
30. Castiglione, F., Tieri, P., Palma, A., Jarrah, A.S.: Statistical ensemble of gene regulatory networks of macrophage differentiation. BMC Bioinf. **17**(19), 506 (2016)
31. Madonia, A., Melchiorri, C., Bonamano, S., Marcelli, M., Bulfon, C., Castiglione, F., et al.: Computational modeling of immune system of the fish for a more effective vaccination in aquaculture. Bioinformatics **33**(19), 3065–3071 (2017). https://doi.org/10.1093/bioinformatics/btx341
32. Heyward, V.H., Gibson, A.L.: Advance Fitness Assessment and Exercise Prescription. Human Kinetics, Champain (2014)
33. Saunders, P.T., Koeslag, J.H., Wessels, J.A.: Integral rein control in physiology. J. Theor. Biol. **194**(2), 163–173 (1998). https://doi.org/10.1006/jtbi.1998.0746
34. Lenart, P.J., Parker, R.S.: Modeling exercise effects in type 1 diabetic patients. In: IFAC Proceedings of the World Congress on Automatic Control (2002)
35. Roy, A., Parker, R.S.: Dynamic modeling of exercise effects on plasma glucose and insulin levels. J. Diabetes Sci. Technol. **1**(3), 338–347 (2007). https://doi.org/10.1177/193229680700100305
36. C. Dalla Man, Camilleri, M., Cobelli, C.: A system model of oral glucose absorption: validation on gold standard data. IEEE Trans. Biomed. Eng. **53**(12 Pt 1), 2472–2478 (2006). https://doi.org/10.1109/TBME.2006.883792
37. Elashoff, J.D., Reedy, T.J., Meyer, J.H.: Analysis of gastric emptying data. Gastroenterology **83**(6), 1306–1312 (1982)
38. Atwater, W., Woods, C.: The chemical composition of american food materials. Farmers' Bulletin No. 28. US Department of Agriculture Washington (1896)
39. Melanson, E.L., Keadle, S.K., Donnelly, J.E., Braun, B., King, N.A.: Resistance to exercise-induced weight loss: compensatory behavioral adaptations. Med. Sci. Sports Exerc. **45**(8), 1600–1609 (2013)
40. Mifflin, M.D., St Jeor, S.T., Hill, L.A., Scott, B.J., Daugherty, S.A., Koh, Y.O.: A new predictive equation for resting energy expenditure in healthy individuals. Am. J. Clin. Nutr. **51**(2), 241–247 (1990)
41. McArdle, W.D., Katch, F.I., Katch, V.L.: Essentials of Exercise Physiology. Lippincott Williams & Wilkins, Philadelphia (2006)
42. Kohlstadt, I.: Scientific Evidence for Musculoskeletal, Bariatric, and Sports Nutrition. CRC Press, Boca Raton (2006)
43. Acheson, K.J.: Influence of autonomic nervous system on nutrient-induced thermogenesis in humans. Nutrition **9**(4), 373–380 (1993)
44. Westerterp, K.R.: Diet induced thermogenesis. Nutr. Metab. (Lond.) **1**(1), 5 (2004). https://doi.org/10.1186/1743-7075-1-5
45. Westerterp, K.R., Donkers, J.H., Fredrix, E.W., Boekhoudt, P.: Energy intake, physical activity and body weight: a simulation model. Br. J. Nutr. **73**(3), 337–347 (1995)

46. Wang, Z., Heshka, S., Pierson, R., Heymsfield, S.: Systematic organization of body composition methodology: an overview with emphasis on component-based methods. Am. J. Clin. Nutr. **61**, 457–465 (1995)
47. Tchernof, A., Després, J.P.: Pathophysiology of human visceral obesity: an update. Physiol. Rev. **93**, 359–404 (2013)
48. Spalding, K.L., Arner, E., Westermark, P.O., Bernard, S., Buchholz, B.A., Bergmann, O., et al.: Dynamics of fat cell turnover in humans. Nature. **453**(7196), 783–787 (2008)
49. Jo, J., Gavrilova, O., Pack, S., Jou, W., Mullen, S., Sumner, A.E., et al.: Hypertrophy and/or hyperplasia: dynamics of adipose tissue growth. PLOS Computat. Biol. **5**(3), 1–11 (2009). https://doi.org/10.1371/journal.pcbi.1000324
50. Jo, J., Shreif, Z., Periwal, V.: Quantitative dynamics of adipose cells. Adipocyte **1**(2):80–88 (2012). https://doi.org/10.4161/adip.19705
51. Parlee, S.D., Lentz, S.I., Mori, H., MacDougald, O.A.: Chapter six - quantifying size and number of adipocytes in adipose tissue. In: Macdougald, O.A. (ed.) Methods of Adipose Tissue Biology, Part A. Methods in Enzymology, vol. 537, pp. 93–122. Academic Press, Cambridge (2014). Available from: http://www.sciencedirect.com/science/article/pii/B9780124116191000069
52. Ostrowski, K., Schjerling, P., Pedersen, B.K.: Physical activity and plasma interleukin-6 in humans–effect of intensity of exercise. Eur. J. Appl. Physiol. **83**(6), 512–515 (2000)
53. Castiglione, F., Celada, F.: Immune System Modelling and Simulation. CRC Press, Boca Raton (2015)
54. Wolfram, S.: A New Kind of Science, vol. 5. Wolfram Media, Champaign (2002)
55. Castiglione, F.: Agent-based modeling and simulation, introduction to. In: Marilda S., David P.-C., Filippo C. (eds.) Complex Social and Behavioral Systems: Game Theory and Agent-Based Models (Encyclopedia of Complexity and Systems Science Series) 1st ed., pp 661–665 (2020)
56. Chen, X., Wang, S., Huang, Y., Zhao, X., Jia, X., Meng, G., et al.: Obesity reshapes visceral fat-derived MHC I associated-immunopeptidomes and generates antigenic peptides to drive CD8+ T cell responses. Iscience **23**(4), 100977 (2020)
57. Hirsch, I.B., Marker, J.C., Smith, L.J., Spina, R.J., Parvin, C.A., Holloszy, J.O., et al.: Insulin and glucagon in prevention of hypoglycemia during exercise in humans. Am. J. Physiol. **260**(5 Pt 1), E695–E704 (1991)
58. Knop, F.K., Vilsboll, T., Madsbad, J.J., Krarup, T.: Inappropriate suppression of glucagon during OGTT but not during isoglycaemic i.v. glucose infusion contributes to the reduced incretin effect in type 2 diabetes mellitus. Diabetologia **50**, 797–805 (2007)
59. Lockard, B., Mardock, M., Oliver, J.M., Byrd, M., Simbo, S., Jagim, A.R., et al.: Comparison of two diet and exercise approaches on weight loss and health outcomes in obese women. Int. J. Environ. Res. Public Health **19**(8), 4877 (2022)
60. Volek, J.S., VanHeest, J.L., Forsythe, C.E.: Diet and exercise for weight loss. Sports Med. **35**(1), 1–9 (2005)
61. Flores, M., Glusman, G., Brogaard, K., Price, N.D., Hood, L.: P4 medicine: how systems medicine will transform the healthcare sector and society. Pers. Med. **10**(6), 565–576 (2013)

Molecular Fingerprint Based and Machine Learning Driven QSAR for Bioconcentration Pathways Determination

Mauro Nascimben, Silvia Spriano, Lia Rimondini, and Manolo Venturin

Abstract Quantitative structure-activity relationship associates molecules' structural characteristics to their bio-activity, and it can be performed via machine learning to find risky chemicals that accumulate in living organisms. The present analysis focused on investigating how structural information of molecules encoded by molecular fingerprints can be used to predict bioaccumulation pathways. Numerical experiments involved extreme gradient boosting, support vector machines, and neural networks, including spiking neural networks. This investigation might be the first attempt to apply this particular kind of biologically inspired neural network for predicting molecules' functions from their fingerprints. The computational models forecasted three possible bioaccumulation processes, with support vector machines obtaining the mean peak accuracy and the spiking neural network architectures achieving satisfactory results: the leaky neuron spiking neural network range of outcomes was not statistically different from the accuracies of the support vector machine algorithms. In addition, an algorithm broadly found in chemoinformatics literature as the extreme gradient boosting algorithm fulfilled compatible accuracies with spiking neural networks and support vector machines. This three-class machine learning-driven bioconcentration modeling of chemicals established a foundation

M. Nascimben (✉)
Department of Health Sciences, Università del Piemonte Orientale, Novara, Italy
Enginsoft SpA, Padua, Italy
e-mail: m.nascimben@enginsoft.com

S. Spriano
Department of Applied Science and Technology, Politecnico di Torino, Turin, Italy
e-mail: silvia.spriano@polito.it

L. Rimondini
Department of Health Sciences, Università del Piemonte Orientale, Novara, Italy
e-mail: lia.rimondini@med.uniupo.it

M. Venturin
Enginsoft SpA, Padua, Italy
e-mail: m.venturin@enginsoft.com

G. Bretti et al. (eds.), *Mathematical Models and Computer Simulations for Biomedical Applications*, SEMA SIMAI Springer Series 33,
https://doi.org/10.1007/978-3-031-35715-2_7

for future analysis pipelines focusing on predicting bioaccumulation in biological tissues by investigating molecules' structures.

1 Introduction

The processing of chemical substances in living organisms requires an equilibrium between accumulation and elimination: when this equilibrium is disrupted because the uptake rate overcomes the elimination phase, the risk of chronic poisoning increases. Accumulation in biological tissues involves substances taken with diet, by respiration, or through the skin, while elimination is mediated by catabolism or excretion. Bioaccumulation is mainly adopted when referring to nonresident chemicals that are imported into the organism; xenobiotic substances retained in the bodies may be organic, metals, or radioactive elements (radionuclides) [1]. The intermediate procedure between absorption and elimination exhibited by the living creatures is the biotransformation of the exogenous chemicals by enzymatic action into compounds more easily excretable [2]. However, in humans, this cleanup ability is influenced by factors like age, gender, health status, and genetic diversity with the addition of environmental and living style variables. Storage sites of exogenous substances vary: hydrophobic organic chemicals are mainly stored in lipid tissues through the passive exchange with water, while metal concentration is associated with several tissues like the liver, kidney, spleen, lung, and muscles. Understanding bioaccumulation mechanisms is essential to protect humans from the adverse effects of exposure to risky elements. A historically famous example of bioaccumulation in workplaces was the mercury poisoning of hatters. The poisoning derived from the felt manipulation was diluted over time and led to long-term lipid accumulation affecting the brain. More recently, heavy metal accumulation in tissues of backyard chickens raised near industrial sites provided a case of possible chemicals absorption involving the human food chain [3]. Another kind of chemical that may persist in the organism and cause side effects are therapeutic drugs [4]. Bacteria may accumulate exceeding molecules into the intracellular space as a partial protection mechanism to diminish the on-site concentration. This particular accumulation strategy of gut microorganisms prevents biotransformation through the metabolism and alters pharmacokinetics.

Computer modeling of kinetic processes causing bioaccumulation should consider the different phases of absorption, distribution, metabolism, and elimination. Mathematical models of bioconcentration may involve the quantitative structure-activity relationships (i.e., QSAR) to find linear or non-linear analogies between measured biological or physicochemical factors and the molecular structure. Alternatively, the mass-balance approach divides the body into compartments to simulate their interplay during the accumulation process. In compartmental models, the metabolism is the stage that encloses significant difficulties for correctly representing bioaccumulation pathways. Over the past, different techniques like explicit single-reaction enzyme models, genetic algorithms, neural networks, hierarchical rule-based simulations, or probabilistic approaches have been proposed to study the kinetics during this phase. Rather than mechanistic modeling, QSAR studies could be addressed by employing machine learning algorithms (i.e., ML), a group of data-driven methods that gather information from the experimental variables and learn data recognition through supervised or unsupervised processes. Machine learning establishes a predictive model that connects input factors to an outcome, providing in silico drivers for experimental design choices [5]. In-vivo studies are resource-intensive and pose ethical concerns; thus, in-silico methods provide alternative ways to direct research hypotheses and support experimental decisions. The advantage of superseding animal testing for the safe use of chemical substances with QSAR analysis was recognized in the European legislative framework REACH (Registration, Evaluation and Authorisation of Chemicals—Regulation (EC) No 1907/2006). Because of their nature, ML outcomes should be explained by in vitro or in vivo biological evidence, establishing a bi-directional relationship with laboratory experiments. They support decision-making, and indirectly, experimental results confirm ML findings by unveiling the "black box". Another practical benefit of ML is that input features can be tested exhaustively without the risk of combinatorial explosion during large-scale computations [6]. In addition, ML could be integrated into composite 3D or more advanced multidimensional QSAR pipelines to enhance drug discovery [7, 8].

The features of chemical compounds could be encoded in binary vectors, also called molecular fingerprints (i.e., MF). In MF sequences, the molecular structure is represented as 0 or 1 digits, meaning the presence or absence of pre-defined substructures, and offering a convenient data transformation to speed up cheminformatics calculations for virtual screening and molecules grouping by similarity searching [9]. The MF sequence's length depends on the type of encoding: for example, the "Molecular ACCess System" keys (i.e., MAACS) are available in 166 or 960 bits, while PubChem fingerprints are 881 binary digits long. The shorter MAACS type is freely available and, for this reason, often included in open-source cheminformatics software. Hashed fingerprints are a particular type of MF that describes molecules without using pre-defined substructures but applying the hash function to map arbitrary-sized structural data to a specific length vector. Recently, researchers started to explore the possibility of exploiting the structural properties of molecules encoded by MF with ML models for chemometric analysis. For example, in [10] the authors used Morgan and atom-pair MFs to predict ionic liquids' refractive index and viscosity. Their QSAR experiment found that Xgboost-based regression on MFs showed consistent performance in mapping the target variables, and their outcomes were compatible with those found employing traditional ML methods relying on molecular descriptors as input features. The authors of [11] tested different MFs in a QSAR framework based on artificial neural networks (i.e., NN) for ligand binding affinities: with high-throughput virtual screening, scholars could check 200,000 chemical structures to search for those with high binding potential with a target receptor. Three molecules with good affinity emerged from the investigation, demonstrating how structural information alone might be sufficient for QSAR analysis aiming at drug discovery. A study of the genetic toxicity based on compounds' structural information of MFs was carried out in [12]. Random Forest (i.e., RF) and non-linear Support Vector Machines (i.e., SVM) performed well on MF data, including MAACS (AUC, area under the receiver operating characteristic curve was 0.947 and 0.931, respectively). In another manuscript targeting bioconcentration prediction in aquatic species, researchers paired an SVM regressor with a preliminary recursive feature elimination of molecule's structural (from MFs) and physicochemical properties [13]. Feature elimination in ML facilitates learning because it reduces the impact of correlated and less informative features. The authors stated that regression outcomes were in line with results from previously published models. Moreover, the researchers also used the same mixed structural and functional features to classify the toxicity or not of the compounds in the same work (binary classification). After training, the accuracy on unseen data was 87.3% using an SVM classifier, but the different accuracy found on the validation set suggests the chance of overfitting.

Besides being the computational unit in informatics, binary structures are also used to represent neural cell activity. Biological neurons have a concentration of different ions across the plasma membrane (i.e., concentration gradient), inducing a voltage difference between outer and inner membrane compartments (i.e., membrane potential). This baseline electrical state is kept constant thanks to active ionic channels, part of the basal metabolism, compensating for passive flux across the membrane that follows the electrochemical gradient. In general, the resting electrical potential of the membrane is constant along time, and it could be portrayed as a vector of zeros until the neuron does not receive external excitatory or inhibitory stimuli.The incoming stimulus alters the membrane potential by opening the voltage-gated cationic-specific channels when the neuron is activated through one of its afferents. The rapid influx of sodium generates a membrane depolarization (aka action potential or simply "firing") that is suddenly compensated by potassium efflux restoring the membrane's resting potential. The quick depolarization of the neuron is marked by inserting a 1 in the binary sparse vector showing neuron activity. The dynamics of voltage-gated channels leading to membrane charge restoration also include a short refractory period that prevents the neuron from firing again for a certain amount of time. In neurophysiology, the neural activity is approximated by a set of differential equations or modeling the membrane voltage as a resistors-capacitor circuit to show current-voltage relations. Spiking neural networks (i.e., SNN) are algorithms that replace the activation function-based neurons derived from the perceptron [14] employed in ANNs, not accounting for membrane dynamics, with biologically inspired neurons [15]. Moreover, learning is also inspired by the processing principles found in the human brain, allowing higher developed abilities like online learning [16]. Due to their nature, SNNs accept binary vectors as inputs, resembling the neural activity of a pre-synaptic neuron and output spike trains whose information could be decoded as the probability of an outcome.

This study aims to conduct a virtual screening experiment starting from MFs to predict three possible accumulation outcomes: the chemical is retained in the lipidic tissue, particles are biotransformed and consequently eliminated or stored in specific sites like proteins. The in-silico models evaluated during the numerical experiments will be based on distinct machine learning techniques, including spiking neural networks. In particular, the experiment involving SNNs intended to verify if MFs could serve as inputs mimicking spike data. Theoretically, MFs being binary sequences in a certain sense, satisfy the "three S" characteristic of SNNs: spikes, sparsity, and static suppression (a.k.a, event-driven processing). As previously reported, a few ML methodologies have already applied MFs for QSAR, but to the best of our knowledge, this is the first attempt that includes SNNs.

2 Materials and Methods

The authors of [17] prepared and annotated the database under exam that aggregates 779 molecules described in "simplified molecular-input line-entry system" format (i.e., SMILE), an alphanumeric string reproducing the structural patterns. For each molecule, the definition of the outcome class was recorded as the following: class 1 encloses 460 compounds that mainly concentrate in lipidic tissue, class 2 contains 64 chemicals that interact with other tissues, and class 3 reports 255 molecules biotransformed by the organism. The authors also included nine functional descriptors calculated programmatically not considered in the present investigation. Although the dataset investigated aquatic species, the modern approach for the toxicokinetic models is to consider humans together with the environment, identifying similarities between life forms for a unified health risk assessment [18]. Cross-species extrapolation may enhance understanding of chemistry-biology interactions and data from different sources integrated for predictive risk models returning exportable insights.

2.1 Data Processing

Short MAACS fingerprints were generated with the Python RDKit package [19] from the SMILEs after standardization of the molecule structure and normalization of the functional groups (Fig. 1). The database contained an unequal distribution of molecules through classes: without proper treatment, results may be unreliable, with the model being overexposed to the majority class during training. Classes were equalized by oversampling, a straightforward technique that experimental evidence showed superior to undersampling [20]. All models tested underwent two-round cross-validation (i.e., CV): an inner (or out-of-bag) loop with train and validation data splits for hyperparameters tuning, while the final prediction was on a subset of unseen data (20%) in the outer cycle. The two-round (aka nested) CV reduces overfitting because hyperparameter search and model evaluation are two distinct phases [21]. The nested CV configured as a stratified fivefold probed the hyperparameter space to find the best combination of values, and after achievement, the estimator retrained with the optimal parameter set on the test set, as shown in Fig. 3. The whole procedure has been repeated ten times to explore

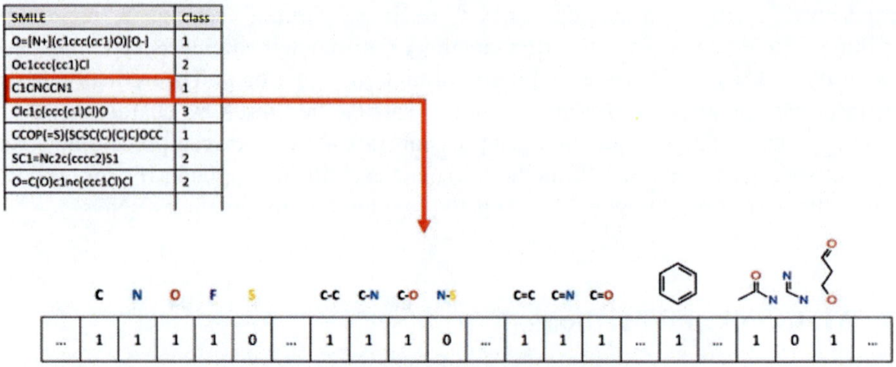

Fig. 1 Exemplification of MF creation from the SMILES present in the dataset

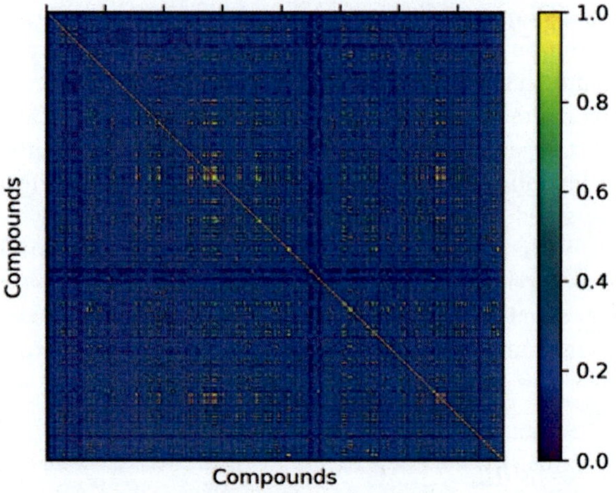

Fig. 2 Gauging the similarity between molecules of the dataset by the Jaccard–Tanimoto index. Each point of the matrix is a coefficient measured between two MFs. In general, the molecular structure of the compounds under analysis is discordant, proving that the ML models will not focus on a family of molecules only

the feature space extensively and evaluate the performance variability of the fully tuned classifiers. The similarity between MAACS MFs of the dataset has been computed by the Jaccard–Tanimoto coefficient, a measure of diversity between binary vectors [22]. It is the proportion of on-bits shared by two fingerprints divided by their union. Figure 2 is an image displaying the symmetric matrix collecting similarity indexes of all compounds under exam transformed into a color scale: yellow values mean high affinity while blue points indicate that molecule structures diverge from another. The averaged Jaccard–Tanimoto coefficient for the dataset was 0.21, assuring the heterogeneity of the molecules under analysis (Fig. 3).

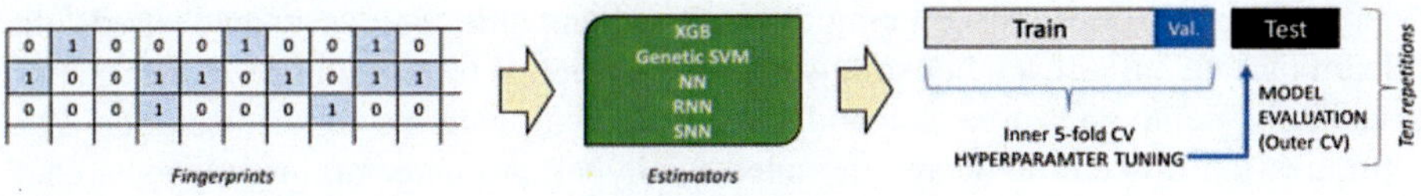

Fig. 3 Experimental procedure during the numerical experiments

2.2 *Machine Learning Models*

The numerical experiments employed the following techniques to predict bioaccumulation:

- Extreme Gradient Boosting (i.e., XGB), inspired by the work of [10] on MFs
- SVM is a technique already employed in [13] and [12]. However, the current work investigated SVM models tuned using genetic algorithms
- ANN shallow or multilayer, similarly to what is shown in [11] with shallow NN
- ANN in recurrent configuration (i.e., RNN)
- SNN

The actual study tested the ability of recurrent networks to model binary sequences, albeit RNNs were previously employed on SMILEs rather than MFs. Furthermore, the SVM models were paired with a genetic algorithm for parameter optimization, while for XGB, the parameters were tested by the grid and Bayesian search. Hyperparameter tuning is an important stage of ML development and one of the methods that help researchers control the model's overfitting [23]. The ANNs and SNNs were verified with gradient clipping to avoid exploding gradients and weight decay regularization as an overfitting prevention strategy (also called L2-norm regularization). For overfitting control, dropout has been applied on RNNs. For all types of neural networks, class membership has been calculated with cross-entropy loss adjusted for multiclass problems. Implementation of algorithms was in Python programming language; in particular, NNs were based on Torch [24, 25], SVM using Scikit–Learn [26], and XGB with XGBoost [27] libraries.

2.2.1 Extreme Gradient Boosting

XGBs are decision tree ensembles, part of the family of nonlinear supervised ML techniques [28], that gather multiple weak classifiers to build a composite classifier for enhanced prediction. In XGBs, the input variables receive a weight, adjusted recursively by the tree sequence, and increased on erroneously classified instances. Weights portray the contribution of every single example to the loss function and consequently to the gradient. The weight adjustment procedure includes their regularization to shrink them by shifting the splits taken. In recent literature, XGBs appeared a popular choice in the field of QSAR targeting the prediction of molecular property or activity [29–31]. For XGBs, the hyperparameters that underwent tuning were "minimal split loss" (pruning the splits when values are less than desired), the "learning rate" after each boosting step, the number of trees included as estimators, maximum depth of a tree, L1 and L2 regularization terms of the weights. Two optimization procedures were attempted for hyperparameter tuning. One searched for the optimal parameter combination over a grid of values, while the other was a sequential design strategy (i.e., Bayesian, through the text abbreviated as Bayes) with the loss function's prior and posterior probability distributions modeled on parameters by a Gaussian process [32]. In the XGBs classifiers, the softmax function has been used to extract the probabilities and accomplish multiclass classification.

2.2.2 Support Vector Machines

Earlier works in cheminformatics put into action SVM for QSAR in drug discovery [33, 34] or to find binding affinities between drugs and proteins [35]. In actual work, the SVM model used a non-linear kernel based on a radial basis function to seek an adequate hyperplane in the feature space, maximizing the boundary between the three categories of bioaccumulation. The hyperplane that maximized the margin between decision boundary and support vectors on the training data was selected iteratively to minimize the classification error. The SVM classifiers may apply soft or hard margin, with the former using a regularization parameter (i.e., usually called "C"), to control the penalty of misclassification during training and deliver a relaxed constrain function that accommodates noisy data. In the current numerical experiment on SVMs, together with regularization, also curvature of the decision boundary has been optimized with a genetic algorithm (i.e., GA). GAs are based on the principle of Darwinists' evolution, where genetic variations provide abilities that allow individuals to survive. In ML, each individual is a solution to a problem, and each individual survives through iterations (also called generations) based on the selection of the best outcomes (i.e., fitness) [36]. Each generation of individuals provides a new set of solutions to the problem, with GA evaluating the performance in regards to the constraints imposed initially. Individuals' fitness to the problem requirements creates a pool of solutions that survive within the next generation and undergo genetic manipulations, subdivided into crossover and mutations: crossover recombines the possible solutions while mutations introduce random variations. The randomly chosen mutations help the algorithm escape local minima, while recombination could be similar to chopping data from two vectors of numbers to generate two new sequences of values. The combination of both crossover and mutation helps the algorithm cover all the search space of optimal solutions. In addition, more sophisticated genetic mechanisms such as inversion, dominance, and genetic edge recombination are available. A prominent individual selection method in GA algorithms is called "tournament selection", where the fittest candidates of the actual generation are selected by comparing subgroups of k individuals. Withal, a standard CV hyperparameter grid search has been evaluated, and in both SVMs, the probabilities of the classifiers were calibrated. The classes were assigned with one-versus-one heuristic.

2.2.3 Neural Networks

Neural networks connect processing units (aka neurons) that assemble weighted inputs and a quantity called bias by summation to produce one output value resulting from a transfer function (also called activation function). This information flow through the NN is commonly called "forward pass", and it can involve neurons arranged in one or more subsequent layers. Once the forward pass ends, the error between computed and desired output is used to update the input weights (i.e., "backpropagation"), aiming to reach a global minimum in the loss function. Weights and biases are adjusted according to an optimization function during the backpropagation. The optimization function gets the partial derivative of the loss function, modifying the initial weight values towards the global loss minima, where the computed and ground truth values are closer to each other. Researchers broadly exploited NN in QSAR by designing complex architectures involving several layers of neurons [37–40]. The number of layers and neurons in each layer have been probed in current numerical experiments. The parameters of the optimizer function underwent analysis likewise. A special kind of NN called recurrent artificial neural network was included in the experiments to verify if their temporal dynamic behavior may retain MF structural information more efficiently. Indeed, the RNNs have an internal state associated with each neuron that allows sequential data processing. The internal state collects information of previous inputs, and the memorized previous states are consolidated in evaluating the current input. Two common types of RNN are the long short-term memory (i.e., LSTM) and gated recurrent unit (i.e., GRU): both use gating mechanisms to regulate the memorization process, and in deep NN designs, they can circumvent the vanishing gradient problem that affects backpropagation. Scrutinizing QSAR literature, recurrent architectures could be found mainly combined with NNs for ensemble learning [41, 42].

2.2.4 Spiking Neural Networks

Spiking neural networks progressed in the last 20 years when scholars applied the neuroscience principles of the timing of action potentials found in human neurons to determine novel information-processing artificial intelligence routines [43]. The research community showed interest in SNNs' capacity to reproduce natural intelligence mechanisms and find alternatives to the computational costs connected with ANNs. Indeed, traditional NNs are resource intensive in terms of memory and calculation speed, as highlighted in EU and US governmental plans for AI, where official documents underline the requirement for faster chips [44, 45]. These regulatory initiatives advertised the problem without providing possible solutions for an issue that should be tackled urgently in the light of Moore's law and von Neumann's bottleneck for computers rooted in the separation between memory and processing unit, demanding a continuous and costly data exchange between them [46]. An answer could be provided by developing SNN-specific computers called neuromorphic calculators. The clock-based traditional hardware is replaced by event-driven devices that reach high-efficiency thanks to accumulation-based alternate computing [47–49]. Indeed, SNNs produce an output like biological neurons do, only when the membrane potential hits a threshold value, rather than NNs neurons that output values at every propagation cycle. In addition, discoveries in neurobiological research are driving the SNNs design toward sophisticated evolving connectionist systems [50]. SNN for bioaccumulation prediction featured both a leaky integrate and fire neuron model (i.e., LIF [51]) and the improved version with synaptic conductance (i.e., SYN [52]). The difference between them is the presence in the latter of a parameter that controls the decay of the synaptic current, while the membrane potential waning parameter exists in both types of neurons. The decays follow an exponential rule. A surrogate fast sigmoid function has been used to smooth the Heaviside step function characterizing the binary vector to overcome the lack of derivability of the discrete values during backpropagation [53]. This workaround also facilitates the adoption of traditional NN optimizers and loss functions. It should be underlined that surrogate gradients match the so-called neo-Hebbian "three-factors" plasticity mechanisms found in brain processing [54], thus maintaining biological plausibility [55]. The spike trains produced by the output layer of the network have been converted into a class label establishing membership according to the neuron of the three with the highest amount of spikes fired. A simulation of the behavior of LIF and SYN neurons when they receive one MF as input is in Fig. 4.

SNN Neuron Models

The LIF neuron model implemented in actual work considers the membrane passive ionic flow V only (Eq. (1), setting $V_{rest} = 0$) and could be illustrated by an electrical circuit, where the resistance and a capacitor determine the membrane voltage changes (Fig. 5). The capacitor describes the membrane insulation, while the resistor stands for the membrane conductance (conductance and resistance have an inverse relation). An alternative LIF design could include a battery to maintain the resting voltage V_{rest}, equivalent to the membrane resting potential given by active ionic transportation. When the LIF receives spike impulses from a pre-synaptic neuron over time t, a current pulse I is injected into the circuit, and the neuron responds by firing if $V(t) > \vartheta$. After each firing, the threshold ϑ is subtracted from the membrane voltage to simulate the discharge during the refractory period. The equations embody passive membrane properties as modeled by passive electronic components:

$$\tau_{RC} \frac{dV}{dt} = V_{rest} - V(t) + RI(t) \tag{1}$$

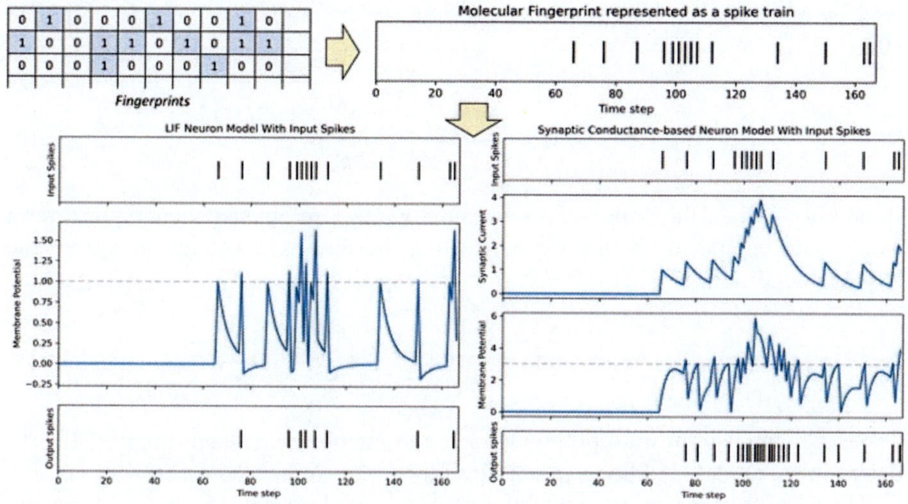

Fig. 4 In the upper part of the figure, one MF from the dataset is mapped as a spike train in a raster plot, with vertical bars representing firing times (the 1 s in the binary MF). In the lower part of the figure, the MF serves as input spike train for two neurons: a LIF (left) and an SYN powered with synaptic conductance (right). The membrane potentials vary over time due to the incoming stimuli from the MF input spike train. The membrane threshold (dashed horizontal line) can be adjusted to control the firing rate. The resulting output spike for the two neurons shows a spike every time the membrane voltage passes the threshold

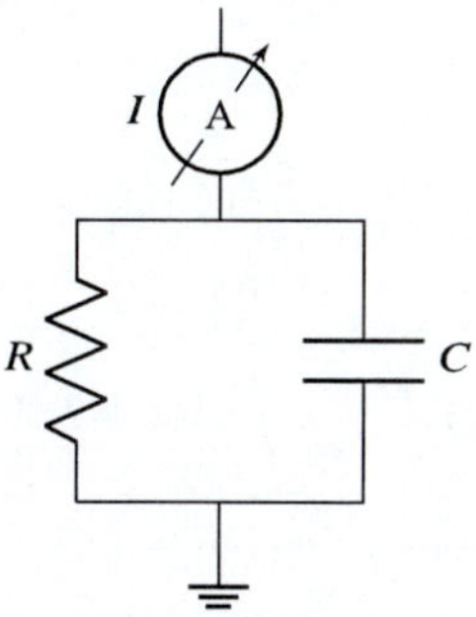

Fig. 5 Membrane behavior in LIF neurons can be interpreted through an electric circuit with one resistance R and a capacitor C in parallel. When the neuron receives stimuli, a current I is injected into the circuit, and it divides into the R and C branches determining membrane potential variations. When the current pulse ends, the membrane voltage returns to zero according to the time constant of the circuit τ_{RC}

And by applying the forward Euler method, the membrane voltage a short time afterward is

$$V(t + \Delta t) = V(t) + \frac{\Delta t\Big(V_{rest} - V(t) + RI(t)\Big)}{\tau_{RC}} \tag{2}$$

In addition, when the neuron fires because $V(t) > \vartheta$, instantaneously the reset mechanism enters into action by subtracting the threshold voltage to restore the baseline conditions (Eq. (3)).

$$V(t + \Delta t) = \exp\left(\frac{\Delta t}{\tau_{RC}}\right) V(t) + I(t + \Delta t) - \vartheta \tag{3}$$

The exponential term multiplying $V(t)$ is the membrane voltage decay. LIF is a deterministc model, and firing rate follows the input current increases.

The SYN neuron simulates the interaction between presynaptic and postsynaptic neurons: in the LIF neuron, the membrane voltage jumps immediately when a current is received while the membrane potential raises gradually in the SYN neuron. It simulates the slow gathering of neurotransmitters in the postsynaptic neuron. Indeed, SYN contains an additional term called "synaptic current" I_{syn} slightly different from the current input I: the synaptic input is more precisely described by a change in conductance rather than a current injection. The reaction of a postsynaptic neuron to a presynaptic stimulus can be magnified or reduced by the difference between I_{syn} generated by a time-dependent synaptic conductance and membrane potential [56].

$$I_{syn} = \exp\left(\frac{\Delta t}{\tau_{syn}}\right) I_{syn}(t) + I(t + \Delta t) \tag{4}$$

$$V(t + \Delta t) = \exp\left(\frac{\Delta t}{\tau_{RC}}\right) V(t) + I_{syn}(t + \Delta t) - \vartheta \tag{5}$$

Likewise, in Eq. (3), in the SYN the threshold value is subtracted only when the neuron fires (Eq. (5)). The conductance time constant τ_{syn} reflects changes caused by the neurotransmitter in the neuron synapse [57]. For example, the synaptic profiles of neurons with different time constants are illustrated in Fig. 6 when one afferent presynaptic stimulus reaches the postsynaptic SYN. The LIF neuron has the τ_{syn} value in Eq. (4) equal to zero, and biologically it reflects only the fast excitatory transmission of AMPA receptors (α-amino-3-hydroxyl-5-methyl-4-isoxazole-propionate) commonly found in the central nervous system [58]. Instead, the synaptic time constant adds the NMDA (N-methyl-D-aspartate) receptors' contribution to the postsynaptic neuron, whose time course depends on sodium, calcium, and potassium influx. In the examples in Fig. 4, the exponential terms have been set as the following equations and abbreviated as "Beta" and "Alpha"

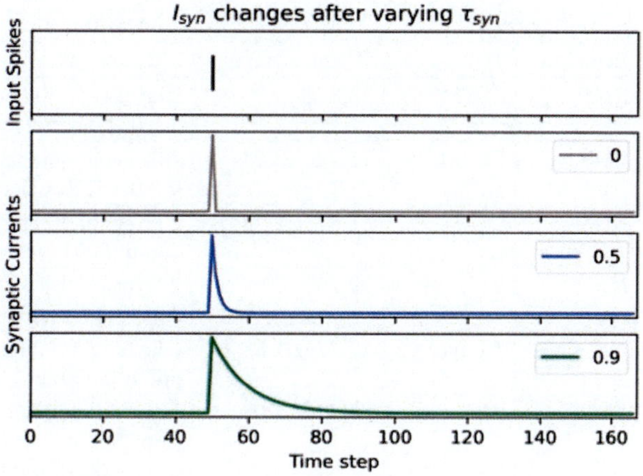

Fig. 6 A single spike at $t = 50$ is delivered to a resting SYN neuron. If τ_{syn} is set to zero, the synaptic current raises and decays immediately (this is what happens in LIF neurons). In the presence of a synaptic time constant set to 0.5 or 0.9 the decay over time simulates the two types of excitatory receptors, AMPA and NMDA

in Tables 1 and 2.

$$\exp\left(\frac{\Delta t}{\tau_{RC}}\right) = 0.9 \quad \exp\left(\frac{\Delta t}{\tau_{syn}}\right) = 0.8$$

3 Results

The balanced accuracy has been selected as an evaluation metric [59] to compare results between estimators on the unseen series of MFs examples (aka test set). It is calculated as the arithmetic mean of sensitivity and specificity. Table 1 describes the performance of all classifiers and their tuned parameters (the hyper-parameter search range is in Table 2 of the Appendix). The standard deviation values are similar, ranging between 2% to 3%, except for the RNN LSTM with ReLU activation. Accuracy values range from 71% to 87%. Accuracy distribution is delineated in the boxplots of Fig. 7. Estimators' performance levels on the test sets showed identical ranges in the performance established with SVM classifiers (mean accuracy around 87%). The extents enclosed by the boxplots of XGBs and SNN LIF (mean accuracy around 83%) were consistent and appeared to overlap SVMs' ranges partially. The SNN SYN and RNN GRU values were closer to each other (mean accuracy of approximately 81%)but departed from the limits of the GA SVM boxplot. NNs and sigmoid LSTM (78 to 79%) ranges were similar but at the last places of the overall accuracies, while LSTM with ReLU showed inefficient results (71%). However, a visual inspection of the results might lead to incorrect considerations if not assisted by a statistical assessment of the performance of each learning algorithm [60]. Each estimator's test set accuracies underwent pairwise statistical analysis based on the Mann-Whitney U independent samples two-tailed

Table 1 Balanced accuracy on test set data

Estimator	Type	% Bal. accuracy	Characteristics[b]
RNN	LSTM	78.8 ± 2.63	2 layers of 150 neurons (2000 epochs) Adam opt. lr 0.001, sigmoid activation
RNN	LSTM	71.7 ± 13.94	2 layers of 400 neurons (1000 epochs) Adam opt. lr 0.0001, ReLU activation
RNN	GRU	81.47 ± 2.78	2 layers of 100 neurons (4000 epochs) RMSProp opt. lr 0.001, weight decay 0.001
NN	2 Layers	79.76 ± 2.98	2 layers with 1000 and 600 neurons (4000 epochs). Adam opt. lr 0.0005, Tanh[a]
NN	1 Layer	79.77 ± 1.88	1 layer of 1000 neurons (4000 epochs) AdamW opt. lr 0.0001, weight decay 0.01, Tanh[a]
SNN	LIF neur.	83.77 ± 2.07	1 layer of 1500 neurons (300 epochs), surr. grad. slope 50, Stochastic Gradient opt. lr 0.0005, weight decay 0.005, Beta 0.95
SNN	SYN neur.	81.96 ± 2.57	1 layer of 1000 neurons (400 epochs), surr. grad. slope 75, Adam opt. lr 0.0001, Beta 0.85, Alpha 0.9
SVM	GA	87.85 ± 1.64	Genetic alg. crossover prob. 0.3, mutation prob. 0.5, generations 100, pop. size 1000
SVM	Grid	86.9 ± 1.9	C=1, kernel coefficient 5.62
XGB	Grid	83.32 ± 2.19	Reg. lagrangian mult. 0.1, lr 0.4, max tree depth 9, trees number 50, L1 0, L2 0.1
XGB	Bayes opt.	83.39 ± 2.03	Reg. lagrangian mult. 0.1, lr 0.3, max tree depth 12, trees number 115, L1 0.1, L2 0.4

[a] Tanh: Hyperbolic tangent sigmoid activation function
[b] lr: Learning rate, opt: Optimizer

test with posthoc Bonferroni corrections (Fig. 8). The alternative hypothesis was that the accuracies coming from the test sets of two classifiers were not equal for at least one value. Based on statistical analysis, the SVM classifiers do not perform significantly differently from the SNN Leaky and the XGB with parameters optimized by the Bayesian probabilistic expectation scheme. Conversely, there is a significant difference between SVMs and NNs expressed by the adjusted 95% or 99% chance probabilities. Moreover, SVM tuned by genetic algorithms performs significantly differently from the SNN SYN neuron model, while this difference is not present when considering SVM optimized by grid search. In past literature, RNNs were employed over SMILEs and not MF; indeed, their performance during the current investigation did not reach the outcomes of XGB, SVMs, or SNN LIF.

4 Discussion

The three accumulation mechanisms of the molecules examined in the present investigation provide an exposure benchmark applicable to forecast biosorption pathways in extant creatures. The top results pertained to the SVM classifiers,

Table 2 Hyperparameter search range during the inner CV loop

Classifier	Search type	Parameter extents
SVM	Grid	C [1e−9;1e9], Kernel coefficient [1e−9;1e9]
SVM	GA	Population size [100;2000], Gene mutation prob. [0.1;0.69, Gene crossover prob. [0.1;0.9], Tournament size [3;9], Num. of generations [100;500]
SNN LIF	Randomized	Beta [0.5;0.95], Surrogate gradient slope [25;100], Num. of hidden neurons [300;1500] Number of epochs [100;500], Optimizers ['Adam','Adamax','Stochastic Gradient Optimizer','Adaptive Gradient Algorithm', 'Adadelta','AdamW','RMSProp'][a]
SNN SYN	Randomized	Alpha [0.5;0.95], Beta [0.5;0.95], Surrogate gradient slope [25;100], Num. of hidden neurons [300;1500] Number of epochs [100;500], Optimizers ['Adam','Adamax','Stochastic Gradient Optimizer','Adaptive Gradient Algorithm', 'Adadelta','AdamW','RMSProp'][a]
XGB	Grid, Bayes	Gamma [0;200], Learning rate [0.01;0.7], Max depth [5;14], Number of estimators [50;150], Reg. alpha [0;200], Reg. lambda [0;200]
RNN LSTM[b]	Randomized	Number of hidden neurons [100;1500], Number of epochs [500;4000], Optimizers ['Adam','Adamax','Stochastic Gradient Optimizer','Adaptive Gradient Algorithm', 'Adadelta','AdamW','RMSProp'][a]
NN[b]	Randomized	Number of hidden neurons [100;1500], Number of epochs [500;4000], Optimizers ['Adam','Adamax','Stochastic Gradient Optimizer','Adaptive Gradient Algorithm', 'Adadelta','AdamW','RMSProp'][a]
RNN GRU[b]	Randomized	Number of hidden neurons [100;1500], Number of epochs [500;4000], Optimizers ['Adam','Adamax','Stochastic Gradient Optimizer','Adaptive Gradient Algorithm', 'Adadelta','AdamW','RMSProp'][a]

[a] Parameters in common were learning rate and weight decay plus optimizer-specific parameters
[b] Search included tests for 1 or 2 layers structure

with little difference between grid search or GA hyperparameter optimization. Results achieved by SVM are compatible with the findings of other scholars. On a dataset addressing mutagenic toxicity, the authors of [61] applied SVM to MAACS fingerprints for a two classes prediction task, obtaining 84.1% and 92.7% accuracy on validation and test set, respectively. Another SVM estimator touched 90.7% accuracy in forecasting toxic bioaccumulation or not on a mixed dataset with structural (from MFs) and functional features [13]. Other work employing structural information encoded in MFs with a length of 1024 bits and functional attributes arrived at 88.7% accuracy with the SVM classifier predicting the bioactivity of two families of enzymes (histone acetyltransferases or histone deacetylases) [62]. The previous binary classification tasks employing SVMs obtained results compatible with what emerges from the actual investigation but hereby solving a three-class problem. The analysis procedures in the mentioned articles included

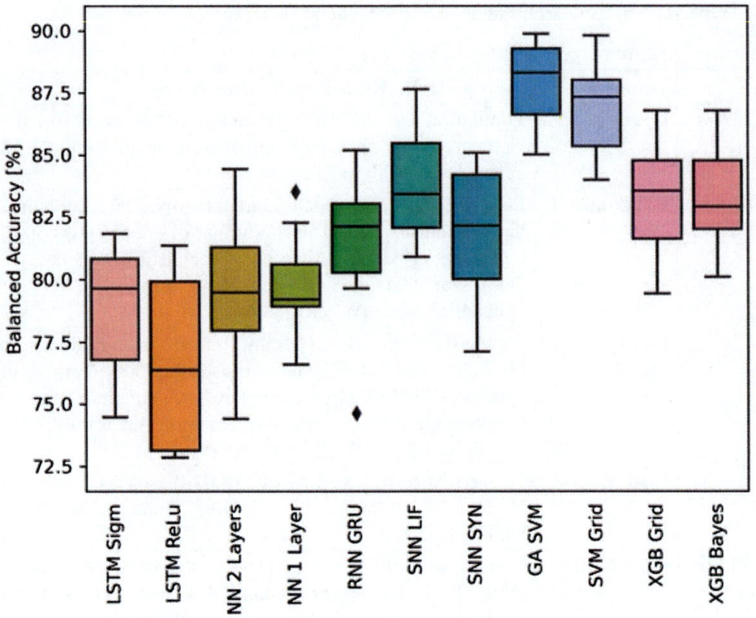

Fig. 7 Results of the numerical experiments for each classifier tested

a fivefold CV, likewise in our numerical experiments. The other classifiers that reached a satisfactory accomplishment were the XGBs and the SNN LIF. While XGB can be found extensively in the QSAR modeling literature [63–65], SNNs were first related to QSAR in the present manuscript. In a machine learning experiment for drug discovery involving XGB as a binary classifier and targeting the mutagenicity of the compounds, the authors of [66] achieved 75% and 74% balanced accuracy on 2048 bits extended connectivity or functional fingerprints. Another work aiming to classify compounds that inhibit β-secretase, an enzyme involved in the accumulation of amyloid-β in the brain, reported an XGB accuracy of 88.1% in categorizing MAACS fingerprints into two classes [67].

SNNs differ significantly from the standard operating modes of the NNs but proved efficient MFs handling for classification purposes. Among the two configurations checked, simpler LIF neurons acted slightly better than the SNN utilizing SYN neurons. This outcome will be verified on more sophisticated datasets in subsequent studies for structure-based virtual screening. In particular, the possibility of invoking learnable parameters in self-adaptation fashion, both for the time constants and the membrane threshold, might improve existing results. The GRU RNN demonstrated accuracy close to the SYN SNN, but the presence of one outlier underlines some difficulties in attaining a stable working performance. Both NNs configurations were inspired by the architecture pursued in [11] without attempting deep designs. In [68], the authors did not identify advantages in increasing the depth of the neural network above three layers, suggesting that two or three should be sufficient for

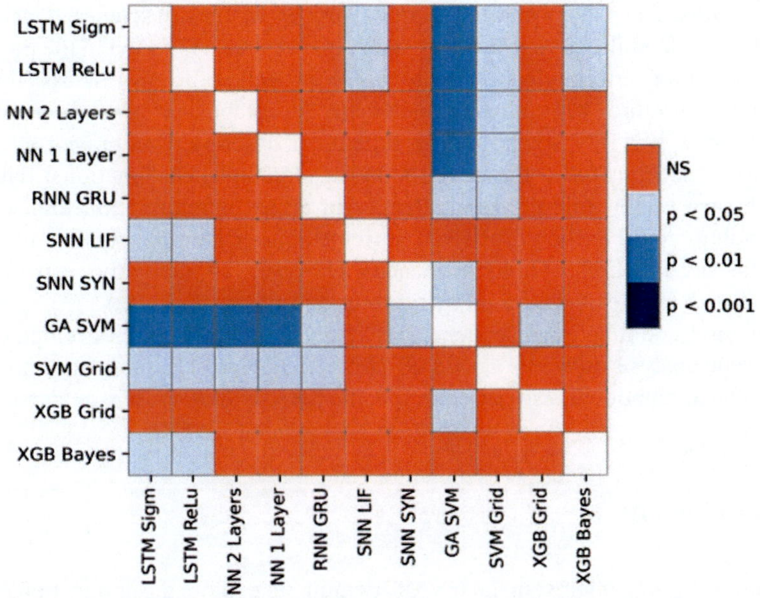

Fig. 8 Bonferroni-corrected posthoc p values of Mann-Whitney U rank test. The term "NS" means "Not Significant"

QSAR bioactivity analysis. The present work confirms this observation; indeed, one or two layers could equally model bioaccumulation from molecule structures. Moreover, for NNs, the same authors suggested applying a large number of neurons in the hidden layers. Indeed, in the numerical experiments involving the NNs, the best results were with 1000 neurons, and for the second layer, 600 units. Another aspect emerging from Table 1 involving the neural networks is the requirement of more training epochs for NNs compared to SNNs. Both one or two layers NNs architectures required 4000 training epochs to reach the best performance, while SNN LIF and SNN SYN learned in 300 and 400 epochs, respectively.

Regarding the compound list under investigation, previous works limited its analysis to a binary categorization of the bioconcentration factor from functional attributes, not evaluating the chemical structure. In [17, 69], the authors of the dataset reported that the accuracy for classifying two classes of bioaccumulation was 73.5% using functional features and classification trees. From this point of view, the current multi-class analysis is more comprehensive than previously published papers on the same database of molecules and enhances the prediction of bioaccumulation pathways using structural rather than functional attributes. In addition, the analysis pipelines studied in the present work were tested on commodity hardware to ensure the broad application of the findings. The application of the proposed analysis pipelines on a public domain database of molecules and the usage of reduced computational resources might facilitate the reproducibility of current results. This latter aspect is of critical interest in modern science, considering the lack of replicability of published mathematical models reported in the past [70].

The structural information encoded by MF proved sufficient to classify bioaccumulation into three physiological pathways. As established in other cheminformatics papers [71, 72], incorporating functional descriptors improved structural ML outcomes. For SNNs, the commixture of structural and functional data in a single binary vector might pose challenges in terms of information loss. Ideally, the paradigm could encode functional attributes into spike trains and combine the structural and functional binary sequences. However, an alternative exists, and it could take advantage of studies on spatio-temporal spiking machines [73], providing the structural and functional sequences as separate inputs. Future developments of the present analysis that introduced SNN for QSAR will explore the molecular structural and functional association in specific SNN-QSAR consensus frameworks.

5 Conclusions

There is an ongoing interest in the QSAR literature to explore the application of MFs for bioactivity prediction. The present investigation proved the possibility of using molecular fingerprint structural information for bioaccumulation modes forecasting in life forms. Using a database of 779 molecules, three accumulation routes were detected from MAACS MFs, with the peak performance achieved by the SVM classifiers. SNNs were tested for the first time on MFs and obtained 83.77% and 81.96% prediction accuracy on unseen data. The outcomes of SNN with leaky neurons and XGB paired with Bayesian optimization did not differ significantly from the SVMs accuracies. While SVM and XGB were already employed as structure-based virtual screening tools from MFs, SNN ended up being a credible alternative for this type of task. Future works will furtherly develop the SNN-QSAR framework and export it to other domains of cheminformatics, also testing the possibility of an integrated structural-functional approach.

Appendix

Table 2 contains the search limits for tuning each hyperparameter during the CV procedure. A standard approach to hyper-parameter tuning is given by Grid search, where all combinations of a set of values for each parameter were tested. Randomized search is an alternative method, mainly employed to overcome the drawback of Grid search that might explore the hyper-parameter space in not influent areas. Randomly sampling from a specified continuous distribution during the numerical experiments offered a more scalable approach than Grid search. The Bayesian optimization procedure models the hyper-parameters search space using a Gaussian Process to minimize an objective function representing the model with the best accuracy given a set of parameters. The advantage is that over time, regions of hyper-parameter space not performing adequately are discarded while the search focuses on desirable areas. In the current investigation, Bayesian optimization looked for parameters over a dense grid of values.

Author contributions

MN (conceptualization, methodology, numerical experiments, analysis, manuscript draft writing, manuscript editing and revision), SS (project supervision, manuscript revision), LR (funding, project administration, project supervision, manuscript revision), MV (manuscript revision). All authors discussed the results and commented on the manuscript.

Acknowledgments This project has received funding from the European Union's Horizon 2020 research and innovation program under the Marie Sklodowska Curie grant agreement No 860462. This manuscript respects only the author's view, and the Agency and the Commission are not responsible for any use that may be made of the information it contains.

References

1. Connell, D.W.: Bioaccumulation of Xenobiotic Compounds. CRC Press, Boca Raton (1989)
2. Ali, H., Khan, E.: Trophic transfer, bioaccumulation, and biomagnification of non-essential hazardous heavy metals and metalloids in food chains/webs-concepts and implications for wildlife and human health. Human Ecol. Risk Assess. Int. J. **25**(6), 1353–1376 (2019)
3. Kar, I., Mukhopadhayay, S.K., Patra, A.K., Pradhan, S.: Bioaccumulation of selected heavy metals and histopathological and hematobiochemical alterations in backyard chickens reared in an industrial area, india. Environ. Sci. Pollut. Res. **25**(4), 3905–3912 (2018)
4. Klünemann, M., Andrejev, S., Blasche, S., Mateus, A., Phapale, P., Devendran, S., Vappiani, J., Simon, B., Scott, T.A., Kafkia, E., et al.: Bioaccumulation of therapeutic drugs by human gut bacteria. Nature **597**(7877), 533–538 (2021)
5. Qiao, Z., Li, L., Li, S., Liang, H., Zhou, J., Snurr, R.Q.: Molecular fingerprint and machine learning to accelerate design of high-performance homochiral metal–organic frameworks. AIChE J. **67**(10), e17352 (2021)
6. Tang, J., Aittokallio, T.: Network pharmacology strategies toward multi-target anticancer therapies: from computational models to experimental design principles. Curr. Pharm. Des. **20**(1), 23–36 (2014)
7. Durai, P., Ko, Y.-J., Pan, C.-H., Park, K.: Evolutionary chemical binding similarity approach integrated with 3d-qsar method for effective virtual screening. BMC Bioinf. **21**(1), 1–18 (2020)
8. Chen, Y.-N., Qian, J., Liang, R., Zeng, W.-B., Dong, J., Lin, Q.-L.: Precise hapten design of sulfonamides by combining machine learning and 3d-qsar approaches. Food Anal. Methods 1–13 (2022)
9. Yin, Z., Ai, H., Zhang, L., Ren, G., Wang, Y., Zhao, Q., Liu, Q.: Predicting the cytotoxicity of chemicals using ensemble learning methods and molecular fingerprints. J. Appl. Toxicol. **39**(10), 1366–1377 (2019)

10. Ding, Y., Chen, M., Guo, C., Zhang, P., Wang, J.: Molecular fingerprint-based machine learning assisted qsar model development for prediction of ionic liquid properties. J. Mol. Liq. **326**, 115212 (2021)
11. Myint, K.-Z., Wang, L., Tong, Q., Xie, X.-Q.: Molecular fingerprint-based artificial neural networks qsar for ligand biological activity predictions. Mol. Pharm. **9**(10), 2912–2923 (2012)
12. Fan, D., Yang, H., Li, F., Sun, L., Di, P., Li, W., Tang, Y., Liu, G.: In silico prediction of chemical genotoxicity using machine learning methods and structural alerts. Toxicol. Res. **7**(2), 211–220 (2018)
13. Ai, H., Wu, X., Zhang, L., Qi, M., Zhao, Y., Zhao, Q., Zhao, J., Liu, H.: Qsar modelling study of the bioconcentration factor and toxicity of organic compounds to aquatic organisms using machine learning and ensemble methods. Ecotoxicol. Environ. Saf. **179**, 71–78 (2019)
14. Rosenblatt, F.: The perceptron: a probabilistic model for information storage and organization in the brain. Psychol. Rev. **65**(6), 386 (1958)
15. Taherkhani, A., Belatreche, A., Li, Y., Cosma, G., Maguire, L.P., McGinnity, T.M.: A review of learning in biologically plausible spiking neural networks. Neural Netw. **122**, 253–272 (2020)
16. Lobo, J.L., Del Ser, J., Bifet, A., Kasabov, N.: Spiking neural networks and online learning: An overview and perspectives. Neural Netw. **121**, 88–100 (2020)
17. Grisoni, F., Consonni, V., Vighi, M., Villa, S., Todeschini, R.: Investigating the mechanisms of bioconcentration through qsar classification trees. Environ. Int. **88**, 198–205 (2016)
18. Rivetti, C., Allen, T.E.H., Brown, J.B., Butler, E., Carmichael, P.L., Colbourne, J.K., Dent, M., Falciani, F., Gunnarsson, L., Gutsell, S., et al.: Vision of a near future: Bridging the human health–environment divide. toward an integrated strategy to understand mechanisms across species for chemical safety assessment. Toxicol. Vitro **62**, 104692 (2020)
19. Landrum, G., et al.: Rdkit: a software suite for cheminformatics, computational chemistry, and predictive modeling (2013). https://doi.org/10.5281/zenodo.591637
20. Mohammed, R., Rawashdeh, J., Abdullah, M.: Machine learning with oversampling and undersampling techniques: overview study and experimental results. In: 2020 11th International Conference on Information and Communication Systems (ICICS), pp. 243–248. IEEE (2020)
21. Cawley, G.C., Talbot, N.L.C.: On over-fitting in model selection and subsequent selection bias in performance evaluation. J. Mach. Learn. Res. **11**, 2079–2107 (2010)
22. Todeschini, R., Consonni, V., Xiang, H., Holliday, J., Buscema, M., Willett, P.: Similarity coefficients for binary chemoinformatics data: overview and extended comparison using simulated and real data sets. J. Chem. Inf. Model. **52**(11), 2884–2901 (2012)
23. Ying, X.: An overview of overfitting and its solutions. J. Phys. Conf. Series **1168**, 022022 (2019)
24. Paszke, A., Gross, S., Massa, F., et al.: Pytorch: An imperative style, high-performance deep learning library. In: H. Wallach, H. Larochelle, A. Beygelzimer, F. d'Alché-Buc, E. Fox, R. Garnett (eds.) Advances in Neural Information Processing Systems, vol. 32, pp. 8024–8035. Curran Associates, Inc, Red Hook (2019)
25. Eshraghian, J.K., Ward, M., Neftci, E., Wang, X., Lenz, G., Dwivedi, G., Bennamoun, M., Jeong, D.S., Lu, W.D.: Training spiking neural networks using lessons from deep learning. Preprint (2021). arXiv:2109.12894
26. Pedregosa, F., Varoquaux, G., Gramfort, A., Michel, V., Thirion, B., Grisel, O., Blondel, M., Prettenhofer, P., Weiss, R., Dubourg, V., Vanderplas, J., Passos, A., Cournapeau, D., Brucher, M., Perrot, M., Duchesnay, E.: Scikit-learn: machine learning in Python. J. Mach. Learn. Res. **12**, 2825–2830 (2011)
27. Chen, T., Guestrin, C.: Xgboost: a scalable tree boosting system. CoRR (2016). abs/1603.02754
28. Chen, T., Guestrin, C.: Xgboost: a scalable tree boosting system. In: Proceedings of the 22nd ACM SIGKDD International Conference on Knowledge Discovery and Data Mining, pp. 785–794 (2016)
29. Jiang, Z., Xu, J., Yan, A., Wang, L.: A comprehensive comparative assessment of 3d molecular similarity tools in ligand-based virtual screening. Brief. Bioinf. **22**(6), bbab231 (2021)

30. Li, S., Ding, Y., Chen, M., Chen, Y., Kirchmair, J., Zhu, Z., Wu, S., Xia, J.: Hdac3i-finder: a machine learning-based computational tool to screen for hdac3 inhibitors. Mol. Inf. **40**(3), 2000105 (2021)
31. Ye, Q., Chai, X., Jiang, D., Yang, L., Shen, C., Zhang, X., Li, D., Cao, D., Hou, T.: Identification of active molecules against mycobacterium tuberculosis through machine learning. Brief. Bioinf. **22**(5), bbab068 (2021)
32. Turner, R., Eriksson, D., McCourt, M., Kiili, J., Laaksonen, E., Xu, Z., Guyon, I.: Bayesian optimization is superior to random search for machine learning hyperparameter tuning: Analysis of the black-box optimization challenge 2020. In: NeurIPS 2020 Competition and Demonstration Track, pp. 3–26. Proceedings of Machine Learning Research (2021)
33. Zernov, V.V., Balakin, K.V., Ivaschenko, A.A., Savchuk, N.P., Pletnev, I.V.: Drug discovery using support vector machines. the case studies of drug-likeness, agrochemical-likeness, and enzyme inhibition predictions. J. Chem. Inf. Comput. Sci. **43**(6), 2048–2056 (2003)
34. Heikamp, K., Bajorath, J.: Support vector machines for drug discovery. Expert Opin. Drug Discov. **9**(1), 93–104 (2014)
35. Xue, C.X., Zhang, R.S., Liu, H.X., Yao, X.J., Liu, M.C., Hu, Z.D., Fan, B.T.: Qsar models for the prediction of binding affinities to human serum albumin using the heuristic method and a support vector machine. J. Chem. Inf. Comput. Sci. **44**(5), 1693–1700 (2004)
36. Filho, J.L.R., Treleaven, P.C., Alippi, C.: Genetic-algorithm programming environments. Computer **27**(6), 28–43 (1994)
37. Goh, G.B., Siegel, C., Vishnu, A., Hodas, N.O., Baker, N.: Chemception: a deep neural network with minimal chemistry knowledge matches the performance of expert-developed qsar/qspr models. Preprint (2017). arXiv:1706.06689
38. Ghasemi, F., Mehridehnavi, A., Fassihi, A., Pérez-Sánchez, H.: Deep neural network in qsar studies using deep belief network. Appl. Soft Comput. **62**, 251–258 (2018)
39. Ding, Q., Hou, S., Zu, S., Zhang, Y., Li, S.: Visar: an interactive tool for dissecting chemical features learned by deep neural network qsar models. Bioinformatics **36**(11), 3610–3612 (2020)
40. Mozafari, Z., Chamjangali, M.A., Arashi, M., Goudarzi, N.: Performance of smoothly clipped absolute deviation as a variable selection method in the artificial neural network-based qsar studies. J. Chemom. **35**(5), e3338 (2021)
41. Kwon, S., Bae, H., Jo, J., Yoon, S.: Comprehensive ensemble in qsar prediction for drug discovery. BMC Bioinf. **20**(1), 1–12 (2019)
42. Li, Y., Xu, Y., Yu, Y.: Crnntl: Convolutional recurrent neural network and transfer learning for qsar modeling in organic drug and material discovery. Molecules **26**(23), 7257 (2021)
43. Ponulak, F., Kasinski, A.: Introduction to spiking neural networks: information processing, learning and applications. Acta Neurobiol. Exp. **71**(4), 409–433 (2011)
44. Molyneux, C.G., Oyarzabal, R.: What is a robot (under eu law). RAIL **1**, 11 (2018)
45. OSTP White House: Preparing for the future of artificial intelligence. Technical Report, Executive Office of the President-National Science and Technology (2016)
46. Backus, J.,: Can programming be liberated from the von neumann style? a functional style and its algebra of programs. Commun. ACM **21**(8), 613–641 (1978)
47. Akopyan, F., Sawada, J., Cassidy, A., Alvarez-Icaza, R., Arthur, J., Merolla, P., Imam, N., Nakamura, Y., Datta, P., Nam, G.-J., et al.: Truenorth: Design and tool flow of a 65 mw 1 million neuron programmable neurosynaptic chip. IEEE Trans. Comput.-Aided Des. Integr. Circuits Syst. **34**(10), 1537–1557 (2015)
48. Furber, S.B., Lester, D.R., Plana, L.A., Garside, J.D., Painkras, E., Temple, S., Brown, A.D.: Overview of the spinnaker system architecture. IEEE Trans. Comput. **62**(12), 2454–2467 (2012)
49. Davies, M., Srinivasa, N., Lin, T.-H., Chinya, G., Cao, Y., Choday, S.H., Dimou, G., Joshi, P., Imam, N., Jain, S., et al.: Loihi: a neuromorphic manycore processor with on-chip learning. IEEE Micro **38**(1), 82–99 (2018)
50. Kasabov, N.K.: Time-Space, Spiking Neural Networks and Brain-Inspired Artificial Intelligence. Springer, Berlin (2019)

51. Tal, T., Schwartz, E.L.: Computing with the leaky integrate-and-fire neuron: logarithmic computation and multiplication. Neural Comput. **9**(2), 305–318 (1997)
52. Destexhe, A.: Conductance-based integrate-and-fire models. Neural Comput. **9**(3), 503–514 (1997)
53. Zenke, F., Ganguli, S.: Superspike: supervised learning in multilayer spiking neural networks. Neural Comput. **30**(6), 1514–1541 (2018)
54. Kuśmierz, Ł., Isomura, T., Toyoizumi, T.: Learning with three factors: modulating hebbian plasticity with errors. Curr. Opin. Neurobiol. **46**, 170–177 (2017)
55. Neftci, E.O., Mostafa, H., Zenke, F.: Surrogate gradient learning in spiking neural networks: Bringing the power of gradient-based optimization to spiking neural networks. IEEE Signal Process. Mag. **36**(6), 51–63 (2019)
56. Frolov, A.A., Medvedev, A.V.: Substantiation of the "point approximation" for describing the total electrical activity of the brain with use of a simulation model. Biophysics **31**(2), 332–337 (1986)
57. Amit, D.J., Tsodyks, M.V.: Quantitative study of attractor neural network retrieving at low spike rates. i. substrate-spikes, rates and neuronal gain. Network: Comput. Neural Syst. **2**(3), 259 (1991)
58. Kelly, J.B., Zhang, H.: Contribution of ampa and nmda receptors to excitatory responses in the inferior colliculus. Hearing Res. **168**(1–2), 35–42 (2002)
59. Brodersen, K.H., Ong, C.S., Stephan, K.E., Buhmann, J.M.: The balanced accuracy and its posterior distribution. In: 2010 20th International Conference on Pattern Recognition, pp. 3121–3124. IEEE (2010)
60. Bouckaert, R.R., Frank, E.: Evaluating the replicability of significance tests for comparing learning algorithms. In: Pacific-Asia Conference on Knowledge Discovery and Data Mining, pp. 3–12. Springer, Berlin (2004)
61. Xu, C., Cheng, F., Chen, L., Du, Z., Li, W., Liu, G., Lee, P.W., Tang, Y.: In silico prediction of chemical ames mutagenicity. J. Chem. Inf. Model. **52**(11), 2840–2847 (2012)
62. Shi, J., Zhao, G., Wei, Y.: Computational QSAR model combined molecular descriptors and fingerprints to predict HDAC1 inhibitors. Médecine/Sciences **34**, 52–58 (2018)
63. Yang, M., Tao, B., Chen, C., Jia, W., Sun, S., Zhang, T., Wang, X.: Machine learning models based on molecular fingerprints and an extreme gradient boosting method lead to the discovery of jak2 inhibitors. J. Chem. Inf. Model. **59**(12), 5002–5012 (2019)
64. Baidya, A.T.K., Ghosh, K., Amin, S.K.A., Adhikari, N., Nirmal, J., Jha, T., Gayen, S.: In silico modelling, identification of crucial molecular fingerprints, and prediction of new possible substrates of human organic cationic transporters 1 and 2. New J. Chem. **44**(10), 4129–4143 (2020)
65. Khalifa, N., Konda, L.S.K., Kristam, R.: Machine learning-based qsar models to predict sodium ion channel (Nav 1.5) blockers. Future Med. Chem. **12**(20), 1829–1843 (2020)
66. Chu, C.S.M., Simpson, J.D., O'Neill, P.M., Berry, N.G.: Machine learning – predicting ames mutagenicity of small molecules. J. Mol. Graph. Model. **109**, 108011 (2021)
67. Singh, R., Ganeshpurkar, A., Ghosh, P., Pokle, A.V., Kumar, D., Singh, R.B., Singh, S.K., Kumar, A.: Classification of beta-site amyloid precursor protein cleaving enzyme 1 inhibitors by using machine learning methods. Chem. Biol. Drug Des. **98**(6), 1079–1097 (2021)
68. Koutsoukas, A., Monaghan, K.J., Li, X., Huan, J.: Deep-learning: investigating deep neural networks hyper-parameters and comparison of performance to shallow methods for modeling bioactivity data. J. Cheminf. **9**(1), 1–13 (2017)
69. Ballabio, D., Grisoni, F., Todeschini, R.: Multivariate comparison of classification performance measures. Chemometr. Intell. Lab Syst. **174**, 33–44 (2018)
70. Tiwari, K., Kananathan, S., Roberts, M.G., Meyer, J.P., Shohan, M.U.S., Xavier, A., Maire, M., Zyoud, A., Men, J., Ng, S., et al.: Reproducibility in systems biology modelling. Mol. Syst. Biol. **17**(2), e9982 (2021)
71. Guedes, I.A., Pereira, F.S.S., Dardenne, L.E.: Empirical scoring functions for structure-based virtual screening: applications, critical aspects, and challenges. Front. Pharmacol. **9**, 1089 (2018)

72. Maia, E.H.B., Assis, L.C., De Oliveira, T.A., Da Silva, A.M., Taranto, A.G.: Structure-based virtual screening: from classical to artificial intelligence. Front. Chem. **8**, 343 (2020)
73. Kasabov, N.K.: Neucube: a spiking neural network architecture for mapping, learning and understanding of spatio-temporal brain data. Neural Netw. **52**, 62–76 (2014)

Advanced Models for COVID-19 Variant Dynamics and Pandemic Waves

Ryan Weightman, Samantha Moroney, Anthony Sbarra, and Benedetto Piccoli

Abstract Identifying the driving forces of COVID-19 case counts can help decision makers predict possible effects of virus on populations. This would allow for more swift and directed mitigation tactics, possibly even before new case waves appear. We analyze the role of virus mutations in the dynamics of infection spread via the comparison of cases-over-time data with variant specific data. What we find is a strong correlation between characteristic waves in cases and the evolution of the variant mutations themselves. Namely, when a new variant becomes dominant, it is usually followed by a local maximum in cases. We then use this information to fit an epidemiological model which couples ordinary differential equations with Markov chain dynamics to allow for viral mutation. We see that variants dynamics in such a model is enough to both elicit the characteristic waves in cases, and estimate when they appear over a time horizon. This study pave the way to the use of epidemiological models with variants dynamics for accurate predictions and to guide interventions.

1 Introduction

The COVID-19 pandemic began for the United States on January 21st, 2020 with its first reported case in the state of Washington [21]. Within 10 days, the United States had declared a state of emergency, marking the beginning of the fight to keep the virus under control. Early on in pandemic mitigation, non-pharmaceutical

R. Weightman (✉) · A. Sbarra · B. Piccoli
Center for Computational and Integrative Biology, Rutgers University–Camden, Camden, NJ, USA
e-mail: rjw163@camden.rutgers.edu; ajs686@scarletmail.rutgers.edu; piccoli@camden.rutgers.edu

S. Moroney
Department of Mathematical Sciences, Rutgers University–Camden, Camden, NJ, USA
e-mail: sfm96@scarletmail.rutgers.edu

G. Bretti et al. (eds.), *Mathematical Models and Computer Simulations for Biomedical Applications*, SEMA SIMAI Springer Series 33,
https://doi.org/10.1007/978-3-031-35715-2_8

interventions (NPIs) such as masking, testing for the virus, and social distancing were implemented to keep cases and deaths at a minimum while governments around the world worked to better understand how to control viral propagation. This cycle continued in the US until the first vaccines were administered in December 2020 [31]. Soon after, the largest vaccination campaign in history ensued [5], seeming to diminish the number of virus related deaths and severe cases. At this point, many countries began to lift NPIs in an effort to ease society back into some level of normalcy and remove economic stress brought on by COVID-19 regulations [6]. Once the reality set in that the vaccine, while effective against severe symptomatic COVID-19, would not completely eradicate the virus, gears shifted to long-term mitigation [35].

At each step in this process mathematical models have been used to answer driving questions about the virus. First, researchers raced to understand what combination of NPIs could minimize death and economic cost to a population [36]. Then, vaccines were released and researchers became interested in understanding how a vaccine affects viral dynamics and what demographic should be vaccinated first to optimize distribution impact [1, 12, 43]. As vaccines lowered mortality rates, the question shifted to how a population should go about easing out of the less sustainable NPIs [42]. At this point in time, the virus has reached an endemic state; understanding long term viral dynamics through the modeling of longer time-spans is essential for decision makers [4, 10]. Such models include time dependent infection rates [22], shifting social dynamics [46], the role of human behavior in viral transmission [2], the kinematic formulations of contacts themselves to study social heterogeneity [17], the use of such social contact insights in driving public policy [47] and more. Each case requires an implementation that allows for typically constant parameters to shift to become more dynamic. As the human population and virus characteristics change, so must the models. In fact, it is becoming increasingly clear that, while many factors contribute to case counts, new mutation seems to be the key in understanding when local maxima will turn up on a given time horizon [40].

In this chapter, we work at furthering the understanding of this last point: By analyzing and visualizing COVID-19 case data along with virus variant information, we hope to gain a deeper understanding of viral trajectories and the underlying driving forces of case extrema. Our approach is based on two main datasets: the John Hopkins University repository of number of cases and the the GISAID database [18], which includes genetic sequencing, thus enabling identification of the virus variant responsible for the infection. Combining both datasets allows us to elucidate the role played by variants in the appearance of cases increases and peaks. We focus primarily on the state of New Jersey, see Fig. 3, but analyze 9 additional US states representative of diverse demographic profile and geographical location. For all the states we observe the phenomenon of increase and peak following the appearance and of a new variant and its dominance over the other. We also show that other factors, such as holidays and NPIs, may affect the dynamics.

To render our claims more quantitative, we focus on three measures. First, we compute the distance in days between the onset of a new variant (accounting for at least 15% of cases) and its dominance (accounting for more than 80% of cases). We call this number the "switching time" and found high consistency for Alpha (except Alaska), Delta and Omicron variants. The first two variants became dominant in around a month in all states, while Omicron took a couple of weeks, confirming the higher infectivity of the latter. Secondly, we compute the time in days between the variant dominance and case peak. This time consistency is less pronounced for the Alpha variant, but this may be due to the fact the GISAID data were available only for part of the time when Alpha was active. Only 5 states out of 10 show a clear peak for Alpha, while all 10 show a peak for Delta and Omicron. Alpha and Delta took more than two months to cause case peaks, while Omicron took only 25 days (on average and only two weeks for Missouri, New Jersey and Pennsylvania). Finally, we analyze the trend similarities between the dynamic of the percentage of cases due to a variant and the total number of cases. We found a reasonably consistent delay between the derivative peak of the first and the derivative peak of the second. Then we use this information to compare the shifted dynamics of percentage of cases and of total number of cases, and found that relative L^1 errors of around 20% in most states for Delta and lower for Omicron. (We excluded Alpha from this analysis for lack of sufficient data.) We conclude that the onset of a new variant is the key factor for the dynamic of cases, and the latter can be predicted using information about the increase in percentage of cases due to the variant becoming dominant.

Next, epidemiological modeling is briefly introduced before we couple our knowledge of COVID-19 trends with an epidemiological model introduced in [45], which couples a classical system of Ordinary Differential Equations (ODEs) with a Markov chain (MC) to capture dynamics of viral mutation. We first describe the model in details, then use the model to reproduce the waves observed in the analysed data. In case of appearance of variants of similar strengths, the dynamics present the typical peak in waves following a new variant dominance, as well as deep in cases when there is switching between variants. We also include the possibility of reinfection, which is dependent on the variant of first infection. To do so, we need to introduce new compartments, which account for the variant of first infection. The resulting dynamics shows the appearance of endemicity. In this case, the population of recovered peaks quickly after the first infection wave, but then the recovered individuals loose immunity to other variants and we observe a wave-like pattern in the recovered population.

Finally, we test our model by fitting specifically to New Jersey dynamics. To do so, we use data to fit the model parameters (see Sect. 5.3) with adapted parameters for New Jersey's four main variants: Alpha, Delta, BA.1, and BA.2. The results show a good match w.r.t. the total observed cases with clear peaks and transition times between variants. Moreover, the peak timing has a good correspondence with data from John Hopkins University repository. This result is encouraging for the future use of the model to understand the dynamics due to virus variant and study optimization problems related to non-pharmaceutical interventions.

2 Description of Data

COVID-19 case, hospitalization, and death data has been meticulously collected since the start of the pandemic. Johns Hopkins University provides a rich public dataset, providing cases-over-time information for each state in the United States through their COVID-19 Data repository [11]. From here we extract case frequencies from 10 states since the onset of the virus: Alaska, California, Florida, Hawaii, Maine, Missouri, New Jersey, New York, Pennsylvania, and Wyoming. For brevity, all example figures here will use New Jersey data unless otherwise specified, but all analyses were conducted on all 10 sample states.

We took cumulative case data and found the difference between dates in order to get a daily case count. To account for reporting errors, a seven day rolling average was taken for each day, smoothing the data.

The data for some states includes days where the cumulative case total decreases, thus giving rise to a negative count for the daily cases. This result is most likely due to states not reporting during certain weekends or holidays, and cases being retroactively tacked onto the Friday before or Monday after. We removed these negative values before the smoothing took place. See Fig. 1 as an example.

The choice of states for our analysis was motivated by their heterogeneous qualities with respect to each other, such as population size, population age, structure (urban, rural etc.) and other characteristics, such as Hawaii and Alaska

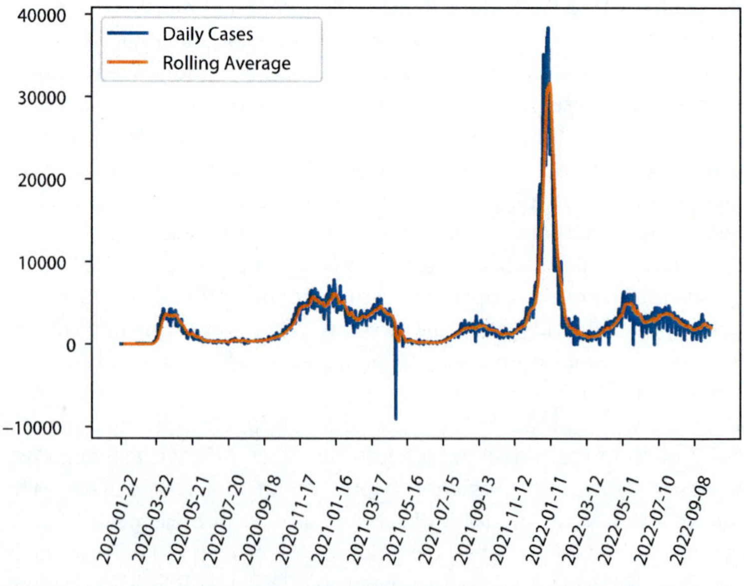

Fig. 1 Cases over time for the state of New Jersey. In blue: raw cases provided by Johns Hopkins University's repository. In orange: smoothed cases with negative values removed and rolling 7 day average taken

Table 1 State demographic information for the states in our sample space. This information is taken from the US census as well as the government websites of the states themselves

State	Pop	Community structure	Average age
Alaska	0.7	80% urban, 20% rural	34.6
California	39.2	94% urban, 6% rural	36.7
Florida	22	92% urban, 8% rural	42.2
Hawaii	1.4	92% urban, 8% rural	39.4
Maine	1.4	60% urban, 40% rural	44.8
Missouri	6.2	63% urban, 37% rural	38.7
New Jersey	8.9	90% urban, 10% rural	40
New York	18.8	95% urban, 5% rural	36.9
Pennsylvania	12.9	79% urban, 31% rural	40.9
Wyoming	0.5	65% urban, 35%$rural$	38

being non-mainland cases. Such characteristics can be major contributors to case count, government policy, public response to mandates, and even virus variant dynamics. While in much of mainland US the dominating variant was fairly homogeneous, there are outliers. For example, while the other states were dealing with the Alpha variant, the dominant variant in Alaska was B.1.1.519, originally detected in Mexico [19]. See Table 1 for a demographic breakdown of each state [33, 41].

We also utilize variant data taken from the GISAID database [18]. GISAID is an initiative that promotes the rapid sharing of data, including genetic sequencing, in order to better understand human viruses. While cases-over-time data has been reported since onset of the pandemic, variant data is a bit harder to come by, especially for early cases (before November 2020). Therefore, we only have significant data starting from November 2020. We acquired all data for our 10 states, focusing on location of patient and variant of COVID-19. An example of the raw variant distribution over time can be seen in Fig. 2 for the state of New Jersey.

The incidence number (y-axis in Fig. 2) of this data is simply the number of cases submitted on a given day. This number does not give information on global case count, but the frequency of each variant in comparison to the other variants should give a good estimate for the times in which new variants of concern (VOC) have become dominant in the infected state-space. For example, in 2021-03, less cases were reported to GISAID than 2021-09, as seen in Fig. 2. However, it is clear from Fig. 1 that there were more new cases in 2021-03 than 2021-09. We took this into consideration by normalizing the variant data to one and considering the distribution of the variants over the infected state-space on any given day. We then smoothed the data to fill in any holes in reporting. The combination of these two data sets gives us our estimates for both the number of people infected and the dominant variant at a given time point. A visual example of how these data-sets, combined, can shed light on interesting patterns can be found in Fig. 3: This already visually suggests a strong correlation between the shifting of variants and local maxima in total cases.

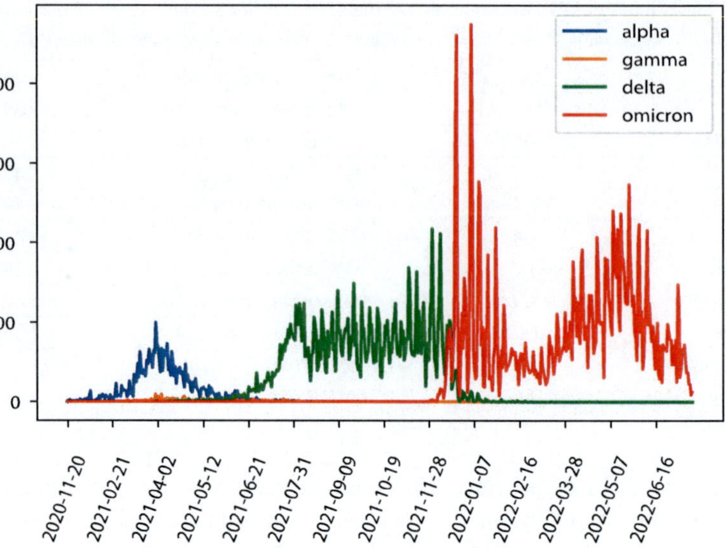

Fig. 2 New Jersey variant data submitted over time through GISAID database

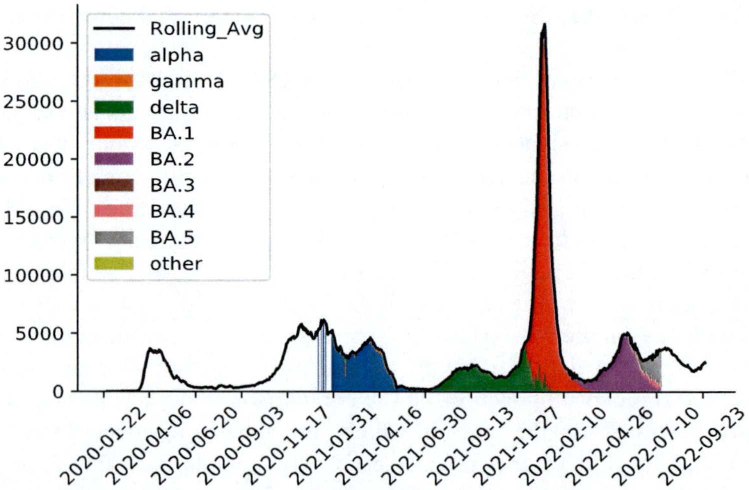

Fig. 3 A combination of GISAID and JHU data. We take our smoothed cases from the JHU data and superimpose a stacked variant percent plot underneath. Note that during the early pandemic, there is reported case data but not reported variant data

Certain states have more sparse reporting in terms of cases, as seen in Fig. 4, and certain states have less data reported in terms of variants, as seen in Fig. 5. We assume here that GISAID reported variant incidence is an adequate random sampling to represent the population. It is possible that there are unforeseen biases skewing the data, such as error due to smaller sample size, or even a severity

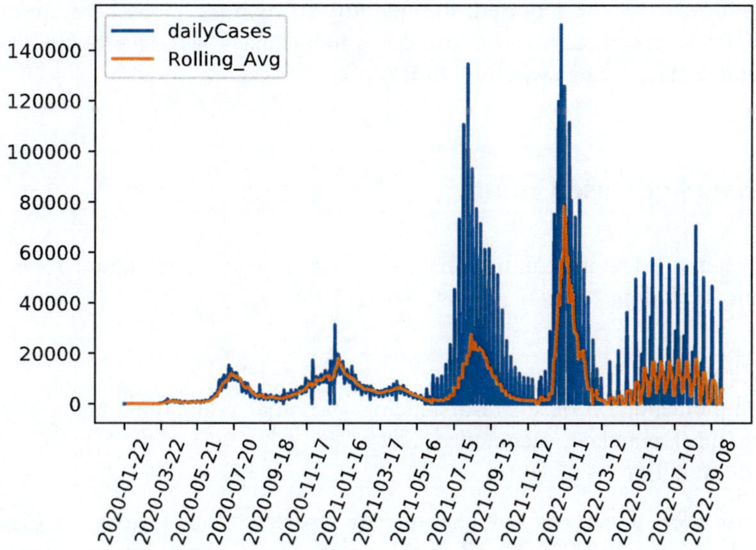

Fig. 4 Florida COVID-19 daily cases over time provided by Johns Hopkins University's repository

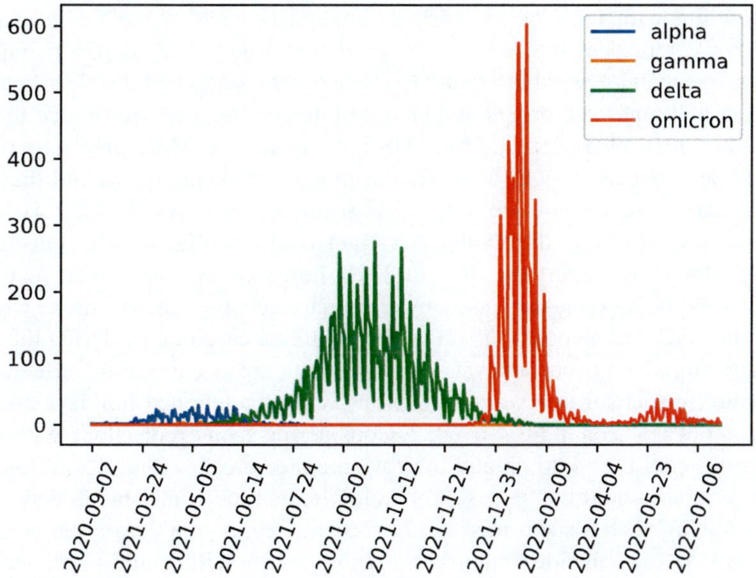

Fig. 5 Wyoming variant data over time provided by the GISAID initiative

factor in how cases are reported; the majority of reported cases here are cases of COVID-19 severe enough to be tested, meaning that a less severe strain may be underrepresented during switching times.

3 Drivers of Case Count

There are many factors contributing to the variations in case count for a global pandemic, including:

1. holidays;
2. public events;
3. the implementation of social-distancing and other regulations (such as a lock-down, mask mandate, vaccination, etc.);
4. variant mutation.

We use New Jersey as our example to visualize such factors. In Fig. 3 we see the daily cases of COVID-19 in the state of New Jersey over the timeline of the virus. Plotted beneath is a stacked percentage plot denoting which variants were represented in the system each day and what percent of the total infected they encompassed. There seems to be a strong correlation between a new variant becoming dominant in a system and peaks in cases. In fact, there is a local maximum corresponding with each of the variants studied in each of the states analyzed. In Fig. 6 we represent the same case count by variants and highlight important holidays. It is well known that around the time of Christmas and the New Year there are family and other gatherings heightening interactions. One can notice that case maxima are often also preceded by a US holiday season. More precisely, the first Omicron peak occurred right after Thanksgiving. Interestingly, around the winter holiday season we do not see such a drastic peak in cases in 2020 as we see in 2021, suggesting that the Omicron variant could be much more transmissible than the Alpha variant. At the time in 2020, however, as seen in Fig. 7, the first vaccines were being released and many non-pharmaceutical interventions were still being followed, which could explain the less extreme case impact during the winter months. Summarizing, our analysis confirms that case counts can be affected on a large scale by societal intervention, such as widely celebrated holidays and strict mitigation laws as seen in Figs. 6 and 7. Considering such events along with variant distributions could serve to greatly improve the predictive capabilities of long term models. While it could be the case, for example, that the winter break was a factor in BA.1 having such a steep increase in cases, it seems true across the board that local case maxima coincide with new mutations, as we will detail in next section.

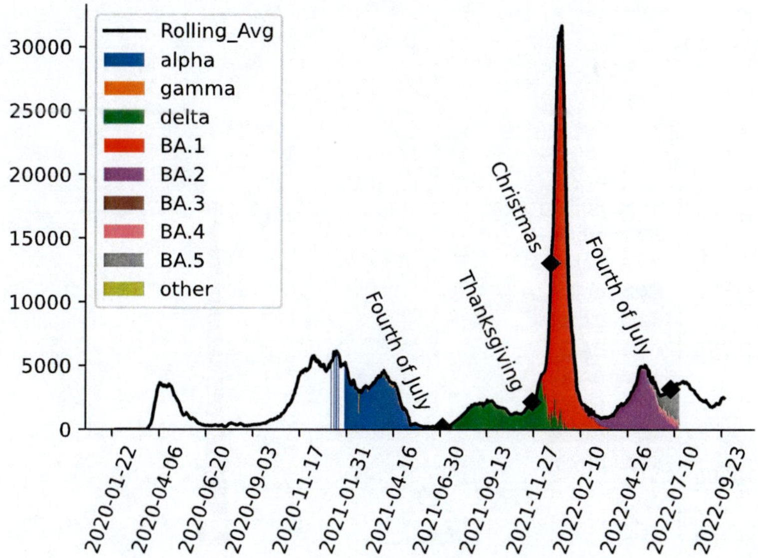

Fig. 6 Widely celebrated US holidays plotted over daily case data for the state of new Jersey

4 Data Analysis

Across all states in our testing sample we see strong correlations between heightened case rates and new variants. We provide quantitative evidence for such observations in the following way. First, we calculate a "switching time" for each dominant variant in each state, then we find the distance in days between new variants taking hold and local case maxima, and lastly we compare the derivative of the normalized variant space to the derivative of the case data over time.

4.1 Computation of "Switching Time"

Let us start with computing the time for a variant to become dominant. First, we consider the percentage of cases due to a specific variant, with 0% meaning no reported cases have the given variant and 100% meaning all cases have that variant. Then, "switching time" for a variant becoming dominant is defined to be the time it takes to go from infecting 15% or less of the infected population to infecting 80% or more. In simple words, this represents the number of days where a variant has gone from appearing in a non negligible number of cases to having the highest frequency by a large margin. Due to limited data from GISAID, we must account for outlier cases. For example, in the state of New Jersey there is one early report of the Delta variant that shifts the Delta variant's "switching time" from around 30

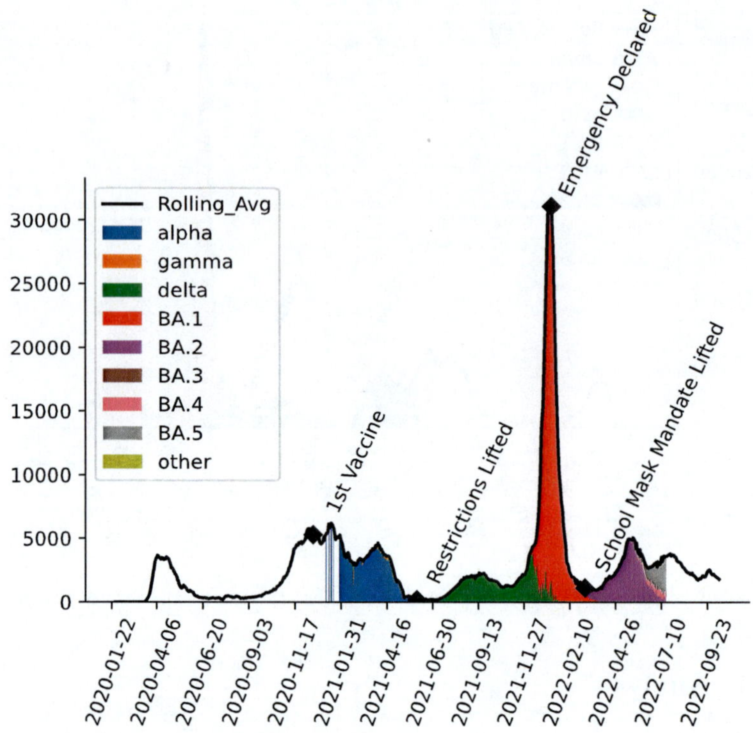

Fig. 7 Government mandates which could affect case count plotted over daily case data for New Jersey

days to around 230 days. While this data point may be valid, we wish to lessen the impact of such outliers on our analysis. To remedy this we take a rolling average of 10 days to smooth the data. This smoothing can be seen in Figs. 8 and 9.

In formulas, first, we set:

$$N_{ij} = \frac{I_{ij}}{\sum_k I_{ik}} \tag{1}$$

where i is a given day in our time horizon, j is a variant represented in the system, I_{ij} is the total reported infected on day i with variant j. Therefore, N_{ij} is the normalized incidence number of each variant for each day where the sum of variant contributors on a given day is 1. Then the switching time, S, is defined as follows. First set:

$$t^j_{app} = \inf\{i : N_{ij} \geq 15\}, \; t^j_{dom} = \sup\{i : N_{ij} \leq 80\} + 1.$$

Thus t^j_{app} is the first time in which the variant j represents at least 15% of the cases and t^j_{dom} is the first time in which the variant j is responsible for more than 80% of

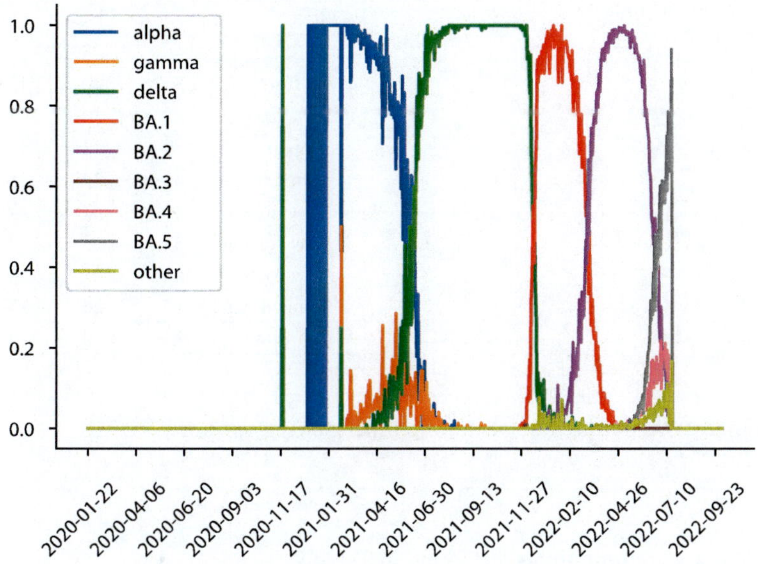

Fig. 8 Raw normalized variant data for New Jersey

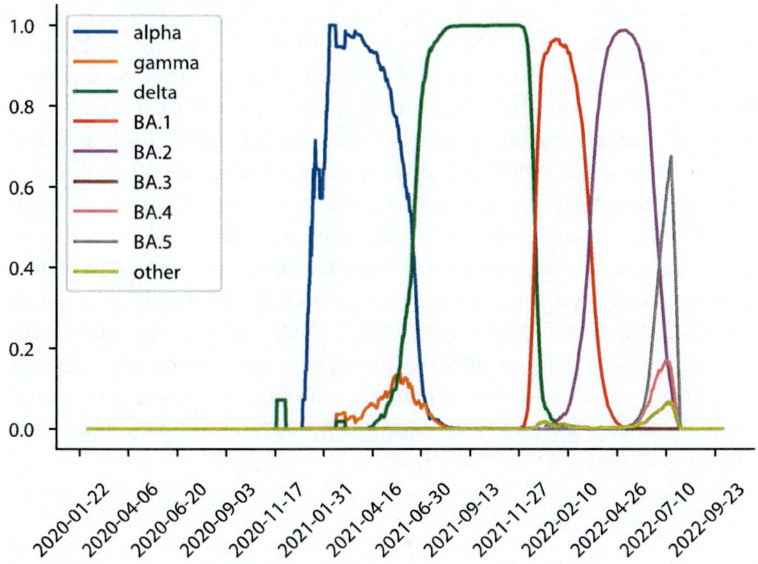

Fig. 9 Smoothed normalized variant data for New Jersey

Table 2 Switching time of variants in days

State	Alpha	Delta	Omicron
Alaska	–	29	16
California	35	40	12
Florida	20	30	11
Hawaii	59	41	12
Maine	43	23	15
Missouri	20	27	17
New Jersey	33	33	13
New York	25	27	14
Pennsylvania	19	24	13
Wyoming	35	30	13

cases. Finally, we define the "switching time" for the variant j as:

$$S_j = t_{dom}^j - t_{app}^j. \tag{2}$$

Results of switching times are found in Table 2.

The Alpha variant never took hold in Alaska, as it was overshadowed by variant B.1.1.519. Variant B.1.1.519 had a switching time of 19 days. The average switching time for the Alpha variant is 32.1, Delta variant is 30.4, and Omicron variant is 13.6, with standard deviations of 12.4, 6.1, 1.9, respectively. Alpha and Delta have similar switching times, while Omicron is less than half of both. This discrepancy could be due to a host of factors, like those described in the previous section, such as a heightened infection rate, beginning its ascent at the start of the winter holidays, or more likely, a combination of case driving factors. In [32] a wave analysis is completed to understand whether the third wave of COVID-19 in India had any heterogeneous characteristics. It is estimated that the reproduction number during the peak of Omicron (third wave) was more than double the reproduction numbers of the previous two waves, suggesting that Omicron has more than double the replication rate of the previous two variants, which is consistent with our findings. An important distinction to make here is that there are multiple factors that drive a reproduction number. Basic infection rate is the most commonly considered, but in this case, there are observable adaptations unique to Omicron that neutralize the effects of both natural immunity brought on by previous infection and immunity brought on by previous vaccination [9, 40].

4.2 Days Between Variants Dominance and Cases Peak

As observed above, after a new variant takes over it is often followed by a local maximum in total cases. This happens across many variants and many of the examined states like New Jersey. One example of a state with no noticeable peak

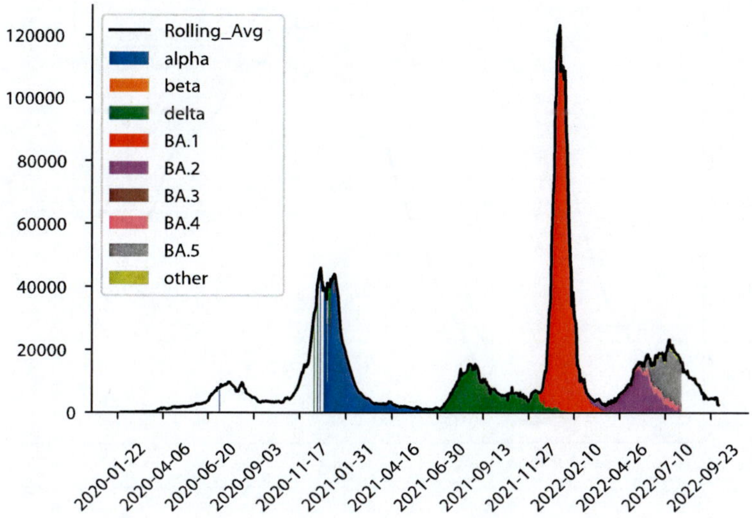

Fig. 10 A combination of GISAID and JHU data visualizing cases and variants over time for the state of California. We take our smoothed cases from the JHU data and superimpose a stacked variant percent plot underneath

in cases in the chosen range is California, which can be seen in Fig. 10 to be on a steady decrease despite the Alpha variant having taken hold. It is possible that the second peak, where the Alpha variant begins, could be the peak associated with Alpha, but variant level data does not go back far enough to be sure, so we omit this.

To quantify the observation of cases peak following a variant becoming dominant, we proceed as follows. First, we normalize the variant distribution as found in (1). We define a variant dominant if $N_{ij} > 0.8$, i.e. it accounts for at least 80% of the cases, and compute the number of days for which the variant is dominant. We also normalize the number of cases during the same period. Let C be the vector containing the daily cases, thus C_i is the cases on day i. Then, the normalized number of cases NC_i on day i in our date range is given by

$$NC_i = \frac{C_i}{\max_i(C_i)}. \tag{3}$$

Then we compute the number of days between the event of the variant becoming dominant, i.e. representing 80% of the cases reported to GISAID, and the next local maximum in total number of cases. So we define

$$c^j = \inf\{i : NC_i = 1\}, \tag{4}$$

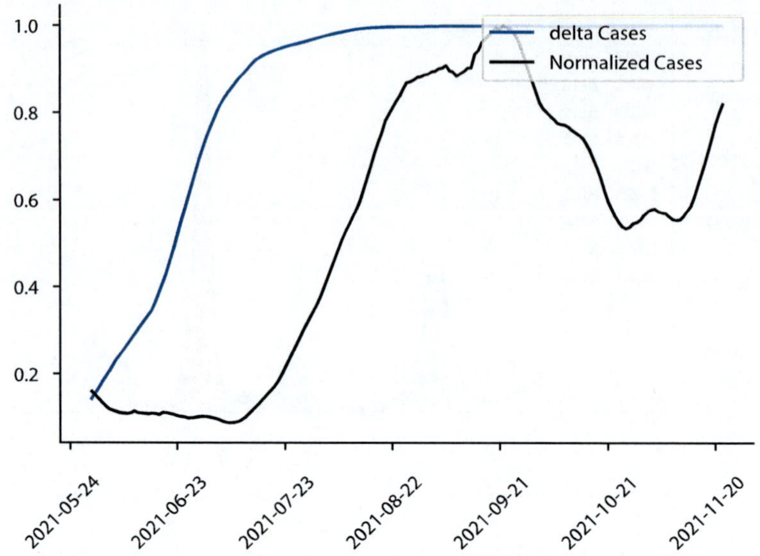

Fig. 11 Percent of cases due to Delta variant and the daily cases normalized to one

Table 3 Distance from variant peak to case peak in days

State	Alpha	Delta	Omicron
Alaska	–	82	29
California	–	62	22
Florida	82	48	23
Hawaii	–	54	34
Maine	50	123	46
Missouri	–	59	14
New Jersey	75	85	15
New York	73	83	28
Pennsylvania	72	86	15
Wyoming	–	85	26

then set

$$D_j = c^j - t^j_{dom} \tag{5}$$

which is the number of days that it takes from a new variant taking over dominance for there to be a local maximum in cases. An example of this procedure can be seen in Fig. 11 where the Delta variant became dominant in early July of 2021 and a local maximum in cases occurred in mid September 2021.

We report the computed values of D_j for the three main variants (Alpha, Delta, Omicron) and the 10 states in Table 3.

Note that when there is no value listed, there is no noticeable peak in cases following a variant takeover. For the Alpha variant, for example, we see that it took on average 70.4 days for a local maximum in cases to appear with standard deviation of 10.78; this is with only five of the states in our sample considered as the other five had no noticeable peak in cases within the range considered.

The Delta variant had an average of 76.7 with a higher standard deviation of 20.7. Once again, the characteristics of the Alpha and Delta variants are very similar. Lastly, the Omicron variant had an average value of 25.1 with standard deviation of 9.51, almost one third of the time taken during both the Alpha waves and the Delta waves. This is once again consistent with the idea that the Omicron variants have heightened reproduction numbers.

4.3 *Comparing the Trend of Variant Progression with Cases Progression*

Let us pass now to consider the trend in variants becoming dominant and the increase of total number of cases. More precisely, we compare how the derivative of the variants over time aligns with the derivative of cases over time. An example of this can be found in Fig. 12, where we look at New Jersey's delta variant compared with daily case data.

Notice that 58 days after the delta variant has a maximum derivative, the daily cases see a similar rate of increase.

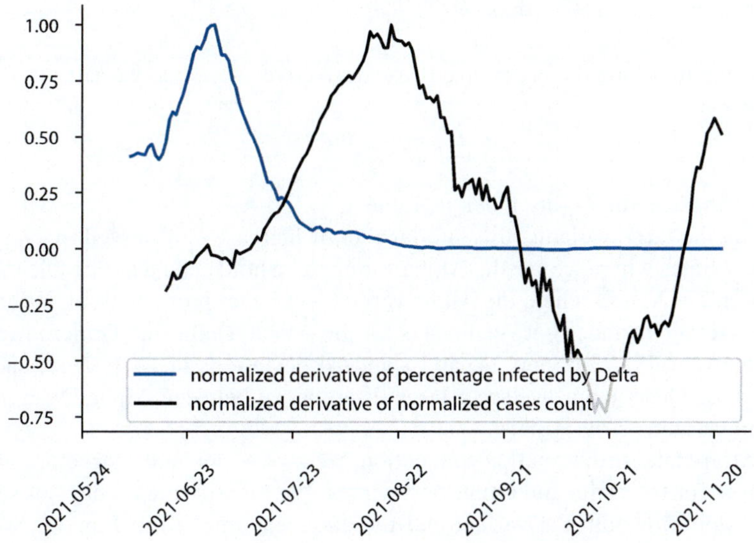

Fig. 12 Comparison of normalized derivatives of percent of Delta cases and normalized daily cases

Table 4 Distance of derivative of variant peak to derivative of case peak in days

State	Alpha	Delta	Omicron
Alaska	–	80	30
California	–	154	20
Florida	101	43	20
Hawaii	–	53	36
Maine	31	69	47
Missouri	–	58	17
New Jersey	79	58	15
New York	85	56	19
Pennsylvania	89	80	13
Wyoming	–	85	23

First, we consider the point of maximum growth of the percentage of cases due to a specific variant and that of total cases. Using the symbol f' to indicate the incremental ratio over a day for a function f defined in the range of interest, we set:

$$M'_{ij} = \frac{N'_{ij}}{\max_i N'_{ij}}, \quad v'^j = \inf\{i : M'_{ij} = 1\}$$

which is the first day in which the normalized derivative of the variant is 1 and

$$MC'_{ij} = \frac{NC'_{ij}}{\max_i NC'_{ij}}, \quad c'^j = \inf\{i : MC'_{ij} = 1\}$$

which is the first time the normalized case derivative is 1. Then we set:

$$D'_j = v'^j - c'^j. \tag{6}$$

The found values for D'_j are given in Table 4.

Across the three variants, this similar rate of increase is observed in every case besides in states where we see the Alpha variant take hold and no noticeable increase in cases and in Alaska where the Alpha variant was never prominent. In this case we have an average distance between peaks for the Alpha, Delta, and Omicron variants respectively of 77, 73.6, and 24, and standard deviations of 24.1, 29.8, and 10.1 respectively. Once again, we see a large discrepancy between Alpha/Delta and the Omicron variant.

To extrapolate further on this connection, we take a look at the structure of these derivatives for the Delta and Omicron variants. We identify the 20 days of steepest positive slope for both the variant and the cases exemplified in Fig. 12. We then superimpose both plots for those 20 days over each other and calculate the L_1 distance between the two, as seen in Fig. 13 for New Jersey's Delta variant. We

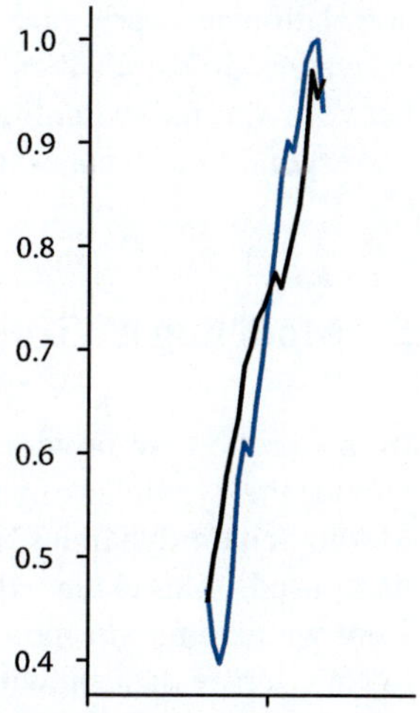

Fig. 13 Comparison after shifting of 58 days

Table 5 L1 distances and relative error (RE) between derivatives of cases and variant distribution

State	Delta	Omicron	RE Delta	RE Omicron
Alaska	3.27	2.34	0.21	0.17
California	2.13	2.63	0.13	0.07
Florida	3.11	2.25	0.23	0.14
Hawaii	1.58	0.81	0.11	0.05
Maine	8	1.64	0.77	0.08
Missouri	1.82	2.13	0.08	0.12
New Jersey	1.52	1.43	0.06	0.13
New York	2.17	2.21	0.21	0.18
Pennsylvania	3.21	0.41	0.22	0.01
Wyoming	1.62	1.3	0.12	0.14

take the cases C' and variant distribution V', and compute

$$L_1 = \sum_{n=1}^{n=20} |C'_n - V'_n|. \tag{7}$$

We then define the relative error as

$$RE = \frac{L_1}{\max(\sum_{n=1}^{n=20} |C'_n|, \sum_{n=1}^{n=20} |V'_n|)}. \tag{8}$$

The results of this can be seen in Table 5.

For reference, we see that Fig. 13 has an L_1 value of 1.52 and relative error of 0.05. Maine had an extremely gradual climb in cases for the Delta variant causing the derivative of cases to be much less pronounced than the derivative of the percentage of Delta variants. Besides this outlier, we see an extremely strong correlation in every case, with Pennsylvania's Omicron variant having a variant derivative almost identical to the derivative of cases. While there is an obvious delay between variant switching and case maxima, a faster switching time does seem to correspond to a steeper climb in cases.

5 Modeling a Virus with Mutation

In this section we provide a method to extend classical epidemiological models to include the dynamics of virus mutations. The model is based on coupling ODEs with Markov chain dynamics and was already proposed in [45]. Moreover, reinfections were used in an extended model in [44] to include the case of endemic diseases. Here we examine in more details the wave dynamics and then we tune the model to New Jersey data showing how it can reproduce correctly the observed waves and their timing.

5.1 Epidemiological Modeling

A classical method used to model a virus in a population is a compartmental model. These models were first proposed by Kermack and McKendrick [23] to tackle virus dynamics. For infectious diseases (such as COVID-19), the population is divided into compartments: Susceptible, Infected, and Removed. The most basic model (SIR) has the following structure:

$$\begin{cases} \dot{S} = -\frac{\beta S I}{N}, \\ \dot{I} = \frac{\beta S I}{N} - \gamma I, \\ \dot{R} = \gamma I, \end{cases} \tag{9}$$

where β and γ are the infection rate and recovery rate respectively. The allure of such a model is its ability to capture information about a population with very little input data required (parameters such as β and γ weigh heavily on the trajectories of the model). These models have been effectively used in research across various traditional infectious diseases including pertussis [14, 34], measles [7], Ebola [24, 28], etc. Due to their predictive capabilities and low computational cost, they continue to be used in modern epidemics such as HIV [15, 30], Lyme, etc., as well as reemerging diseases (tuberculosis [8, 38], cholera [13], etc.). A model developed with a certain population in mind can be tailored to another with little change in computational cost, regardless of the difference in population sizes.

Various extensions of the SIR model have been used to measure and predict COVID-19: including vulnerable groups [48]; considering the economic cost of non-pharmaceutical intervention [29] and, as vaccines became more readily available, optimization of vaccination schedule [26, 39].

To assist policy makers, epidemiological models must aim for high predictive power. However, for a study spanning various months, the mutability of the virus itself can skew model results. Indeed, a mutation of the virus may result in higher viral load, increased viral transmissibility, higher infectiousness, immune escape, increased resistance to monoclonal/polyclonal antibodies from vaccine, and a change in virulence [16, 37]. In the case of COVID-19, the Omicron variant contains mutations that confer resistance to at least four classes of monoclonal antibodies targeting the receptor binding domain of the spike protein. This reality both sheds light on the rise in cases seen in Fig. 3 when the Omicron variant (BA.1) took hold, and suggests a need for a new vaccine tailored to Omicron and its sub-variants.

To model such sudden changes in the characteristics of the virus, the inclusion of discrete changes to the parameter space of a SIR model can be built in via viral mutation where each variant has its own unique parameter space.

5.2 *Definition of MC-ODE System*

We use a SIR model coupled with a nonlinear discrete-time Markov chain (briefly NDMC) which is used to govern emerging virus variants. The system is fully coupled, but the discrete-time nature of the Markov chain lends itself simply to the simulations to follow. Once this is established, we include the possibility of reinfection with immunity to the previously contracted variant.

Assume the infected population is represented by a sequence of random variables $I_i : \Omega \to \mathrm{i}$, with $(\Omega, \mathcal{F}, P)$ measure space and $\mathrm{i} = \{I^1, \ldots, I^p\}$ representing the distribution over the space representing the set of significant virus mutations. The meaning is the following: the total population of infected people $\sum_i I_i$ will follow a standard compartmental model, with $I_i / \sum_i I_i$ representing the probability of the generic infected person to have variant i. Alternatively, we can think of i as a vector representing the infected population distinguished in sub-populations via the virus variants. Therefore, we are left with the following SIR model with multiple variants:

$$\begin{cases} \dot{S} = -\sum_{i=1}^{p} \beta_i \frac{S(t)}{N} I_i(t), \\ \dot{I}_i = \beta_i \frac{S(t)}{N} I_i(t) - \gamma_i I_i(t), \\ \dot{R} = \sum_{i=1}^{p} \gamma_i I_i(t). \end{cases} \tag{10}$$

Here S is the susceptible population, $I_i(t)$ is the population infected by the i-th variant, R recovered population, N the total population, β_i, γ_i, represents the infection rate and recovery rate of the i-th variant. We assume that the evolution of i is given by a NDMC associated to a transition matrix $T = \{t(i, j)\}_{i,j=1,\ldots,p}$. Then, the discrete-time evolution of the MC governs the infected distribution with time-varying mass coupled with the SIR dynamics. Assuming the time step of the Markov chain is given by Δt, the updating term is:

$$\mathcal{I}^+(k\Delta t) = \mathcal{I}(k\Delta t) \cdot T(\mathcal{I}(k\Delta t)),\ k \in \mathbb{N}. \tag{11}$$

This step redistributes the infected population among the different variants every Δt time units, allowing for mutation. At time $k\Delta t$, if $\mathcal{I}$ is the distribution of infected before the mutation step, $\mathcal{I}^+$ is the distribution of infected after the mutation step. $T(\mathcal{I})$ represents a transition matrix for the Markov chain that governs the mutations, and the dependence on $\mathcal{I}$ makes it nonlinear. The state of this MC is dynamic and is tied to both the number of infected people with the given variant and the total infected. We choose $T_{ii}(\mathcal{I}) = \widetilde{T}_{ii} \cdot \psi(I_i)$, where $\psi(I_i) = 1 - \eta[I_i - \bar{I}_i]_+$. Here, $\bar{I}_i$ is a threshold of total infected and η depends on both the total infected and amount of infected with infection i. The presence of the Markov chain term is the key difference between this model and the usual SIR, allowing for the mutation of infected populations seen strongly in the wide field of variants of Sars-CoV-2.

Lastly, we address natural immunity to reinfection. We know that natural immunity is dependent on both the previous variant of infection and the new variant of exposure. For example, after first infection with the Delta variant, one has a certain amount of natural immunity to the same variant. However, it is well documented now that Omicron has a level of immune escape from both the initial vaccine and the previous COVID-19 variants. We accomplish this with the following term:

$$\dot{S}_{R_i} = \sigma_i R_i - \sum_{\hat{i}} \beta(i, \hat{i}) * \frac{S_{R_i}}{N} * I_{\hat{i}}, \tag{12}$$

where $\beta(i, \hat{i})$ is the infection rate of variant $\hat{i}$ among patients that recovered from variant i and σ is a loss of immunity rate. After infection a person retains some immunity to reinfection from similar variants, thus if variant i is sufficiently close to variant $\hat{i}$ from a genetic point of view, then the value of the infection rate $\beta(i, \hat{i})$ is smaller than the original infection rate $\beta_{\hat{i}}$ of variant $\hat{i}$. Lastly, we include a compartment for mortality and a death rate to keep track of mortality. Therefore, our fully coupled model has the following structure:

$$\begin{cases} \dot{S} = -\sum_{i=1}^{p} \beta_i \frac{S(t)}{N} I_i(t), \\ \dot{I}_i = \beta_i \frac{S(t)}{N} I_i(t) - \sum \beta(i, \hat{i}) * \frac{S_{R_i}}{N} * I - \gamma_i I_i(t), \\ \dot{R}_i = \gamma_i I_i(t) - \sigma_i R_i - d * R, \\ \dot{S}_{R_i} = \sigma_i R_i - \sum \beta(i, \hat{i}) * \frac{S_{R_i}}{N} * I, \\ \dot{D} = \sum d * R_i \end{cases} \tag{13}$$

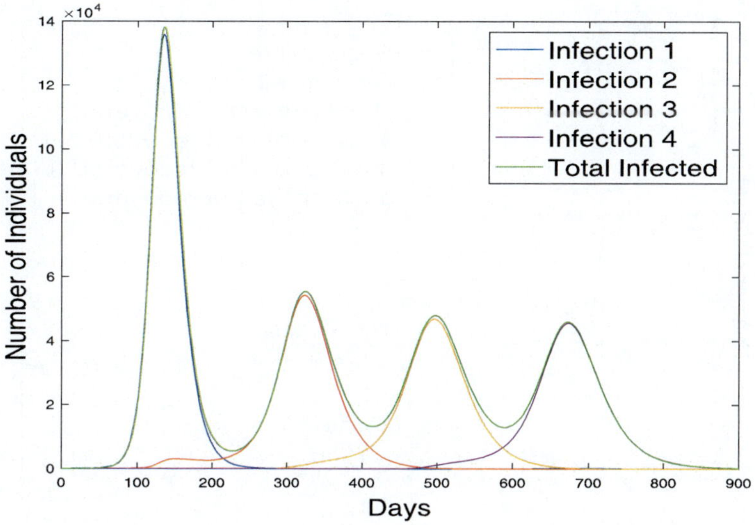

Fig. 14 Infected dynamics for a simulation of (13) with homogeneous variant parameters

5.3 *Simulations*

Using the above described equations, a Markov chain coupled SIRS model is able to simulate the dynamics of a virus spreading through a population while capturing the changing characteristics of the disease due to new variants appearing and taking hold over the majority of the field of infections. Here, we solve our system of ODEs via a Runge Kutta method over a chosen time horizon of 900 days. Every two days, we allow for the infected space to be mutated via the pre-defined Markov chain mutation matrix. We see in Fig. 14 that as two variants have an "exchange of power" there is a dip in cases just as observed previously in the daily case count in Fig. 3. In Fig. 14, we use homogeneous variants with replication rates of 2.5, σ of 0.01, and identical infection and recovery rates.

In Fig. 15, the characteristics of reinfection can be observed. The susceptible compartment quickly empties as much of the population is infected with at least one variant. The total recovered reaches a peak in a short period of time, but then the recovered individuals begin returning to susceptible compartments with immunity to the virus that originally infected them, highlighting the wave-like pattern in the recovered population. Lastly, as each variant has its peak in Fig. 14, there is a delayed peak in the corresponding susceptible but recovered compartment in Fig. 15 followed by a slow walk towards reinfection with another variant. We see that by the end of our time horizon, the first S_R compartment has almost completely been reinfected.

While the ability to capture the shape of the viral dynamics found in real data with this model adaptation is very exciting, it is unlikely biologically that a virus

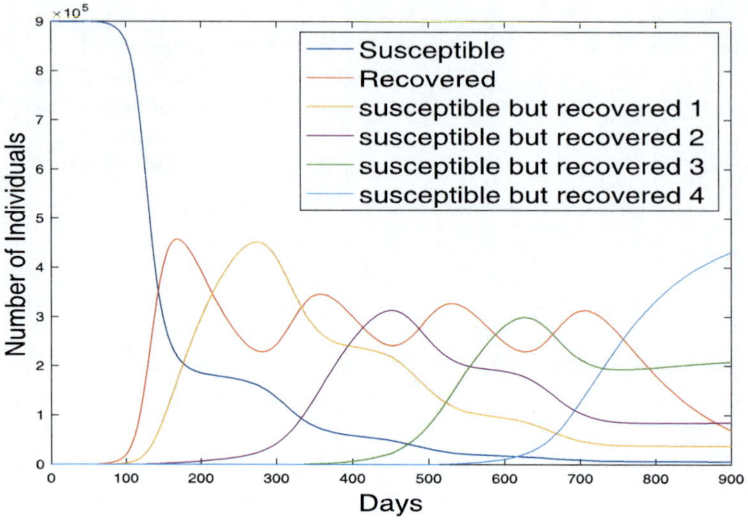

Fig. 15 Susceptible dynamics for a simulation of (13) with homogeneous variant parameters

with solely neutral mutations will mutate so quickly to a new variant. Breakthrough variants should be expected to have some new beneficial trait like heightened infectivity or heightened immune escape in the case of a virus for which a vaccine exists [25].

To test our model's ability to fit a real population, in Fig. 17 we use the population of New Jersey and parameter estimates that match the progressions of Alpha, Delta, BA.1, and BA.2 in New Jersey. We begin with our initial conditions, choosing the total population to be nine million. We are only interested in the dates starting where we have variant level data, so we begin with data from 2021-01-31, including initial infected from this day. We choose the recovery rate to be one over the number of days it takes to recover from the virus. According to the World Health Organization, people with COVID-19 generally have an infectious/symptomatic period of about two weeks, so we choose γ for each variant to be a constant $1/14$ [20]. For the reproduction number, we calculate a simple reproduction number as the new cases on each day divided by the new cases from 14 days prior, then average over the days in which the given variant was dominant ($<80\%$). This method, while simple, provides a good estimate when compared with more complicated algorithms [3]. This gave us R values of 1.12 for Alpha, 1.33 for Delta, 1.72 for Omicron BA.1, and 1.41 for Omicron BA.2. The visual of $R(t)$ can be found in Fig. 16. We are limited to only reported case data, so not only is there a question of severity affecting reporting, but also an acknowledgement that a large majority of COVID-19 cases could be going unreported [27].

For simplicity, to account for such things, we use our results as proportions. If Alpha is normalized to 1, we re-scale such that the Alpha variant begins at an R_0 of 1.5. We do so as this is still below the peak R_t in the Alpha time horizon, and a

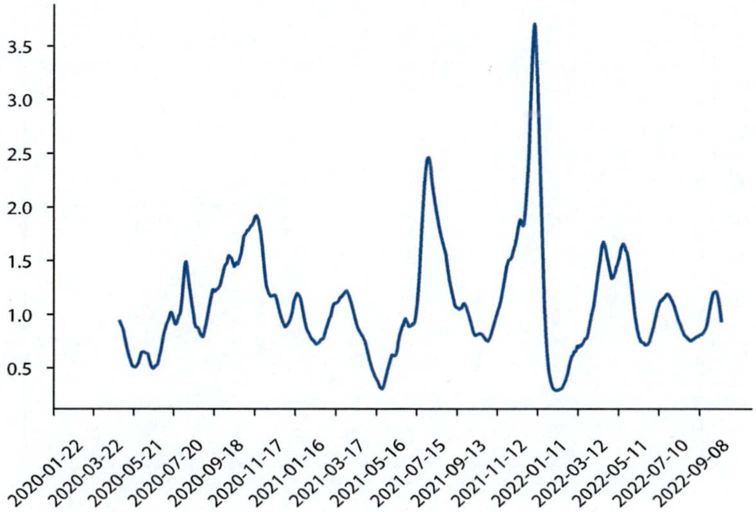

Fig. 16 Replication rate over time calculated from JHU case data

reliable estimate for the Alpha variant globally when compared to other choices of serial intervals (number of days back to divide by) [3]. Thus, the replication rates used in this model are 1.50 for Alpha, 1.78 for Delta, 2.30 for Omicron BA.1, and 1.88 for Omicron BA.2. We calculate the death rate simply as total deaths/cases, and leave it constant across all variants as we are only concerned with the wave dynamics. Lastly, σ is chosen to be .01 (lose immunity in 100 days) for all variants besides the Delta variant. We know that the Omicron variants, which are next to take over after Delta, have a certain level of immune escape [16], so to capture this, we scale up σ for the Delta variant by a factor of three. Note here while the underlying causes of the case spikes are the driving force of the replication rate estimates, we do not explicitly include characteristics such as holidays events or non-pharmaceutical interventions in this model. This ability to generalize a complex system into parameter estimations is a benefit of working with SIR models.

In Fig. 17, we see that the characteristic peaks in cases once again correspond with new variants. We also see that the shape of the waves match closely with recorded data for the state of New Jersey from Fig. 2 where Alpha has a peak, followed by a lower peak for Delta, followed by the Omicron variants having first a large peak, then a lower peak. Further, even the distance between comparable peaks in the two figures is very similar. In the real New Jersey data, peak to peak from Alpha to Gamma it took 165 days, Gamma to BA.1 was 119 days, and BA.1 to BA.2 was 134 days. In our simulation, while slightly different since we are considering all infected at a given point rather than new cases, we still have similar results: peak to peak from Alpha to Gamma it took 171 days, Gamma to BA.1 was 108 days, and BA.1 to BA.2 was the outlier taking 205 days.

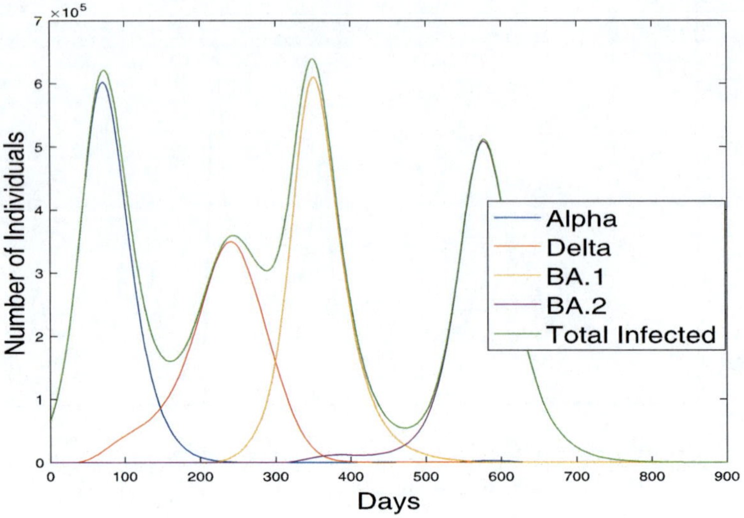

Fig. 17 Infected dynamics for a simulation of (13) with parameter estimates matching the state of New Jersey

6 Discussion

In this chapter we analyzed COVID-19 case and mutation data in order to gain a better understanding of pandemic dynamics and, in particular, local maxima in reported cases. What we found is a strong correlation between variant dynamics and case dynamics. During the exchange of power between variants local case minima occur which, in most instances, are then followed by a peak shortly after the new variant has taken over. Finally another drop in cases follows, restarting the cycle. Across the board, we see that the Omicron variant affected the system much quicker than Alpha and Delta, resulting in short switching times, steeper case climbs and more cases overall. Without a dominant variant after Omicron (but few sub variants) having yet taken over, it is unclear whether this is a long term shift in the parameter space, or if Omicron is an outlier case.

We then introduced advanced epidemiological models, which include variants dynamics through the coupling of a Markov chain to a system of ODEs. The model was able to reproduce qualitative behavior of the waves (peaks and valleys) and, for the state of New Jersey, also the timing of the waves after tuning the parameters appropriately. With COVID-19 cases fluctuating at predictable time points with respect to its mutations, these models can provide more accurate forecasts and allow decision makers to address pandemic management effectively.

Acknowledgments The authors would like to acknowledge the support of the NSF CMMI project # 2033580 "Managing pandemic by managing mobility". The authors also gratefully acknowledge all data contributors including the Johns Hopkins public data repository through the Pandemic

Data Initiative, and the Authors and their Originating laboratories responsible for obtaining the specimens, and their Submitting laboratories for generating the genetic sequence and metadata and sharing via the GISAID Initiative.

References

1. Abbasi, B., Fadaki, M., Kokshagina, O., Saeed, N., Chhetri, P.: Modeling vaccine allocations in the covid-19 pandemic: a case study in Australia. Available at SSRN 3744520 (2020)
2. Allred, S., Chyba, M., Hyman, J.M., Mileyko, Y., Piccoli, B.: The covid-19 pandemic evolution in Hawai 'i and New Jersey: a lesson on infection transmissibility and the role of human behavior. In Predicting Pandemics in a Globally Connected World, vol. 1, pp. 109–140. Springer (2022)
3. Arroyo-Marioli, F., Bullano, F., Kucinskas, S., Rondón-Moreno, C.: Tracking r of covid-19: a new real-time estimation using the Kalman filter. PloS One **16**(1), e0244474 (2021)
4. Badoual, M., Grammaticos, B., Nakamura, G.: Vaccination strategies for a seasonal epidemic: a simple sir model. Open Commun. Nonlinear Math. Phys. **1**, 20 (2021)
5. Bagcchi, S.: The world's largest covid-19 vaccination campaign. Lancet Infect. Dis. **21**(3), 323 (2021)
6. Béland, L., Brodeur, A., Wright, T.: The short-term economic consequences of covid-19: exposure to disease, remote work and government response. Plos one **18**(3), e0270341 (2023)
7. Bokler, B.: Chaos and complexity in measles models: a comparative numerical study. Math. Med. Biol. A J. IMA **10**(2), 83–95 (1993)
8. Bowong, S., Tewa, J.J.: Mathematical analysis of a tuberculosis model with differential infectivity. Commun. Nonlinear Sci. Numer. Simul. **14**(11), 4010–4021 (2009)
9. Brüssow, H.: Covid-19: Omicron–the latest, the least virulent, but probably not the last variant of concern of SARS-CoV-2. Microbial Biotechnol. **15**(7), 1927–1939 (2022)
10. Celani, A., Giudici, P.: Endemic–epidemic models to understand covid-19 spatio-temporal evolution. Spatial Stat. **49**, 100528 (2022)
11. Clarke, J., Majeed, A., Beaney, T.: Measuring the impact of covid-19. bmj **373**, (2021)
12. Cot, C., Cacciapaglia, G., Islind, A.S., Óskarsdóttir, M., Sannino, F.: Impact of us vaccination strategy on covid-19 wave dynamics. Sci. Rep. **11**(1), 1–11 (2021)
13. Crooks, A.T., Hailegiorgis, A.B.: An agent-based modeling approach applied to the spread of cholera. Environ. Model. Softw. **62**, 164–177 (2014)
14. de Greeff, S.C., de Melker, H.E., Westerhof, A., Schellekens, J.F., Mooi, F.R., van Boven, M.: Estimation of household transmission rates of pertussis and the effect of cocooning vaccination strategies on infant pertussis. Epidemiology, **23**, 852–860 (2012)
15. Demirci, E., Unal, A., Ozalp, N.: A fractional order SEIR model with density dependent death rate. Hacettepe J. Math. Stat. **40**(2), 287–295 (2011)
16. Dholariya, S., Parchwani, D.N., Singh, R., Sonagra, A., Motiani, A., Patel, D.: Notable and emerging variants of SARS-CoV-2 virus: a quick glance. Indian J. Clin. Biochem. **36**(4), 451–458 (2021)
17. Dimarco, G., Perthame, B., Toscani, G., Zanella, M.: Kinetic models for epidemic dynamics with social heterogeneity. J. Math. Biol. **83**(1), 1–32 (2021)
18. Elbe, S., Buckland-Merrett, G.: Data, disease and diplomacy: Gisaid's innovative contribution to global health. Glob. Challenge **1**(1), 33–46
19. European Centre for Disease Prevention and Control. SARS-CoV-2 variants of concern as of 23 September 2021 (2021)
20. Gomes, C.: Report of the WHO-China joint mission on coronavirus disease 2019 (covid-19). Braz. J. Implantol. Health Sci. **2**(3) (2020)

21. Haynes, B., Messonnier, N.E., Cetron, M.S.: First travel-related case of 2019 novel coronavirus detected in united states: press release, Tuesday, January 21, 2020 (2020)
22. Huang, D., Tao, H., Wu, Q., Huang, S., Xiao, Y.: Modeling of the long-term epidemic dynamics of covid-19 in the united states. Int. J. Environ. Res. Public Health **18**(14), 7594 (2021)
23. Kermack, W.O., McKendrick, A.G.: Contributions to the mathematical theory of epidemics. II.—the problem of endemicity. Proc. R. Soc. Lond. A, containing papers of a mathematical and physical character **138**(834), 55–83 (1932)
24. Lekone, P.E., Finkenstädt, B.F.: Statistical inference in a stochastic epidemic SEIR model with control intervention: Ebola as a case study. Biometrics **62**(4), 1170–1177 (2006)
25. Li, Q., Nie, J., Wu, J., Zhang, L., Ding, R., Wang, H., Zhang, Y., Li, T., Liu, S., Zhang, M., et al.: SARS-CoV-2 501y. v2 variants lack higher infectivity but do have immune escape. Cell **184**(9), 2362–2371 (2021)
26. Luo, Q., Weightman, R., McQuade, S.T., Díaz, M., Trélat, E., Barbour, W., Work, D., Samaranayake, S., Piccoli, B.: Optimization of vaccination for covid-19 in the midst of a pandemic. Netw. Heterogeneous Media **17**, 443 (2022)
27. Magal, P., Webb, G.: Predicting the number of reported and unreported cases for the covid-19 epidemic in South Korea, Italy, France and Germany. Medrxiv (2020)
28. Mamo, D.K., Koya, P.R.: Mathematical modeling and simulation study of SEIR disease and data fitting of Ebola epidemic spreading in West Africa. J. Multidiscip. Eng. Sci. Technol. **2**(1), 106–114 (2015)
29. McQuade, S.T., Weightman, R., Merrill, N.J., Yadav, A., Trélat, E., Allred, S., Piccoli, B.: Control of covid-19 outbreak using an extended SEIR model. Math. Models Methods Appl. Sci., **31**, 1–26 (2021)
30. ÖZalp, N., Demirci, E.: A fractional order SEIR model with vertical transmission. Math. Comput. Model. **54**(1–2), 1–6 (2011)
31. Painter, E.M., Ussery, E.N., Patel, A., Hughes, M.M., Zell, E., et al.: Demographic characteristics of persons vaccinated during the first month of the covid-19 vaccination program—United States, December 14, 2020–January 14, 2021. Morb. Mortal. Wkly. Rep. **70**(5), 174 (2021)
32. Ranjan, R.: Omicron impact in India: An early analysis of the ongoing covid-19 third wave. MedRxiv (2022)
33. Ratcliffe, M., Burd, C., Holder, K., Fields, A.: Defining rural at the us census bureau. Am. Community Surv. Geography Brief **1**(8), 1–8 (2016)
34. Rohani, P., Zhong, X., King, A.A.: Contact network structure explains the changing epidemiology of pertussis. Science **330**(6006), 982–985 (2010)
35. Rouillard, S., Liu, V.X., Corley, D.A.: Covid-19 and long-term planning for procedure-based specialties during extended mitigation and suppression strategies. Gastroenterology **160**(1), 4–9 (2021)
36. Rowan, N.J., Moral, R.A.: Disposable face masks and reusable face coverings as non-pharmaceutical interventions (NPIs) to prevent transmission of SARS-CoV-2 variants that cause coronavirus disease (covid-19): role of new sustainable NPI design innovations and predictive mathematical modelling. Sci. Total Environ. **772**, 145530 (2021)
37. Saha, P., Banerjee, A.K., Tripathi, P.P., Srivastava, A.K. and Ray, U.: A virus that has gone viral: amino acid mutation in s protein of Indian isolate of coronavirus covid-19 might impact receptor binding, and thus, infectivity. Biosci. Rep. **40**(5) (2020)
38. Side, S., Sanusi, W., Aidid, M.K., Sidjara, S.: Global stability of SIR and SEIR model for tuberculosis disease transmission with Lyapunov function method. Asian J. Appl. Sci. **9**(3), 87–96 (2016)
39. Subbaraman, N.: Who gets a covid vaccine first? Access plans are taking shape. Nature **585**(7826), 492–493 (2020)
40. Tegally, H., San, J.E., Cotten, M., Tegomoh, B., et al.: The evolving SARS-CoV-2 epidemic in Africa: Insights from rapidly expanding genomic surveillance. Science **378**(6615), eabq5358 (2022)
41. The United States Census Bureau. U.S. and world population clock (2022)
42. Vardavas, R., de Lima, P.N., Baker, L.: Modeling covid-19 nonpharmaceutical interventions: exploring periodic NPI strategies. MedRxiv (2021)

43. Wang, X., Wu, H., Tang, S.: Assessing age-specific vaccination strategies and post-vaccination reopening policies for covid-19 control using SEIR modeling approach. Bull. Math. Biol. **84**(10), 1–23 (2022)
44. Weightman, R., Piccoli, B.: Optimization of non-pharmaceutical interventions for a mutating virus [submitted to IEEE control systems letters]. In: American Control Conference (2022)
45. Weightman, R., Sbarra, A., Piccoli, B.: Coupling compartmental models with Markov chains and measure evolution equations to capture virus mutability. Math. Models Methods Appl. Sci. **32**, 2099–2119 (2022)
46. Wirtz, K.: Changing readiness to mitigate SARS-CoV-2 steered long-term epidemic and social trajectories. Sci. Rep. **11**(1), 1–11 (2021)
47. Zanella, M., Bardelli, C., Dimarco, G., Deandrea, S., Perotti, P., Azzi, M., Figini, S., Toscani, G.: A data-driven epidemic model with social structure for understanding the covid-19 infection on a heavily affected Italian Province. Math. Models Methods Appl. Sci. **31**(12), 2533–2570 (2021)
48. Zhou, F., Yu, T., Du, R., et al.: Clinical course and risk factors for mortality of adult inpatients with covid-19 in Wuhan, China: a retrospective cohort study. Lancet **395**(10229), 1054–1062 (2020)

Multifractal Spectrum Based Classification for Breast Cancer

Alex Saúl Salas Tlapaya, Julio César Pérez-Sansalvador, and Noureddine Lakouari

Abstract Fractals are naturally found on complex biological structures such as snowflakes, pulmonary arteries and veins, and anatomic structures at the microscopic scale. Biomedical 1D and 2D signals are extensively analyzed using fractal geometry to identify textures and patterns. In particular, the fractal geometry of cells has been previously used to identify cancer cells during the histological evaluation of sample tissues. However, results are similar to those presented by the combination of correlation and entropy, two of the most valuable features for cancer cell classification in classical machine learning strategies. This work presents a novel approach based on the multifractal spectrum at the patient level to identify breast cancer from microscopic tissue images at different magnification factors stained with Hemotoxylin and Eosin. The main steps involve a preprocessing stage to enhance image quality, followed by the computation of fractal dimensions and then a multifractal spectrum, last used by supervised-based classifiers. We explore the multifractal spectrum as a set of features that lead to a high confidence level in classification results. The reported results show an improvement in the accuracy values reported by state-of-the-art strategies.

A. S. S. Tlapaya
Biomedical Science and Technology Department, Instituto Nacional de Astrofísica Óptica y Electrónica, Puebla, Mexico

J. C. Pérez-Sansalvador (✉)
Biomedical Science and Technology Department, Instituto Nacional de Astrofísica Óptica y Electrónica, Puebla, Mexico

Computer Science Department, Instituto Nacional de Astrofísica Óptica y Electrónica, Puebla, Mexico

CONACyT - Consejo Nacional de Ciencia y Tecnología, CDMX, Mexico
e-mail: jcp.sansalvador@inaoep.mx

N. Lakouari
Computer Science Department, Instituto Nacional de Astrofísica Óptica y Electrónica, Puebla, Mexico

CONACyT - Consejo Nacional de Ciencia y Tecnología, CDMX, Mexico

G. Bretti et al. (eds.), *Mathematical Models and Computer Simulations for Biomedical Applications*, SEMA SIMAI Springer Series 33,
https://doi.org/10.1007/978-3-031-35715-2_9

1 Introduction

Fractals are mathematical objects describing complex structures presenting self-repeating patterns [13]. Fractals are found in nature, such as snowflakes, the branches on trees, the leaves on some plants, the anatomy of the vascular and pulmonary systems, and the complex anatomic structure at the microscopic scale [14].

The application of fractal analysis, primarily based on the approximation of the fractal dimension, has shown its efficacy in the analysis of medical images presenting discontinuous, complex, and fragmented objects [13]. Fractal analysis is extensively applied to life-science applications such as biomedical signals segmentation, pattern recognition, texture analysis, and classification [2, 6, 13, 15].

The fractal dimension is an estimation of a self-similarity pattern, a quantitative measure of the structural complexity of the studied object [22]. Fractal dimension analysis can be applied to tumor tissue fragments to measure the clustering of cancer cells over varying patient samples [10]. Moreover, morphological characteristics relevant to pathologists as the shape of individual cells, also quantified by the fractal dimension, have proved importance of describing malignancies, disease staging, and progression [3, 4].

Recently, the fractal dimension has been used to assess breast cancer risk from mammograms [12], grade tumor malignancy from microscopic tissue analysis [21], classify malignant and benign breast cancer histopathological images within a single magnification [7], quantify the irregularity and complexity of the cancer-stromal cell interfaces [1], investigate lung tumor growth based on its geometry complexity, irregular shapes and patterns [11], grading of prostate adenocarcinoma [16], and describe the phenotypic plasticity or aggressivity on pediatric and adult glioblastoma [25].

Breast cancer is the most common cancer among women in the United States [17]. Detection procedures involve mammograms, magnetic resonance imaging, ultrasound, and thermography [20]. However, a biopsy that leads to the analysis of a stained tissue sample under microscopy by a histopathologist is the final breast cancer diagnostic procedure [9]. The final diagnostic includes grading and staging from the analyzed samples by the histopathologist [20]; a time-consuming task that may lead to false negatives or false positives with a significant impact on treatment strategy, patient outcome, and quality of life [21].

The development of computational tools aiding the specialists in the diagnostic by automating the classification task increases the confidence level of the final results [4, 23]. Additionally, these automated tools may suggest prognosis and even therapy strategies. Furthermore, specialists may be guided in selecting the best therapy strategy to maximize patient outcomes and minimize unwanted therapy effects and risks [22].

Even though the success of fractal dimension on biosignal analysis, it only describes an object by a single value, this may limit the complete analysis of objects exhibiting multifractal features at different space-time scales [13]. In this work, we further extend the application of fractal analysis by computing the multifractal spectrum of breast tissue images captured at different magnification factors to identify patients with a breast cancer diagnosis.

The document is organized as follows: in Sect. 2, we present a review of works applying fractal dimension to the detection and diagnosis of cancer disease. In Sect. 3, we detail the images dataset used in this work, then the proposed classification method is presented in Sect. 4. The results and a detailed discussion is presented in Sect. 5. Finally, we present the conclusions and current limitations in Sect. 6.

2 Related Work

This section presents a selection of research works on the classification of biomedical images involving fractal analysis. For the reader interested in additional applications of fractal analysis, we suggest reviewing [14], where the author presents a general review on the application of fractal analysis for the study of cancer prognosis and the organization of the genome and the epigenome.

Fractal dimension has been used as a unique feature to classify histopathological single magnification images from breast cancer [7], colon cancer [8], and leukemia [4, 23]. Additionally, fractal dimension has shown similar results as those provided by correlation and entropy, two of the previously identified most useful features [8].

In [8], the authors include fractal dimension to sets of features involving correlation, entropy, inverse difference momentum, contrast, angular second momentum, and dissimilarity. Their results show a slight improvement in sensitivity and specificity metrics; relevant for automated techniques in clinical practice.

The authors in [6, 21, 22] use fractal dimension to grade the breast's ductal carcinoma. They performed experiments on a private dataset of tissue images at 100× magnification stained with pan-cytokeratin. The fractal dimension quantifies the deviation from normal epithelial breast morphology; therefore, it is presented as an indicator of abnormal aggressive tumor growth. Moreover, in [22] fractal dimension shows its importance as a disease-specific quantitative and reproducible image-based prognosis marker for invasive ductal carcinoma of the breast.

The risk of developing breast cancer is estimated through fractal analysis of mammographic parenchymal patterns [12]. The authors presented an evaluation of different fractal dimension approximation strategies. They concluded that a modified box-counting method using linear discriminant analysis (LDA) showed better classification results than the conventional box-counting method.

In [7], a fractal dimension based classification for histopathological images at 40× magnification factor is presented. The authors train an SVM on a random half of the BreasKHis public dataset [20] and obtain outstanding results (0.979 F1 score). However, results at 100×, 200× and 400× magnification factors are poor (below 0.26 F1 score). Given the 40× magnification factor results, the authors tested sixteen different SVMs for subtype classification and obtained a maximum F1 score of

0.977. However, whether the fractal dimensions of the other magnifications factors may improve or worsen the classification results remain unanswered.

Regarding multifractal analysis, the authors in [2] compute the multifractal spectrum for the segmentation and classification of biomedical images of healthy liver and fatty liver diseases. Similarly, in [27], the authors present a set of software tools to apply multifractal analysis for the identification of colon cancer. In those works, the authors approximate the multifractal dimension as the average of the fractal dimension of several images of the same class. The accuracy results reach their peak at about 85%.

Regarding the application of fractal analysis to identify other types of cancer, in [4], the authors present a cancer detection method based on fractal dimension for the identification of hairy-cell lymphocytic leukemia from cytological preparations. Lately, in [23], the authors present a software implementation of the previous work to ease tuning threshold parameters.

3 Dataset

In this work, we use the freely available Breast Cancer Histopathological (BreaKHis) dataset [20] to ease validation, reproducibility, and direct comparison of experimental results. The BreaKHis dataset comprises 7,909 microscopic images of breast tumor tissue stained with hematoxylin and eosin (H&E), corresponding to 82 patients. The dataset includes benign and malignant microscopic tissue color images obtained at different magnifications: 40×, 100×, 200×, and 400×. An experienced pathologist establishes the final classification for each image once confirmed by complementary exams such as immunohistochemistry analysis. There are about 24 images per patient per magnification factor. Each classified image as a benign tumor corresponds to one of the following subtypes: adenosis (A), fibroadenoma(F), phyllodes tumor (PT), or tubular adenoma(TA). Each classified image as a malignant tumor type corresponds to one of the following subtypes: carcinoma(DC), lobular carcinoma (LC), mucinous carcinoma (MC), or papillary carcinoma (PC). The images distribution in the dataset is shown in Table 1. In Fig. 1 we present two sample microscopic images from the dataset at 40× and 100× magnification factors. This work will focus on patient-level results based on their associated images.

Observe that previous works have used the same dataset; for example, the authors in [7, 9] concluded that the images with a 40× magnification factor

Table 1 Images distribution by magnification factor, class and number of patients

	Magnification				
Class	40×	100×	200×	400×	# Patients
Benign	625	644	623	588	24
Malignant	1370	1437	1390	1232	58

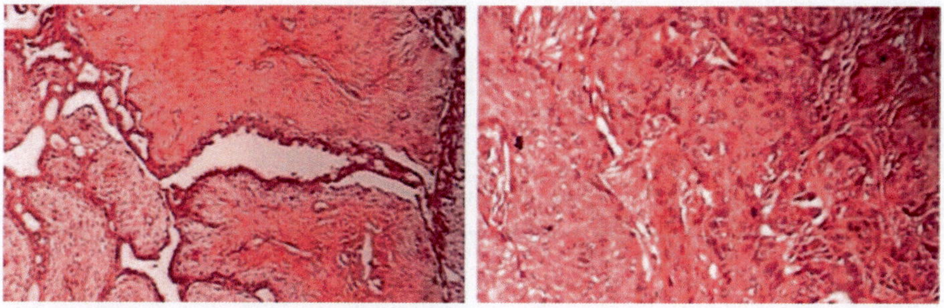

Fig. 1 BreaKHis microscopic images; to the left a $40\times$ magnification factor for fibroadenosis, to the right a $100\times$ magnification factor for ductal carcinoma

have enough information to identify malignant and benign images. Related works implementing a Convolutional Neural Network (CNN) [19] and reusing state-of-the-art neural network architecture for image classification [18] (with about 85% accuracy results) suggest that the $40\times$ and $200\times$ magnification factors present most of the information required by the classification task. However, none has explored the multifractal spectrum of the images on the BreaKHis dataset and presented whether this information improves or worsens the classification results.

4 Patient-Based Breast Cancer Identification

The proposed method considers four main stages for identifying breast cancer at the patient level from the BreaKHis dataset: image processing, fractal dimension computation, multifractal spectrum computation, and patient-based classification, see Fig. 2. The first stage processes each image in the BreaKHis dataset and computes its edges. The second stage takes each edge image from the previous step and computes its fractal dimension. The third stage calculates the multifractal spectrum for each patient by taking all patients' associated images (including all magnifications). The last stage performs the patient-level classification based on the multifractal spectrum of each patient.

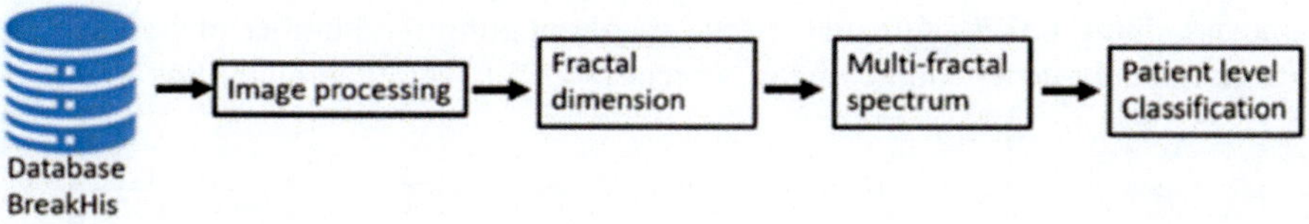

Fig. 2 Main stages for patient based classification

Table 2 Canny filter parameters for edge detection from binary images

Parameter	Value	Value
Aperture size	3	5

We provide a `GitHub` repository[1] with the source code and the fractal dimensions of the dataset to ease reproducibility, validation, and comparison of results.

4.1 Image Processing

This stage computes the edges of each image of the BreaKHis dataset. This involves a set of image processing steps going through image grayscaling, histogram equalization, binarization, and edge detection.

The first step transforms each color image to its grayscale representation, and then we apply histogram equalization to normalize all images; this step also improves the image's contrast. The we perform a thresholding from this new set of images to generate binary images. The *threshold* value was set to 50. Finally, we apply a Canny edge detection filter to compute the edges from the resulting images. Note that we compute two different edge images for each binary image, each generated from a different value for the kernel size parameter for the Canny filter; see Table 2. We use the `OpenCV` [5] `Python 4.5.1.48` version library and its implementation of the Canny filter for image processing.

4.2 Fractal Dimension

The second stage is the fractal dimension estimation based on the edges representation of the microscopic images. There are several algorithms to approximate the fractal dimension from digital images; we use the box-counting method to estimate the fractal dimension of the edges images computed in the previous stage. The box-counting method is the most widely used for approximating the fractal dimension on digital images mainly due to its simplicity and effectiveness [13, 28].

The box-counting method involves covering the geometric object of interest with nonoverlapping boxes with side length ϵ and counting the number of boxes $N(\epsilon)$ that completely cover the geometric object [26, 28]. The theoretical box-counting

[1] https://github.com/tachidok/multifractal_breaKHist.

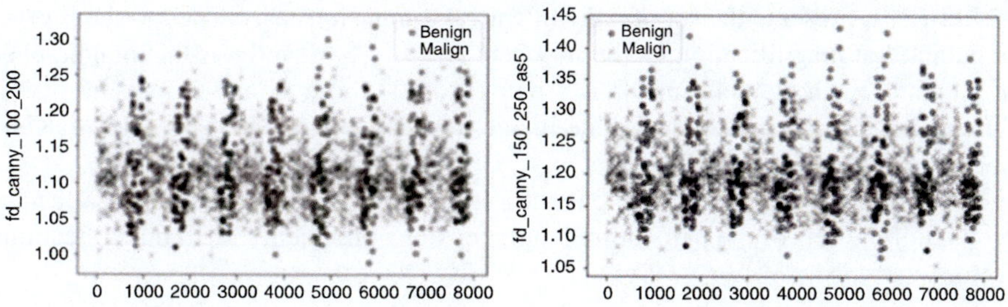

Fig. 3 Fractal dimensions for the 40× magnification factor images. The image a the left shows the fractal dimension with Canny filter (parameters values set 1). The image at the right shows the fractal dimensions with Canny filter (parameters values set 2)

dimension is defined by Eq. (1).

$$d_{\text{box}} = \lim_{\epsilon \to 0} \frac{\log N}{\log \frac{1}{\epsilon}} \tag{1}$$

We use the box-counting method implementation by Francesco Turci in [24]; this involves an iterative process to track $N(\epsilon)$ as it tests different values for ϵ to cover the input image. The absolute value of the slope of the straight line that fits the log-log plot of ϵ vs $N(\epsilon)$ is the approximated fractal dimension of the input image.

Note that at the end of this stage, we have two fractal dimensions associated with each microscopic image on the BreaKhis dataset. Recall that at the image processing stage, we generated two edges images representations for each original BreaKHis microscopic image; therefore, we calculated a fractal dimension for each edge image.

In Fig. 3, we present the fractal dimensions associated with each microscopic image of the BreaKHis dataset at the 40× magnification factor. Observe that the fractal dimension at this magnification factor does not show a clear difference between benign and malignant tissue images. Therefore we take advantage of the multiple magnification factors of the BreaKHis dataset and generate the multifractal spectrum.

4.3 *Multifractal Spectrum*

The third stage involves generating the multifractal spectrum for each patient. Given that the BreakHis dataset provides multiple magnifications of its microscopic tissue images, then the multifractal spectrum of each patient results from the fractal dimensions of the patient's associated microscopic images at the 40×, 100×, 200× and 400× magnification factors.

Let D_i^m represents the set of pairs of fractal dimensions associated to the images of patient i at magnification factor m, where $i = 1, \ldots, n$, and n is the total number of patients. Each element in D_i^m is a pair (d_1, d_2) where d_1 and d_2 are the fractal dimensions of first and second edge image, respectively, of the original BreaKHis microscopic image.

The multifractal spectrum of patient i is defined by Eq. (2), i.e., the cartesian-product of the sets of the pairs of fractal dimensions of patient i at all magnifications factors.

$$M_i = D_i^{40\times} \times D_i^{100\times} \times D_i^{200\times} \times D_i^{400\times} \tag{2}$$

Observe that the number of elements in M_i is given by Eq. (3).

$$|M_i| = |D_i^{40\times}| \cdot |D_i^{100\times}| \cdot |D_i^{200\times}| \cdot |D_i^{400\times}| \tag{3}$$

5 Experiments and Results

In this section, we present the set of experiments for identifying patients in the benign or malign class based on the multifractal spectrum of its associated microscopic tissue images at all magnification factors.

5.1 The Extended Dataset: Structure and Preprocessing

Our multifractal spectrum dataset is extended, for each element in M_i we add the corresponding benign or malign label based on the pathologist's classification of the images for patient i. An entry structure on the extended dataset for a given patient is shown in Table 3.

As part of the preprocessing stage, we removed duplicated entries, normalized them, and undersampled the dataset to manage unbalanced inherited from the original BreaKHis dataset. Then we randomly sample 70% of the dataset for training the classifier and leave the rest for testing.

Table 3 An entry on the extended dataset showing the multifractal spectrum for patient i

Patient ID	Fractal dimensions 40×	Fractal dimensions 100×	Fractal dimensions 200×	Fractal dimensions 400×	Label

Table 4 Classifiers accuracy

Classifier	Accuracy
Naive-Bayes	0.6936
SVM	0.8084
Multilayer perceptron	0.8177
K-NN	0.9635
Decision tree	0.9899
Random forest	0.9969

5.2 *Classification Results*

We report the accuracy values with k-fold cross-validation ($k = 10$) for patient classification based on its multifractal spectrum. In Table 4, we present the associated accuracy for a selection of machine learning classifiers. We observe that Naive-Bayes performed the worst since it assumes feature independence and the same contribution of each to the final classification result. Note that the obtained SVM results are better than those reported in [7] for 100×, 200×, and 400× magnification factors. However, our approach requires microscopic images at multiple magnification factors to compute the multifractal spectrum. The classifiers with the best results were decision tree and random forest.

In order to gain insight into feature importance for the classification task, we trained a decision tree classifier with all the data (no training or testing sets). We obtained the feature importance for the fractal dimensions associated with each magnification factor from the generated tree. We observed that the fractal dimensions related to the 40× magnification factor got feature importance around 55%, followed by the fractal dimensions associated with the 100× magnification factor with about 30% of feature importance. These observations agree with those presented in [7, 18] regarding the importance of the 40× and 100× magnification factors.

In Fig. 4 we present the confusion matrices for the tested classifiers. Observe that the Multilayer perceptron presents a false-positive rate of about 5%; this would involve starting treatment on disease-free patients with the associated therapy effects. More importantly, the same classifier presents a false-negative rate of 13%; this would involve leaving without treatment to patients with an in-developing tumor. Based on previous results and the significant impact on patient outcomes, this classifier would probably not make it into clinical practice. Similar cases are the SVM and the KNN classifiers, with a false-negative rate of 10% and 3%, respectively. Observe that only the decision tree and the random forest classifiers present false-positive and false-negative rates below 0.006%.

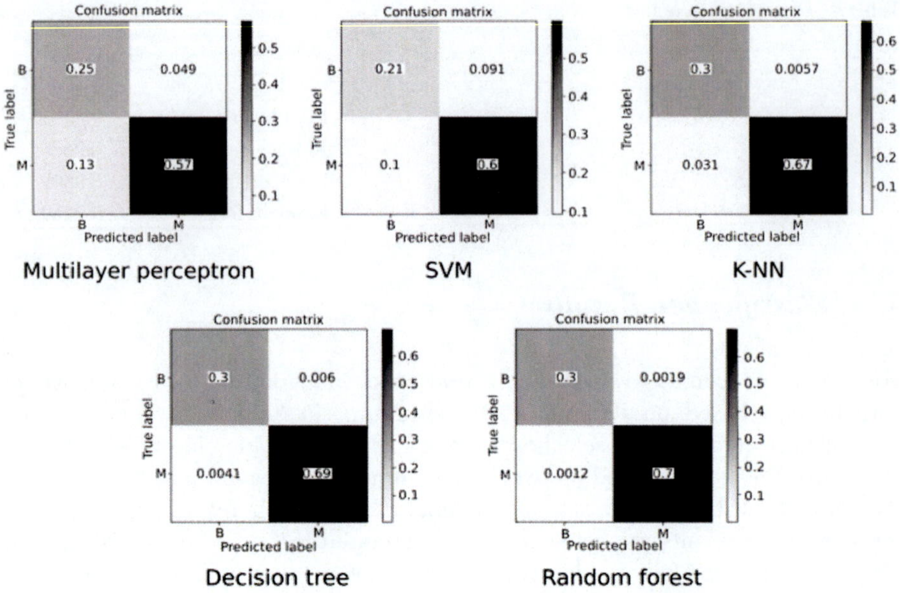

Fig. 4 Confusion matrices

5.3 *Discussion*

Including information associated with 200× and 400× magnification factors significantly improves results compared to works using a single magnification factor [6, 7, 21, 22]. However, those works perform an image-based classification, and our approach performs a patient-level classification. Surprisingly, our accuracy results were better than those based on deep-learning strategies, performing a patient-level classification [18, 19]. This is generally the case when the dataset is small. We have only 82 instances (or patient data) for patient-based classification in this case. The previous suggests that the information provided by the 200× and 400× magnification factors may help decide the classification label for difficult cases at the patient level.

Additionally, in agreement with previous works, we found that the 40× and the 100× magnification factors provide most of the information (about 85% of feature importance) for the final classification decision. The 200× and 400× magnification factors contribute only about 15% to the final classification decision.

It should be noted that a primary constraint of our approach is that it requires microscopic tissue images at different magnifications; otherwise, it would not be possible to compute the multifractal spectrum of each patient.

6 Conclusions

We presented a multifractal spectrum classification strategy for identifying breast cancer at the patient level on the BreaKHis dataset. We took advantage of the multiple magnifications provided by the dataset to generate the multifractal spectrum of each patient. The reported accuracy values show outstanding results for the decision tree and random forest classifiers.

Previous works base their classification results on a single magnification factor, either 40× or 100×, and left open the question of whether including information from other magnification factors would improve or worsen the classification results. In this work, we clearly show that including information about 200× and 400× magnifications significantly improves the final classification results.

Additionally, our results agree with previous works that the fractal dimension of the 40× magnification factor has greater feature importance at the classification stage. However, we noted a significant increase in classification accuracy by including information concerning additional magnification factors, precisely the associated fractal dimension of the 100× magnification factor images. Surprisingly, our results outperformed the deep-learning-based strategies presented in [18, 19] for patient-level classification on the same dataset.

Finally, the fractal dimension contains important information regarding the structure and complexity of microscopic tissues such that it may be helpful to identify benign from malign tumors. Moreover, the multifractal spectrum generated from the fractal dimensions of the microscopic images associated with a given patient improves classification results significantly. The generation of automated tools for breast cancer classification based on the multifractal spectrum would increase the confidence level of the results.

Future work involves the application of the presented strategy to other types of cancer; this would require a dataset with microscopic images at different magnification factors. We also want to test our approach with microscopic images with a different stained technique since related works mention that this may significantly impact the results of fractal dimension approximation. As in reviewed works, we would like to explore the use of the multifractal spectrum for progression prediction, grading, and staging of the disease.

Acknowledgments The first author acknowledges to the "Alianza en Inteligencia Artificial, Proyecto FORDECyT 296737" and INAOE for the financial support to commit this work.

References

1. AbdulJabbar, K., Raza, S.E., Rosenthal, R., Jamal-Hanjani, M., Veeriah, S., et al.: Geospatial immune variability illuminates differential evolution of lung adenocarcinoma. Nat. Med. **26**, 1054–1062 (2020)
2. Ampilova, N., Soloviev, I.: Application of fractal and multifractal analysis algorithms to image segmentation and classification. WSEAS Trans. Biol. Biomed. **13**, 14–21 (2016)

3. Baish, J.W., Jain, R.K.: Perspectives in cancer research fractals and cancer 1. Cancer Res. **60**, 3683–3688 (2000)
4. Bauer, W., Mackenzie, C.D.: Cancer detection on a cell by cell basis using a fractal dimension analysis. Fundam. Issues Elem. Matter **14**, 39–46 (2000)
5. Bradski, G.: The OpenCV Library. Dr. Dobbś J. Softw. Tools (2000)
6. Braverman, B., Tambasco, M.: Scale-specific multifractal medical image analysis. Comput. Math. Methods Med. **2013** (2013)
7. Chan, A., Tuszynski, J.A.: Automatic prediction of tumour malignancy in breast cancer with fractal dimension. R. Soc. Open Sci. **3**, 160558 (2016)
8. Esgiar, A.N., Naguib, R.N., Sharif, B.S., Bennett, M.K., Murray, A.: Fractal analysis in the detection of colonic cancer images. IEEE Trans. Inf. Technol. Biomed. **6**, 54–58 (2002)
9. Gupta, V., Bhavsar, A.: Breast cancer histopathological image classification: is magnification important? In: IEEE Computer Society Conference on Computer Vision and Pattern Recognition Workshops, 2017 July, pp. 769–776 (2017)
10. Jenner, A.L., Smalley, M., Goldman, D., Goins, W.F., Cobbs, C.S., Puchalski, R.B., Chiocca, E.A., Lawler, S., Macklin, P., Goldman, A., Craig, M.: Agent-based computational modeling of glioblastoma predicts that stromal density is central to oncolytic virus efficacy. iScience **25**, 104395 (2022)
11. Lennon, F.E., Cianci, G.C., Cipriani, N.A., Hensing, T.A., Zhang, H.J., Chen, C.T., Murgu, S.D., Vokes, E.E., Vannier, M.W., Salgia, R.: Lung cancer-a fractal viewpoint. Nat. Rev. Clin. Oncol. **12**, 664–675 (2015)
12. Li, H., Giger, M.L., Olopade, O.I., Lan, L.: Fractal analysis of mammographic parenchymal patterns in breast cancer risk assessment. Acad. Radiol. **14**, 513–521 (2007)
13. Lopes, R., Betrouni, N.: Fractal and multifractal analysis: a review. Med. Image Anal. **13**, 634–649 (2009)
14. Metze, K.: Fractal dimension of chromatin and cancer prognosis. Epigenomics **2**, 601–604 (2010)
15. Nayak, S.R., Mishra, J., Khandual, A., Palai, G.: Fractal dimension of rgb color images. Optik **162**, 196–205 (2018)
16. Pleşea, R.M., sebastian Şerbănescu, M., Ciovică, D.V., camelia Roşu, G., Moldovan, V.T., Bungărdean, R.M., Popescu, N.A., Pleşea, I.E.: The study of tumor architecture components in prostate adenocarcinoma using fractal dimension analysis. Rom. J. Morphol. Embryol. **2019**, 501–519 (2019)
17. Siegel, R.L., Miller, K.D., Fuchs, H.E., Jemal, A.: Cancer statistics, 2022. CA: Cancer J. Clin. **72**, 7–33 (2022)
18. Spanhol, F.A., Cavalin, P.R., Oliveira, L.S., Petitjean, C., Heutte, L.: Deep features for breast cancer histopathological image classification. http://web.inf.ufpr.br/vri/breast-cancer-database
19. Spanhol, F.A., Oliveira, L.S., Petitjean, C., Heutte, L.: Breast cancer histopathological image classification using convolutional neural networks. http://web.inf.ufpr.br/vri/breast-cancer-database
20. Spanhol, F.A., Oliveira, L.S., Petitjean, C., Heutte, L.: A dataset for breast cancer histopathological image classification. IEEE Trans. Biomed. Eng. **63**, 1455–1462 (2016)
21. Tambasco, M., Magliocco, A.M.: Relationship between tumor grade and computed architectural complexity in breast cancer specimens. Hum. Pathol. **39**, 740–746 (2008)
22. Tambasco, M., Eliasziw, M., Magliocco, A.M.: Morphologic complexity of epithelial architecture for predicting invasive breast cancer survival. J. Transl. Med. **8**, 1–11 (2010)
23. Timbó, C., da Rosa, L.A., Gonçalves, M., Duarte, S.B.: Computational cancer cells identification by fractal dimension analysis. Comput. Phys. Commun. **180**, 850–853 (2009)
24. Turci, F. Box Counting in Numpy. https://francescoturci.net/2016/03/31/box-counting-in-numpy/
25. Uthamacumaran, A., Craig, M.: Algorithmic reconstruction of glioblastoma network complexity. iScience **25**, 104179 (2022)

26. Vargas, I., Alhallak, K., Kolenc, O.I., Jenkins, S.V., Griffin, R.J., Dings, R.P.M., Rajaram, N., Quinn, K.P.: Rapid quantification of mitochondrial fractal dimension in individual cells. Biomed. Opt. Express **9**, 5269 (2018)
27. Vasiljevic, J., Zivic, N., Reljin, B., Celebic, V.: Application of multifractal analysis on medical images. WSEAS Trans. Inf. Sci. Appl. **5**, 1561–1572 (2008)
28. Wu, J., Jin, X., Mi, S., Tang, J.: An effective method to compute the box-counting dimension based on the mathematical definition and intervals. Results Eng. **6** (2020)